AF566663

The Visual System
from Genesis to Maturity

The Visual System
from Genesis to Maturity

Roberto Lent
Editor

Birkhäuser
Boston • Basel • Berlin

Roberto Lent
Departamento de Neurobiologia
Instituto de Biofísica Carlos Chagas Filho
Universidade Federal do Rio de Janeiro
Centro de Ciências de Saúde, Bloco G
Cidade Universitária
21941 Rio de Janeiro R.J.
Brazil

Library of Congress Cataloging-in-Publication Data

The Visual system from genesis to maturity / edited by Roberto Lent.
p. cm.
Includes bibliographical references and index.
ISBN 0-8176-3598-X (H : acid-free). — ISBN 3-7643-3598-X
1. Eye—Differentiation. 2. Eye—Growth. 3. Visual pathways--Differentiation. 4. Visual pathways—Growth. I. Lent, Roberto.
QP475.V63 1992
599'.01823—dc20 92-21738
CIP

Printed on acid-free paper.

ISBN 0-8176-3598-X
ISBN 3-7643-3598-X

Typeset by ATLIS Graphics
Printed and bound by Edward Bros., Ann Arbor, MI
Printed in the United States of America

9 8 7 6 5 4 3 2 1

Cover: A detail of "Solomon's Justice", by the Brazilian painter Candido Portinari (1903–1962), showing an ambiguous figure.
Art Museum of São Paulo (MASP). Reproduced by permission of the Portinari Project.

Contents

Preface

I must confess openly, before presenting this book to the reader, that the task of producing it was much more enjoyable to me than I had suspected at the project's beginning. First, I had the opportunity to discuss with my colleagues at the Department of Neurobiology of the Institute of Biophysics in Rio (who are internationally known experts in the different areas of visual research), which particular themes would best illustrate the state-of-the-art. Second, after the difficulty of having to select some of these themes and prospective authors out of a very numerous list, I had the pleasure to receive them all in Rio de Janeiro, and to chair a very active, small meeting held at the Brazilian Academy of Sciences. Third, I enjoyed the privilege of reading all the manuscripts containing scholarly, up-to-date reviews of each theme, as well as original data in some cases. Finally, I was deeply moved with the prospect of dedicating the result of this collective endeavor to the three men who initiated visual research in Rio: Professors Carlos Eduardo Rocha-Miranda, and Eduardo Oswaldo-Cruz, and the laboratory technician Mr. Raymundo F. Bernardes.

The book is organized into two broad sections preceded by an overview. The first section—Genesis—covers many of the numerous fronts of research concerning the developmental biology of the visual system, from either a molecular, a cellular, or a systems angle. The second section—Maturity—takes the same broad approach and covers the structure and physiology of different parts of the visual system, "from photoreceptors to the cortical modules," to quote Pasko Rakic from his Overview.

Both the book and the meeting that preceded it were partially supported by funds obtained from the National Research Council (CNPq), the International Brain Research Organization (IBRO/UNESCO), the Advanced Program of Neuroscience (PAN), the Brazilian Academy of Sciences and IBM-Brazil. The book itself has been competently produced by Birkhäuser Boston.

I hope that we have succeeded in extending to the readers—students, neuroscientists, ophthalmologists, neurologists, and others—the pleasure that we had organizing, discussing, and contributing to the book.

Roberto Lent
Instituto de Biofísica Carlos Chagas Filho

List of Contributors

Ruben Adler, Wilmer Ophthalmological Institute, Johns Hopkins University School of Medicine, Baltimore, Maryland, 21205, USA

Michael Ariel, Departments of Behavioral Neuroscience and Psychiatry, University of Pittsburgh, Pittsburgh, Pennsylvania, 15260, USA

Raymundo F. Bernardes, Instituto de Biofísica Carlos Chagas Filho, Universidade Federal do Rio de Janeiro, Centro de Ciências da Saúde, 21941 Rio de Janeiro, Brazil

Richard T. Born, Department of Neurobiology, Harvard Medical School, Boston, Massachusetts, 02115, USA

Hollis T. Cline, Department of Physiology and Biophysics, University of Iowa College of Medicine, Iowa City, Iowa, 52242, USA

Martha Constantine-Paton, Department of Biology, Kline Biology Tower, Yale University, New Haven, Connecticut, 06511, USA

Fernando G. De Mello, Instituto de Biofísica Carlos Chagas Filho, Universidade Federal do Rio de Janeiro, Centro de Ciências da Saúde, 21941 Rio de Janeiro, Brazil

Rodney J. Douglas, Anatomical Neuropharmacology Unit, Medical Research Council, Mansfield Road, Oxford, England

Manuel Esguerra, Department of Brain and Cognitive Sciences, Massachusetts Institute of Technology, Cambridge, Massachusetts, 02139, USA

Rosalia B. Faria, Departamento de Neurobiologia, Universidade Federal Fluminense, Instituto de Biologia, 24001 Niterói, Brazil

Mario Fiorani, Jr., Instituto de Biofísica Carlos Chagas Filho, Universidade Federal do Rio de Janeiro, Centro de Ciências da Saúde, 21941 Rio de Janeiro, Brazil

Patrícia F. Gardino, Instituto de Biofísica Carlos Chagas Filho, Universidade Federal do Rio de Janeiro, Centro de Ciências da Saúde, 21941 Rio de Janeiro, Brazil

Ricardo Gattass, Instituto de Biofísica Carlos Chagas Filho, Universidade Federal do Rio de Janeiro, Centro de Ciências da Saúde, 21941 Rio de Janeiro, Brazil

Luiz Gawryszewski, Departamento de Neurobiologia, Universidade Federal Fluminense, Instituto de Biologia, 24001 Niterói, Brazil

Charles G. Gross, Department of Psychology, Princeton University, Princeton, New Jersey, 08544, USA

Jong-On Hahm, Department of Brain and Cognitive Sciences, Massachusetts Institute of Technology, Cambridge, Massachusetts, 02139, USA

Jan N. Hokoç, Instituto de Biofísica Carlos Chagas Filho, Universidade Federal do Rio de Janeiro, Centro de Ciências da Saúde, 21941 Rio de Janeiro, Brazil

Roberto Lent, Instituto de Biofísica Carlos Chagas Filho, Universidade Federal do Rio de Janeiro, Centro de Ciências da Saúde, 21941 Rio de Janeiro, Brazil

Rafael Linden, Instituto de Biofísica Carlos Chagas Filho, Universidade Federal do Rio de Janeiro, Centro de Ciências da Saúde, 21941 Rio de Janeiro, Brazil

Marla B. Luskin, Department of Anatomy and Cell Biology, Emory University School of Medicine, Atlanta, Georgia, 30322, USA

Andrew P. Mariani, Neurological Sciences 1, Division of Research Grants, National Institutes of Health, Bethesda, Maryland, 20892, USA

Kevan A. C. Martin, Anatomical Neuropharmacology Unit, Medical Research Council, Mansfield Road, Oxford, England

Rosalia Mendez-Otero, Instituto de Biofísica Carlos Chagas Filho, Universidade Federal do Rio de Janeiro, Centro de Ciências da Saúde, 21941 Rio de Janeiro, Brazil

Eduardo Oswaldo-Cruz, Instituto de Biofísica Carlos Chagas Filho, Universidade Federal do Rio de Janeiro, Centro de Ciências da Saúde, 21941 Rio de Janeiro, Brazil

Antônio Pereira, Jr., Instituto de Biofísica Carlos Chagas Filho, Universidade Federal do Rio de Janeiro, Centro de Ciências da Saúde, 21941 Rio de Janeiro, Brazil

Cristovam W. Picanço-Diniz, Departamento de Fisiologia, Centro de Ciências Biológicas, Universidade Federal do Pará, 66059 Pará, Brazil

Walter M. Pinheiro, Departamento de Neurobiologia, Universidade Federal Fluminense, Instituto de Biologia, 24001 Niterói, Brazil

Maria C. G. Piñon, Instituto de Biofísica Carlos Chagas Filho, Universidade Federal do Rio de Janeiro, Centro de Ciências da Saúde, 21941 Rio de Janeiro, Brazil

Pasko Rakic, Section of Neurobiology, Yale University School of Medicine, New Haven, Connecticut, 06510, USA

Ary S. Ramoa, Section of Neurobiology, Yale University School of Medicine, New Haven, CT, 06510, and Instituto de Biofísica Carlos Chagas Filho, Universidade Federal do Rio de Janeiro, Centro de Ciências da Saúde, 21941 Rio de Janeiro, Brazil

Giacomo Rizzolatti, Istituto di Fisiologia Umana, Università di Parma, I-43100 Parma, Italy

Carlos E. Rocha-Miranda, Instituto de Biofísica Carlos Chagas Filho, Universidade Federal do Rio de Janeiro, Centro de Ciências da Saúde, 21941 Rio de Janeiro, Brazil

Hillary R. Rodman, Department of Psychology, Princeton University, Princeton, New Jersey, 08544, USA

Marcelo G. P. Rosa, Instituto de Biofísica Carlos Chagas Filho, Universidade Federal do Rio de Janeiro, Centro de Ciências da Saúde, 21941 Rio de Janeiro, Brazil

Burkhard Schlosshauer, Naturwissenschaftliches und Medizinisches Institut, Der Universität Tubingen in Reutlingen, Reutlingen, Germany

Luiz Carlos L. Silveira, Departmento de Fisiologia, Centro de Ciências Biológicas, Universidade Federal do Pará, 66059 Pará, Brazil

Juliana G. M. Soares, Instituto de Biofísica Carlos Chagas Filho, Universidade Federal do Rio de Janeiro, Centro de Ciências da Saúde, 21941 Rio de Janeiro, Brazil

Aglai P. B. Sousa, Instituto de Biofísica Carlos Chagas Filho, Universidade Federal do Rio de Janeiro, Centro de Ciências de Saúde, 21941 Rio de Janeiro, Brazil

Mriganka Sur, Department of Brain and Cognitive Sciences, Massachusetts Institute of Technology, Cambridge, Massachusetts, 02139, USA

Tania G. Thomaz, Departamento de Neurobiologia, Universidade Federal Fluminense, Instituto de Biologia, 24001 Niterói, Brazil

Roger B. H. Tootell, Department of Neurobiology, Harvard Medical School, Boston, Massachusetts, 02115, USA

Carlo Umiltà, Istituto di Fisiologia Umana, Università di Parma, I-43100 Parma, Italy

Ana L. M. Ventura, Instituto de Biofísica Carlos Chagas Filho, Universidade Federal do Rio de Janeiro, Centro de Ciências da Saúde, 21941 Rio de Janeiro, Brazil

Eliane Volchan, Instituto de Biofísica Carlos Chagas Filho, Universidade Federal do Rio de Janeiro, Centro de Ciências da Saúde, 21941 Rio de Janeiro, Brazil

Edna N. Yamasaki, Instituto de Biofísica Carlos Chagas Filho, Universidade Federal do Rio de Janeiro, Centro de Ciências da Saúde, 21941 Rio de Janeiro, Brazil

An Overview Development of the Primate Visual System: From Photoreceptors to Cortical Modules

Pasko Rakic

This overview is based on a series of longitudinal studies carried out in my laboratory over the last two decades on the developing rhesus monkey. Obviously, most of this work could not have been done without the enormous amount of factual and conceptual advances that have been made during this period in the understanding of visual development in other species. However, a scholarly analysis and comprehensive review of the appropriate literature on this subject is beyond the scope of this communication.

The initial rationale behind the singular effort placed in my laboratory on the analysis of the development of the visual system in the rhesus monkey is that the organization, principles and mechanisms of neural development in this species is, in several major respects, similar to that of the human being. The rhesus monkey has basically the same binocular organization: forwardly placed eyes, fovea, and 10° of central vision that occupies 50% of the surface of the visual cortex. Both rhesus monkey and human beings have almost an identical number and a similar fraction of crossed and uncrossed retinal axons as well as the same number and placement of laminae in the lateral geniculate nucleus. Most importantly, both species share exclusivity of geniculate projections to area 17 of the cerebral cortex, with analogous and uniquely primate ocular dominance, laminar and columnar compartmentalization. Furthermore, both species have trichromatic vision with a comparable photoreceptor mosaic and proportion of short and long wave-sensitive cones. Finally, the divergence factor (defined as the number of higher order cells to which one lower order cell projects) and the convergence factor (defined as the number of lower order cells from which a single higher order cell receives input) is comparable in rhesus monkey and humans. Thus, even numerical relationships and their alteration during development and under pathological conditions in humans can be studied experimentally in the monkey.

On the basis of the similarity in organization, it is also reasonable to expect that the basic principles as well as the cellular and molecular mechanisms involved in the development of the visual system of the macaque monkey are akin to that of human beings. Indeed, both species display a similarly protracted tempo, schedule and sequence of developmental cellular events. Although the duration of gestation in the rhesus monkey is only about two-thirds of that in human beings, one can determine rather precisely the onset and duration of corresponding developmental stages using various cytological and histochemical criteria (e.g., Kostovic and Rakic, 1980, 1984, 1990; Rakic, 1978, 1988; Sidman and Rakic, 1982). Furthermore, the effect of various interventions that produce visual deficits in developing monkeys are similar to those reported in pathological situations in human beings. Thus, with the exception of the great apes, which are inaccessible for experimental research, from a biological standpoint, developmental data on the visual system in macaque monkeys are most likely to be directly applicable to the understanding of normal or diseased states of human vision, and has already inspired strategies for the prevention and treatment of various congenital disorders.

INTEGRAL NATURE OF THE DEVELOPING VISUAL SYSTEM

In spite of the large amount of available data in the literature about the development of the retina, lateral geniculate and visual cortex that could be reasonable subjects for review alone, I will attempt to describe here selected developmental events in the entire visual system, from the photoreceptors to the cerebral cortex. Such an integrative approach may be helpful for understanding the dynamics of the interdependent processes involved in formation of neuronal connections. The integrative approach also can point to the inadequacy of the theories that may provide compelling explanations of the single experiment conducted in one of the visual centers, but fall apart when placed in the context of the visual system as a whole. It seems to me that one of the major tenets of development to come from the study of the mammalian visual system is that the retina, lateral geniculate nucleus and visual cortex are not formed as independent units. Although almost four decades ago, Waddington denoted epigenesis in the broadest sense as a process by which "development is brought about through a series of control interactions between the various parts," only recently has it become fully appreciated that development of the visual system cannot be understood by studying only minute details that take place in its components without considering the properties occurring in the system as a whole. There is little doubt, for example, that the mosaic of photoreceptors and diversity of ganglion cells in the primate retina is related to the magno- (M) and parvocellular (P) subsystems in the lateral geniculate nucleus, and eventually to the specific layers, cellular modules and synaptic circuits that subserve these subsystems within the cerebral cortex. It is also becoming clear that heterogeneous cells communicate during development across various cell-cell junctions and chemical synapses through short-lived diffusible factors and adhesion molecules (e.g., Easter et al., 1985). In the visual system, these interactions act in both directions from the photoreceptors at the periphery towards the central structures in the cortex and back from the cortex toward the retina. Disruption, or even a short delay, in one communication step in either direction can cause a chain reaction affecting heterogeneous cell classes, leading to an abnormal organization of the entire system and, consequently, to abnormal function.

The integrative approach poses extraordinary challenges, as there is no easy and straightforward way to determine how functional circuitry, starting from the photoreceptor and ending in the visual cortex, is formed and maintained in a species with a life span of more than 30 years. To be meaningful, broad multidisciplinary research has to be carried out on the development of the entire visual system, rather than on the narrow aspects that comprise its components. Furthermore, the analysis has to be rigorously quantitative, as many developmental changes may be relatively small and could occur during a protracted time window. Finally, it is also useful if the studies can be carried out in the same laboratory so that the timing of postconceptual age, the criteria of maturation and various technical and procedural aspects are as uniform as possible. We are lucky to live in an era when the introduction of new anatomical tracing methods and production of various cell class-specific markers, neurotransmitter- and receptor-specific ligands as well as computer-aided quantitative methods have opened unprecedented opportunities for probing the principles governing the development of functionally dedicated networks even in the visual system of large and complex primate brain. The refinement of prenatal neurosurgery and the elaboration of methods for interference with mitotic activity, neuronal migration and formation of synaptic connections help to alter the course of development in a way that is useful for testing some basic cellular and molecular mechanisms.

I will include in my presentation our data on the early origin of the visual system, including proliferation kinetics and control of neuronal number. I will follow by an outline of the development of binocular and color-opponent system and the so-called broad-band (P and M) system that forms parallel pathways from their origin in the retina across the lateral geniculate nucleus to the striate cortex. I will also briefly describe formation of cytoarchitectonic maps and intraareal compartmentalization in the occipital lobe, as well as the dynamics of synaptic production and their stabilization during biochemical maturation in the visual system. Finally, I will review some of our recent data on the synaptogenesis and development of neurotransmitter signaling molecules during the lifetime of the rhesus monkey. This global approach will provide a broad picture at the expense of specific details. However, a description of the experimental procedures, factual data, documentation and a large body of additional relevant information can be found in the primary references listed in the bibliography.

NEURONS COMPRISING PRIMATE VISUAL SYSTEM ARE GENERATED PRENATALLY

First, we have to address the question of the time of origin of neurons that comprise the visual system in primates. Studies using ^{3}H thymidine autoradiography in the rhesus monkey revealed that all neurons comprisingthe primate visual system are generated exclusively before birth, during the first half of gestation. This genesis occurs in remarkably precise cell, class-specific and area-specific sequences and in reproducible spatio-temporal gradients. Figure 1A displays graphically the

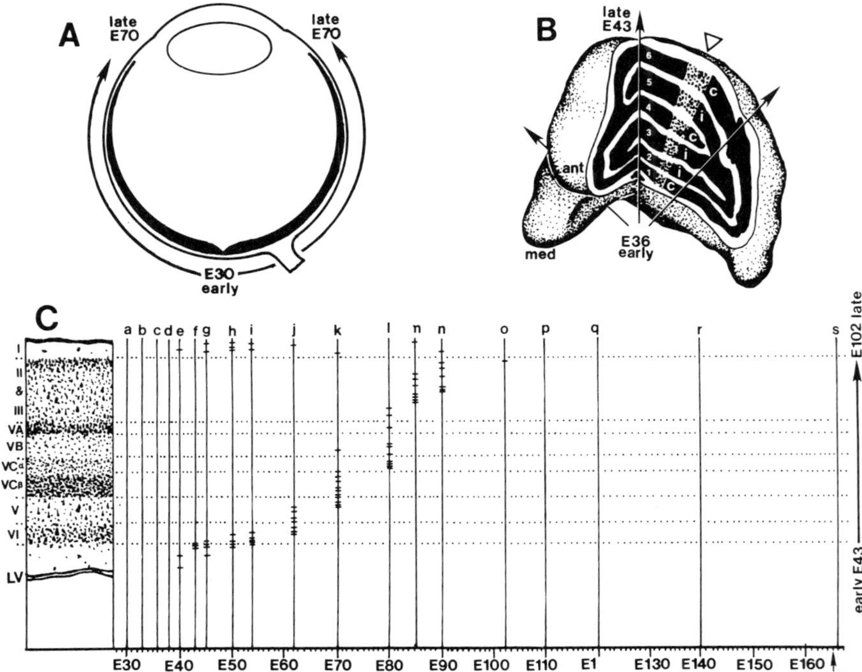

FIGURE 1. Diagrammatic representation of the time of origin and neurogenetic gradients in the retina (A,B), lateral geniculate nucleus (C) and cortical area 17 (D) of the rhesus monkey. The series of over one hundred ^{3}H-TdR labeled animals processed for autoradiography provides the only available data of the time of neuron origin in any primate. The more detailed quantitative data can be found in Rakic, 1974, 1977a,b; La Vail, Rapaport and Rakic, 1991.

onset, course and cessation of the genesis of retinal ganglion cells, as well as all neurons destined for the lateral geniculate nucleus and primary visual cortex. The details of the kinetics of cell proliferation and the rate of daily production of cells can be found in the primary references (Rakic, 1974, 1977a; La Vail et al., 1991). Here, it is sufficient to state that, although neurogenesis of the neurons engaged in the primary visual pathways begins in the retina, followed by onset in the lateral geniculate nucleus and finally the cerebral cortex, the completion of neurogenesis in these three principle structures does not follow this simple hierarchical sequence. For example, photoreceptors situated at the periphery of the retina continue to be generated until birth (La Vail et al., 1991), long after production of neurons has completely stopped in the lateral geniculate nucleus and cerebral cortex (Rakic, 1974, 1977a). Furthermore, within each structure there is a smooth gradient of neurogenesis illustrated graphically by arrows in Figures 1B,C,D.

Understanding the role of neurogenetic gradients is complicated by the fact that all postmitotic cells in the visual system migrate from the place of their origin to their final destinations. This migration may be rather minimal in the retina, but can assume long trajectories in the telencephalon of the large primate brain (e.g., Rakic, 1972). In either case, it is an active, directed and targeted cell movement. Review of this remarkable phenomenon, as observed in the monkey cerebrum, and the discussion of the possible mechanisms of cell translocation and surface-mediated guidance is provided elsewhere (Rakic 1975, 1977b, 1988, 1990). For the present purpose, it is sufficient to state that examination of the diretion of the neurogenetic cell gradients in mature structures, in relation to the pattern of connections made by these same cells, reveals clearly that the data on the time of origin is not sufficient to explain either the laminar or areal topography of primary visual projections in primates. For example, terminal field of cortical input originated from the earlier-generated magnocellular neurons of the lateral geniculate neurons are sandwiched between terminals of input that arise from the later-generated cells of the parvocellular moiety. This arrangement of terminals in the visual cortex cannot be explained simply by the timing of cell origin of projecting and target neurons participating in this synaptic circuitry.

EACH EMBRYONIC STRUCTURE EXPRESSES A SPECIES-SPECIFIC PROTOMAP

The second problem concerns cell determination and formation of topographic maps within each structure of the visual system. This problem is further complicated in the animals where we have to deal also with different parallel subsystems such as M and P pathways, as in the case of the macaque monkey. When and how species specific differences in terms of the number and type of various neuronal classes and their segregation into different subsystems emerge? Contemporary ideas about the development of cortical compartmentalization can be classified into three basic groups: *fate map hypothesis*, which implies that cell phenotypes and their distribution are determined strictly by cell lineage relationships; *tabula rasa hypothesis*, which suggests that all cortical cells are equipotential and their specialization is induced exclusively by contacts with afferents; and the *protomap hypothesis*, which suggests synergistic (reciprocal) interaction between cortical cells and incoming projections (Rakic, 1992). As the prefix *proto* in the latter hypothesis signifies, the details of cell differentiation are not prespecified; rather, cortical cells have a restricted repertoire or a limited genetic potential that can be fully expressed in the area-specific fashion only through interaction with the appropriate set of afferents (Rakic, 1988).

I will elaborate here on the definition of the protomap hypothesis not only because it is based on the work in our laboratory, but also because it has been often misunderstood or confused with the other two hypotheses. Although both the *tabula rasa* and *protomap* proponents recognize the importance of afferents in cortical differentiation, the significant difference between the two views is that the latter presumes that neurons of the cerebrum themselves possess some information about their prospective laminar and areal fate, whereas the former regards cells as totally uncommitted and devoid of any phenotypic commitment or positional information. Thus, unlike the *tabula rasa* hypothesis, which implies that the formation of areal differences is induced entirely by input, the protomap hypothesis postulates two-way cooperation between specific input and target cells, each of which are critical of the performance of certain developmental programs (Rakic, 1988, 1992). On the other hand, the protomap hypothesis is quite different from the fate map model, since the protomap does not imply the existence of an exclusively intrinsic program—on the contrary, it suggests only that cells composing various areas have certain inherited biases. Thus, for example, a portion of the occipital lobe may have "area 17-bias," while the anterior portion of cingulate gyrus has "area 24-bias." This bias may be intrinsically programmed already in the proliferative zone: the part subjacent to the thicker cortex of area 17 produces many more neurons than the portion of the ventricular zone subjacent to area 24 (Rakic, 1974, 1982). The obvious difference between areas of the embryonic cortical plate also exists in their attractiveness to the afferents that originated from the different diencephalic nuclei—area 17 attracting input from the lateral geniculate and area 24 from the nucleus anterior ventralis (Rakic, 1988). Thus, the cortical protomap may be expressed as a simple gradient of certain molecules across the cerebral surface. The crucial question is whether each level of the visual pathway—the retina, lateral geniculate and cortex—has a separate and independent protomap.

We recently found that, after settling in their final positions in the fetal monkey retina, various classes of photoreceptors promptly assume both species-specific proportions as well as typical mosaic-like distribution (Wikler and Rakic, 1990). The emergence of the mosaic of rods and red/green- or blue-sensitive cones, as identified by antibodies to wave-length sensitive opsins, occurs surprisingly early; and we suggested that it may be related to an array of precociously differentiating cones (Wikler and Rakic, 1991). This neuronal protomap appears in the monkey retina in the first half of gestation, before the photoreceptors have established synaptic contacts with either horizontal or bipolar cells within the outer plexiform layer (Nishimura and Rakic, 1987). This finding suggests that the number of photoreceptors to be produced, the basic phenotypic commitment, the species-specific and the positional information all may be initiated without direct connection with neurons situated in the central structures. Our recent study showed that various subtypes of retinal ganglion cells are generated in a sequential order, suggesting that the regularity of their mosaic may also be determined intraretinally (Rapaport et al., 1992). However, in spite of clear signs of species-specific mosaicism, normal differentiation and maintenance of retinal cyto- and synapto-architecture in primates cannot be achieved without interaction with the cerebral visual cortex, since ablation of area 17 leads to the degeneration and eventual elimination of certain classes of retinal cells.

At the opposite end, at the most central part of the primary visual system—in the striate cortex—neurons also display the existence of some species-specific program that seems to prespecify their global cell number as well as their overall genetic potential, such as the ability to form cortical modules that respond predominantly to color or to contrast stimulation. We found that the neurons of the visual cortex in the monkey can form some features of the laminar and modular organization characteristic of macaque cortex even in the absence of any information from the photoreceptors in the retina (Rakic, 1988; Kuljis and Rakic, 1990; Rakic and Lidow, 1992; Rakic et al., 1991). These features include proper lamination, distribution of neurotransmitter receptors and pattern of cytochrome oxidase blobs. Since in these experiments some neurons of the lateral geniculate nucleus survive, though drastically reduced in number, and since they project to the occipital lobe, their role in cortical compartmentalization could not be excluded in any of these experiments.

From the start, it was obvious that geniculocortical input must play an important role in regulating proper cellular differentiation of the primary visual cortex (e.g., Rakic, 1976, 1988). However, initially we could not determine whether cells in the cortex had in their own genome any information concerning their fate and repertoire of their eventual phenotypes. We could address this long-standing problem only after we found that a portion of area 17 in the adult monkey that is deprived of the geniculate afferents in early fetal life nevertheless develops certain cytoarchitectonic features, including formation of a characteristic cytochrome oxidase pattern (Rakic 1992; Rakic et al., 1991). It is, perhaps, significant that cytochrome oxidase blobs in the visual area devoid of the geniculate input were actually more intensely stained and had a somewhat larger diameter. This experiment suggests that the neurons in the restricted portion of the occipital cortex of the macaque monkey, like those of the retina, contain some basic potential for building their species-specific cyto- and chemoarchitecture—the protomap (Rakic, 1988).

The potential for developing the characteristic M and P subsystem, therefore, may be prespecified in the neurons of the entire visual system of the primate brain, but a similar potential may not necessarily exist for the neurons at the corresponding levels in subprimate species. On the basis of results from our experiments, we concluded that neither the retina nor the cerebral cortex can properly differentiate their adult cytoarchitectonic pattern without forming connections with the corresponding neurons in the thalamus. If this is so, do we simply relegate the detailed developmental program of the cerebral cortex to the portion of the dienephalon? Some 15 years ago we observed that, in the early embryonic stages, the primordium of the lateral geniculate nucleus consists of an array of radial units composed of cells that share a common site of origin (Rakic, 1977a). This finding leads to the proposal that such a radial mosaic may represent a protomap of this particular thalamic visual center. Although, experimental evidence for this hypothesis has so far been lacking, the potential for species-specific layering pattern and its subdivision into magnocellular and parvocellular subsystems seems to be intrinsic and species-specific. For example, binocular and chromatic organization of the retinogeniculate systems in new and old world monkeys are remarkably similar, but the cytoarchitectonic organization of the lateral geniculate nucleus in the two species is quite different. The development of these two levels of the visual system are, however, interdependent as the typical layering pattern of the lateral geniculate nucleus would not develop in the absence of retinal input

(Rakic, 1988). The big challenge, therefore, is to find out how the species-specific protomaps—which initially may be established independently in the retina, thalamus and visual cortex—become interconnected and then neurons at each level differentiate into a functionally competent adult cytoarchitectural form.

A COORDINATED DIFFERENTIAL LOSS OF NEURONS AT EACH LEVEL

The third issue that needed to be solved was the control of neuronal number and, most importantly, to determine the proper ratio of various neuronal classes at the retinal, geniculate and cortical level. Our ^{3}H-thymidine autoradiographic analyses described above provided the timetable of neurogenesis in each principle structure in the rhesus monkey visual system but left open the question of the number of cells produced. Therefore, we initiated a series of studies designed to provide detailed quantitative analysis that would provide these data. The results clearly demonstrated that the neurons as well as their axons and synapses are initially overproduced in all visual structures of the macaque where the counts have so far been made (Figure 2). For example, we found that at midgestation the macaque retina contains close to 3,000,000 ganglion cells. Subsequently, about 60% of these cells are eliminated (Rakic and Riley, 1983a). The numbers of other retinal elements in the rhesus monkey, including photoreceptors, horizontal, bipolar and amacrine cells at various embryonic ages are presently not known. The problem with counting these cell classes is that their identification at the early developmental stages, when one could expect them to peak in number, is unreliable with the presently available methods.

Study of the timing, magnitude, and spatial distribution of neuron elimination in the lateral geniculate nucleus shows a loss that is proportionately slightly smaller than in the retina (Figure 2). Thus, before E60, this thalamic nucleus contains about 2,200,000 neurons, 800,000 of which are eliminated over the next 40 to 50 days during the middle third of gestation (Williams and Rakic, 1988). Just to illustrate the magnitude of this loss, it is sufficient to mention that neurons in the lateral geniculate nucleus are eliminated at an average rate of 300 an hour between E48 and E60, and at an average rate of 800 an hour between E60 and E100. Very few neurons are lost after E100, when the population falls to the adult average of 1,400,000 (Figure 2). Degenerating neurons seem to be more common in the magnocellular moiety of the nucleus than in the parvocellular moiety, indicating that these two moieties of this nucleus have a different developmental program. The period of the major wave of cell death occurs before the emergence of the typical six layers of the primate lateral geniculate nucleus, and, importantly, before the establishment of geniculocortical connections and their segregation into ocular dominance columns (Figure 3). Most important, the loss of neurons in the lateral geniculate nucleus begins long before the phase of depletion of retinal axons. This temporal sequence seems to eliminate the hypothesis that cell

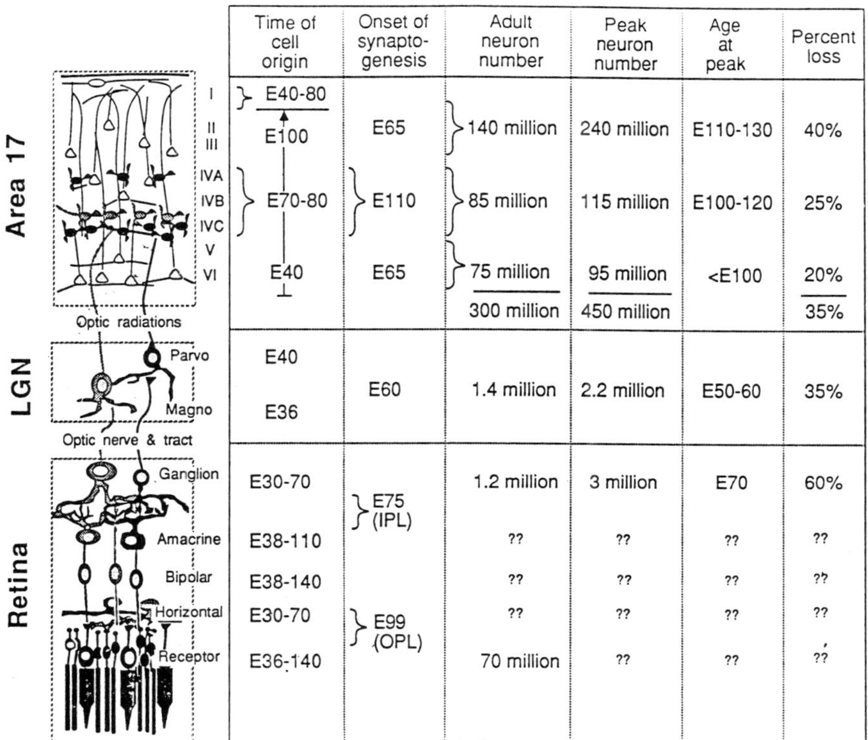

	Time of cell origin	Onset of synapto-genesis	Adult neuron number	Peak neuron number	Age at peak	Percent loss
Area 17: I, II, III	E40-80 / E100	E65	140 million	240 million	E110-130	40%
Area 17: IVA, IVB, IVC	E70-80	E110	85 million	115 million	E100-120	25%
Area 17: V, VI	E40	E65	75 million	95 million	<E100	20%
Area 17: total			300 million	450 million		35%
LGN: Parvo	E40	E60	1.4 million	2.2 million	E50-60	35%
LGN: Magno	E36					
Retina: Ganglion	E30-70	E75 (IPL)	1.2 million	3 million	E70	60%
Retina: Amacrine	E38-110		??	??	??	??
Retina: Bipolar	E38-140		??	??	??	??
Retina: Horizontal	E30-70	E99 (OPL)	??	??	??	??
Retina: Receptor	E36-140		70 million	??	??	??

FIGURE 2. The total number of various neuronal classes present in the primary visual pathway of the developing and adult rhesus monkey. For comparison, the data on the time of origin, onset of synaptogenesis, adult neuron number, peak neuron number, age at the peak and the percent lost are presented for the retina, lateral geniculate nucleus and area 17. For further information see Rakic and Riley, 1983a; Williams and Rakic, 1988; Wikler, Williams and Rakic, 1990. Data on the cell loss in the primary visual cortex are still preliminary. The question marks indicate unavailable data.

death in the retina plays a major role in controlling the size of the entire lateral geniculate or the ratio of neurons in the magno- and parvocellular moiety. Rather, it suggests that both the total number and ratio of magno- and parvocellular components may be regulated to a considerable degree intrinsically and supports the hypothesis that a species-specific protomap may also exist within the lateral geniculate nucleus.

The quantitative study of cell death in the macaque visual cortex is still not fully completed. However, our preliminary results indicate that loss of neurons in this structure ranges between 25% and 40%, depending on the layers (Williams et al., 1987). This magnitude of neuronal loss is in general agreement with the available data for the other mammalian species, including human. However, it should be

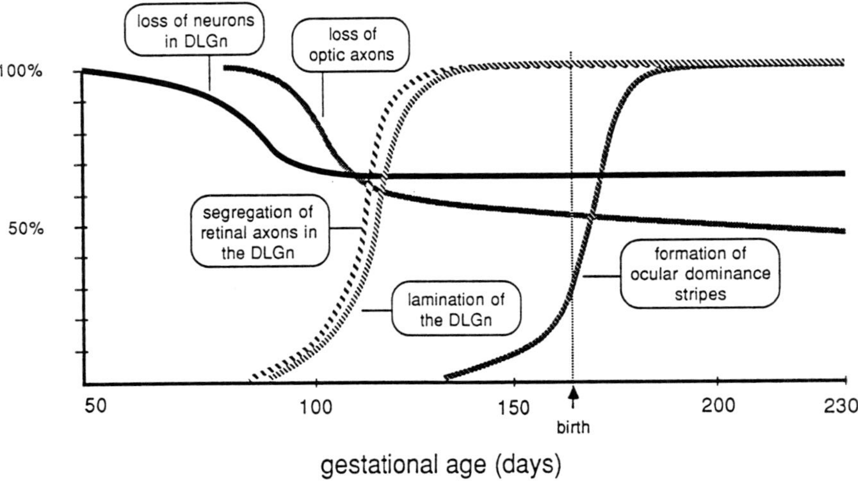

FIGURE 3. Schematic representation of timing of major processes in the development of the primary visual pathway in the rhesus monkey based on the results available in Rakic, 1976; Rakic and Riley, 1983a; and Williams and Rakic, 1988. A particularly important observation is that the loss of neurons in the geniculate nucleus (Black line) is essentially complete by the time the number of optic axons in the nerve begins to fall (gray line).

underscored that, unlike the retina and lateral geniculate nucleus where the direct cells counts were made at each age, in the cortex the estimate of cell elimination depends on the counts of picknotic nuclei, which give less reliable quantitative data for two main reasons. First, we do not know how long each picknotic nucleus remains visible in the fetal tissue; and, second, it is difficult to distinguish between remnants of degenerating neurons and glial cells in such histological material.

TUNING OF SYNAPTIC CONNECTIONS OCCURS THROUGH COMPETITIVE ELIMINATION

The fourth broad area of inquiry in the laboratory concerned the developmental mechanisms involved in the formation of neuronal projections and establishment of their synaptic connections. Injections of the anterograde radioactive axonal tracers into the eyes of monkey fetuses showed that terminations of retinal projections from two eyes in the target structures are initially overlapping and more diffuse than in the adult animals (Rakic, 1976, 1977b, 1986). This phase of intermixed retinogeniculate terminals in rhesus monkey occurs prior to a major wave of synaptogenesis in either structure (Figure 3). Contrary to prevailing views, our electron microscopic results clearly revealed that the pioneer synaptic contacts are formed in the central structures well ahead of the formation of true line projections that proceed from the photoreceptors via bipolars to ganglion cells and further to the cortex via the lateral geniculate (Nishimura and Rakic, 1985, 1987). Furthermore, autoradiographic tracing methods showed that the formation of corticogeniculate projections (Shatz and Rakic, 1981) occur simultaneously with the formation of geniculate input to the cortex (Rakic, 1977), but, most importantly, this reciprocal neuronal system can develop before the establishment of contacts between photoreceptors and bipolar cells in the retina. The topography of the reciprocal geniculocortical system can develop in the absence of both eyes from the early stages (E60) of fetal development (Rakic, 1988).

Subsequent studies revealed that the elimination of the supernumerary axons, synapses and various molecules involved in neural transmission in the visual system occurs during specific time periods but extends up until puberty (Bourgeois and Rakic, 1992; Rakic and Riley, 1983a; Lidow et al., 1991; Rakic et al., 1986). The segregation of synaptic terminals into region-, layer-, column- and cell-specific territories seems to be induced by intracellularly programmed instructions but require further tuning by intercellular competitive interactions (Rakic, 1981; Rakic, 1986; Rakic, 1989; Rakic and Riley, 1983b; Bourgeois et al., 1992). For example, the exponential rate of synaptic production in the cortex starts *in utero* but extends well beyond infancy, for several months after birth (Figure 4). The phase of high synaptic density, with values well above those found in the normal adult, lasts throughout infancy and adolescence and decreases significantly only during sexual maturation, which in this species occurs during the third year of life (Bourgeois and Rakic, 1992; Rakic et al., 1986). We calculated that the number of synapses lost in the striate cortex of a single cerebral hemisphere during this period is about 1.8×10^{11}. The magnitude of this loss is stunning when expressed as loss of about 2500 synapses per each second during a period of about two and one half years (Bourgeois and Rakic, 1992). This phase of primate life obviously may provide the long period of unparalleled opportunity for competitive activity-driven stabilization among initially overproduced various inter- and intracortical connections which comprise the largest fraction of cortical synapses.

The decline in synaptic density in the visual cortex is due primarily to elimination of asymmetrical junctions located on dendritic spines, while synapses on dendritic shafts and cell bodies seems to remain relatively constant during the entire postnatal life (Bourgeois and Rakic, 1992; Rakic et al., 1986). It may be significant that the density of all so-far-examined neurotransmitter receptors in the visual cortex also reaches a maximum level between two and four months of age and then declines to the adult level during the period of sexual maturation (Lidow et al., 1991). For example, this synchronized development of the two dopamine receptor subtypes occurs in the monkey *pari pasu* with the course of synaptogenesis (Figure 5). A similar developmental course was found for all major neurotransmitters so far examined. It may be significant that decrease in receptor

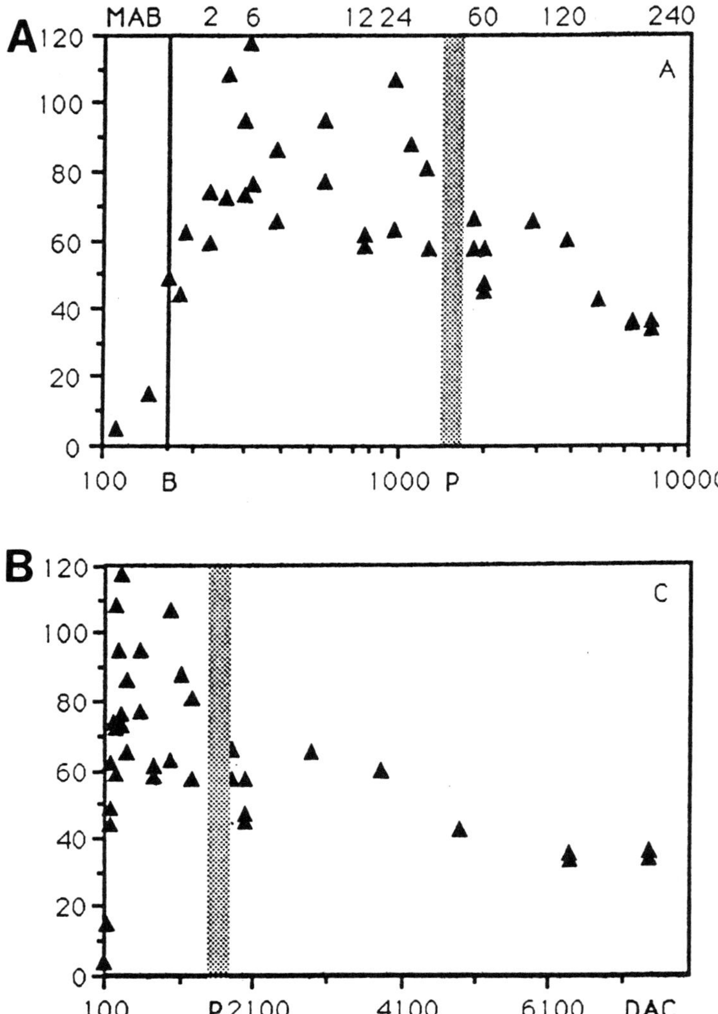

FIGURE 4. Histograms of the density of synapses per 100 cubic microns of *neuropil* in the primary visual cortex of the rhesus monkey at various pre- and postnatal ages. Each triangle represents the value obtained from an uninterrupted electron microscopic montage consisting of over 100 photographs taken across the entire depth of the visual cortex. Times on the upper graph is represented on semi-logarithmic scales. The more rapid decline in synaptic density around the time of puberty (P) is more explicit when the time is given on a linear scale (lower graph). The abbreviation MAB denotes months after birth; DAC denotes days after conception; B denotes birth; and P denotes puberty (From Bourgeois and Rakic, 1992).

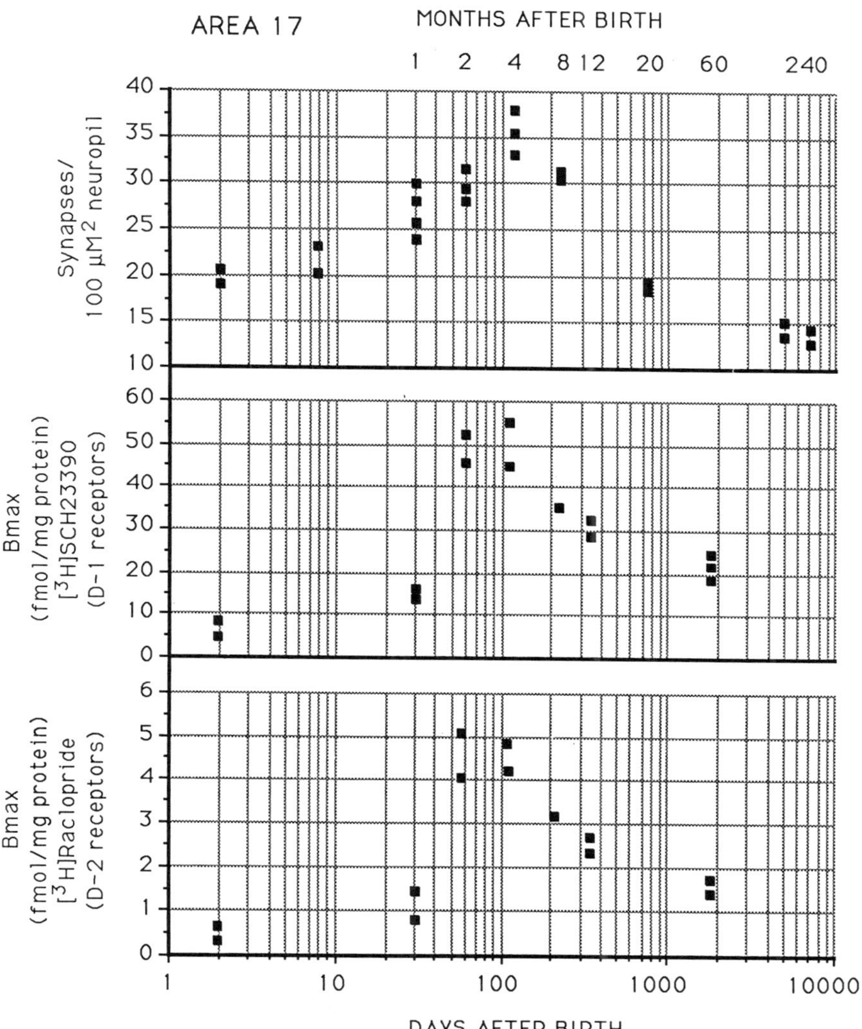

FIGURE 5. Developmental changes in the overall (across all layers) density of the specific binding of radioligands [^{3}H]SCH233890 and [^{3}H]raclopride labeling D_1 and D_2 dopamine receptor subtypes respectively in the primary visual cortical region of the rhesus monkey of different ages. The figure also includes developmental changes in synaptic density in the same regions (adapted from Rakic et al., 1986). For receptor densities, the lines were obtained by locally weighted least square fit with 50% smoothing (KALEIDA GRAPH, Synergy Software, Reading, PA) based on mean B_{max} values obtained from the measurements of the entire cortical thickness in at least two animals at birth, 1, 2, 4, 8, 12, 36 and 60 months of age. Age is presented in postnatal days on a logarithmic scale (adapted from Lidow, Goldman-Rakic and Rakic, 1991).

density during adolescence precedes the phase of synaptic elimination, suggesting that formation of biochemically competent neurotransmission may be a limiting determinant of the final level of synaptic density.

FUNCTIONAL VALIDATION MAY ENHANCE SYNAPTIC STABILIZATION

The findings described above were obtained by different methods, but reveal an unusual degree of coordination between biochemical and structural differentiation, suggesting the possibility that some cellular events may be related to maturation of visual function. The capacity to perform various visual tasks coincides with the time when synaptic density reaches its peak, which suggests that a critical mass of cortical synapses is essential for cognitive functions to emerge. However, the full maturation of such functions may depend upon the elimination of excess synapses that occurs during infancy and adolescence. It seems reasonable to hope that the correlative analysis of the structural-functional relationship after birth may eventually provide instructive insight into cognitive maturation and various genetic and acquired developmental disorders (e.g., Goldman and Rakic, 1979). Considerable progress has been made in the past decade, but here I will try to limit my description to our findings in the visual system.

Contrary to our initial expectation, the phase of ocular-dominance column formation, which begins during the last three weeks of gestation and continues during the first two postnatal months, is associated with an increase rather than decrease in synaptic density in the visual cortex (Bourgeois and Rakic, 1992). Furthermore, premature visual stimulation of the infant monkeys delivered before term does not affect the rate of synaptic accretion or the size, topology, and laminar distribution of synapses in their visual cortex (Bourgeois et al., 1989). Rather, the morphological and ultrastructural parameters of synapses and their density develop according to the time of conception and seem not to be influenced by the time of birth. These results suggest that visual stimulation in infancy may affect the visual cortex predominantly through strengthening, modifying, and eliminating synapses already formed rather than regulating the rate of their production.

Presently, we are studying the plasticity of neural connections and the modifiability of visual cortical maps, as well as the changing pattern of distribution of various molecules by early lesion-induced reduction of specific afferent inputs (e.g., Rakic and Lidow, 1992; Rakic et al., 1991) or by elimination of specific cortical layers using ionizing radiation to eliminate selected neuronal precursors during mitotic division (Algan and Rakic, 1992). The initial results indicate that lesion of a synaptically-related structure or reduced input to a given visual center, occurring during a critical developmental period, may exert changes that in turn influence subsequent developmental events in other related structures, and thereby provide the setting for new patterns of neural and synaptic relationships. However, it appears that specific afferents can induce only those aspects of cell differentiation for which there is a responsive genetic potential. For example, M- and P-subsystem may be induced in cortical neurons in the primate occipital lobe, because cells at each level of the visual system possess potential to develop appropriate circuitry that can subserve these particular systems.

The concept of competitive interactions within each level and between cells at various levels of the visual system need to be verified by experiments involving the selective destruction of visual centers and/or specific pathways in monkey fetuses, neonates and juveniles, and by subsequent analysis of the effects of these procedures on the pattern of axonal pathways, architecture of synaptic contacts and distribution of neurotransmitters and their receptors. We now recognize that these interactions may begin before birth, and therefore may be influenced by spontaneous electrical activity, but that they clearly extend also postnatally and actually continue until the time of puberty when most structural and biochemical properties in primates can be stabilized through visual stimulation. The steady state of synaptic density seems to last throughout more than two decades of adult life in this species, and only at the end of the life span begins to fall again. These quantitative statements are, however, made on the basis of a relatively small number of specimens and their significance need to be confirmed on a larger sample.

The major theme of contemporary research in visual sciences is related to competitive neuronal interactions and to the possible molecular mechanisms underlying these changes. In spite of considerable advances made in the past few years, our understanding of the role of synaptic overproduction in the development of vision at the cellular and molecular level is still rather vague. It seems clear, however, that simple and seemingly straightforward hypotheses will not suffice to explain this enormously complex phenomenon. When one examines temporal sequence and dynamics of developmental events in the entire visual system, from the photoreceptors to the cortical modules, it becomes evident that the overly narrow approach usually falls apart. Ablations of either peripheral or central neuronal structures suggests the existence of complex and interdependent competitive two-way interactions between various levels and between subsystems. Therefore, in-depth analysis of cellular events and the consequences of experimental modification of the visual system from different angles and points of view that are presented at this meeting may set the stage for identification of general biological principles as well as detailed genetic molecular mechanisms involved in synaptic specification of diverse neuronal populations involved in human vision.

Acknowledgments. This series of studies has been supported by the United States Public Health Service. Developing rhesus monkeys were obtained in part from the New England Primate Research Center, Southborough, Massachusetts and the breeding colony at Yale University Medical School.

REFERENCES

Algan O, Goldman-Rakic P S, Rakic P (1992): Selective laminar and areal deletion of cortical neurons by prenatal x-irradiation in macaque monkey. *Abs. Soc. Neurosci.* 18

Bourgeois J.-P., Rakic P (1992): Changing of synoptic density in the primary visual cortex of the rhesus monkey from fetal to adult stage. *J. Neurosci.*, submitted

Bourgeois J.-P., Jastreboff P, Rakic P (1989): Synaptogenesis in the visual cortex of normal and preterm monkeys: Evidence for intrinsic regulation of synaptic overproduction. *Proc. Nat. Acad. Sci. USA* 86:4297–4301

Easter S S, Purves D, Rakic P, Spitzer N C (1985): The changing view of neural specificity. *Science* 230:507–511

Goldman P S, Rakic P (1979): Impact of the outside world upon the developing primate brain. Perspective from neurobiology. *Bull. Menninger Found.* 43:20–28

Kostovic I, Rakic P (1980): Cytology and time of origin of interstitial neurons in the white matter in infant and adult human and monkey telencephalon. *J Neurocytol.* 9:219–242

Kostovic I, Rakic P (1984): Development of prestriate visual projections in the monkey and human fetal cerebrum revealed by transient cholinesterase staining. *J. Neurosci.* 4:25–42

Kostovic I, Rakic P (1990): Developmental history of transient subplate zone in the visual and somatosensory cortex of the macaque monkey and human brain. *J. Comp. Neurol.* 297:441–470

Kuljis RO, Rakic P (1990): Hypercolumns in primate visual cortex develop in the absence of cues from photoreceptors. *Proc. Nat. Acad. Sci. USA* 87:5303–5306

LaVail M M, Rapaport D H, Rakic P (1991): Cytogenesis in the monkey retina. *J. Comp. Neurol.* 309:86–114

Lidow M S, Goldman-Rakic P S, Rakic P (1991): Synchronized overproduction of neurotransmitter receptors in diverse regions of the primate cerebral cortex. *Proc. Nat. Acad. Sci. USA* 88:10218–10221

Nishimura Y, Rakic P (1985): Development of the rhesus monkey retina: I. Emergence of the inner plexiform layer and its synapses. *J. Comp. Neurol.* 241:420–434

Nishimura Y, Rakic P (1987): Development of the rhesus monkey retina: II. A three-dimensional analysis of the sequences of synaptic combinations in the inner plexiform layer. *J. Comp. Neurol.* 262:290–313

Rakic P (1972): Mode of cell migration to the superficial layers of fetal monkey neocortex. *J. Comp. Neurol.* 145:61–84

Rakic P (1974): Neurons in the monkey visual cortex: Systematic relation between time of origin and eventual disposition. *Science* 183:425–427

Rakic P (1975): Timing of major ontogenetic events in the visual cortex of the rhesus monkey. In: *Brain Mechanisms in Mental Retardation*. Buchwald N A, Brazier M, eds. New York: Academic Press

Rakic P (1976): Prenatal genesis of connections subserving ocular dominance in the rhesus monkey. *Nature* 261:467–471

Rakic P (1977a): Genesis of the dorsal lateral geniculate nucleus in the rhesus monkey: site and time of origin, kinetics of proliferation, routes of migration and pattern of distribution of neurons. *J. Comp. Neurol.* 176:23–52

Rakic P (1977b): Prenatal development of the visual system in the rhesus monkey. *Phil. Trans. Roy. Soc. Lond. B.* 278:245–260

Rakic P (1981): Development of visual centers in the primate brain depends on binocular competition before birth. *Science* 214:928–931

Rakic P (1982): Early developmental events: cell lineages, acquisitions of neuronal positions, and areal and laminar development. *Neurosci. Res. Prog. Bull.* 20:439–451
Rakic P (1986): Mechanism of ocular dominance segregation in the lateral geniculate nucleus: competitive elimination hypothesis. *Trends in Neurosci.* 9:11–15
Rakic P (1988): Specification of cerebral cortical areas. *Science* 241:170–176
Rakic P (1989): Competitive interactions during neural and synaptic development. In: *From Neuron to Reading*. Galaburda A M, ed. Cambridge: MIT Press
Rakic P (1990): Principles of neuronal migration. *Experientia* 46:882–891
Rakic P (1992): Determinants of cortical cytoarchitecture. *J. Neurosci.*, in press
Rakic P, Lidow M S (1992): Distribution and density of neurotransmitter receptors in the absence of retinal input from early embryonic stages. *J. Neurosci.*, submitted
Rakic P, Riley K P (1983a): Overproduction and elimination of retinal axons in the fetal rhesus monkey. *Science* 209:1441–1444
Rakic P, Riley K P, (1983b): Regulation of axon numbers in the primate optic nerve by prenatal binocular competition. *Nature* 305:135–137
Rakic P, Bourgeois J-P, Eckenhoff M E, Zecevic N and Goldman-Rakic P S (1986): Concurrent overproduction of synapses in diverse regions of the primate cerebral cortex. *Science* 232:232–235
Rakic P, Gallager D and Goldman-Rakic P S (1988): Areal and laminar distribution of major neurotransmitter receptors in the monkey visual cortex. *J. Neurosci.* 8:3670–3690
Rakic P, Suñer I and Williams R W (1991): A novel cytoarchitectonic area induced experimentally within the primate visual cortex. *Proc. Nat. Acad. Sci. USA* 88:2083–2087
Rapaport D H, Fletcher J T, LaVail M M, Rakic P (1992): Genesis of neurons in the retinal ganglion cell layer of the monkey. *J. Comp. Neurol.*, in press
Shatz C, Rakic P (1981): The genesis of efferent connections from the visual cortex of the fetal rhesus monkey. *J. Comp. Neurol.* 196:287–307
Sidman R L, Rakic P (1982): Development of the human central nervous system. In: *Histology and Histopathology of the Nervous System*. Haymaker W and Adams R D, eds. CC Thomas
Wikler K C, Rakic P (1990): Distribution of photoreceptor subtypes in the retina of diurnal and nocturnal primates. *J. Neurosci.* 10:3390–3400
Wikler K C, Rakic P (1991): Emergence of the photoreceptor mosaic from a protomap of early-differentiating cones in the primate retina. *Nature* 352:397–400
Wikler K C, Williams R W, Rakic P (1990): Photoreceptor mosaic: Number and distribution of rods and cones in the rhesus monkey retina. *J. Comp. Neurol.* 297: 499–508
Williams R W, Rakic P (1988): Elimination of neurons in the rhesus monkey's lateral geniculate nucleus during development. *J. Comp. Neurol.* 272:424–436
Williams R W, Rakic P (1991): Growth cone distribution patterns in the optic nerve of fetal monkeys: Implications for mechanisms of axonal guidance. *J. Neurosci.* 11:1081–1094
Williams R W, Ryder K, Rakic P (1987): Emergence of cytoarchitectonic differences between areas 17 and 18 in the developing rhesus monkey. *Abstr. Soc. Neurosci.* 13: 1044

Section 1

Genesis

1

Cellular and Molecular Mechanisms Regulating Retinal Cell Differentiation

RUBEN ADLER

SOME QUESTIONS ABOUT NEURONAL DIFFERENTIATION IN THE RETINA

The adult neural retina is characterized by both *diversity* and *order*. Diversity is recognized in the presence of different cell types, including the glial cells of Muller, ganglion, horizontal, bipolar, interplexiform, and amacrine neurons, and the highly specialized photoreceptor cells. Several of these cell types can, in turn, be subdivided into two or more subpopulations (i.e., photoreceptors can be rods or cones, and the latter can be subdivided into blue, green, or red based on their visual pigments and sensitivity to light). The complex adult retina is generated, during embryonic development, from a very simple and apparently homogeneous population of mitotically active, neuroepithelial cells. The life history of each adult retinal cell begins with several rounds of mitotic division, the last one of which is considered to represent the "birth" of that cell. After becoming postmitotic, each cell migrates (or translocates) radially to form one of the retinal cell layers, where it undergoes a process of differentiation. The ordered organization of the retina stems from the remarkable correlation between the position occupied by each cell in one of the retinal layers, on the one hand, and the differentiated properties that it expresses, on the other.

One of the goals of contemporary developmental neurobiology is to understand the cellular and molecular mechanisms through which a simple, apparently homogeneous, embryonic neuroepithelium is transformed into a complex, heterogeneous, and orderly tissue such as the adult retina. The nature of the inductive events that determine the fate of each cell remains undisclosed, and it is not even known whether they act before, at the time of, or some time after the cell's terminal mitosis. The transition from the undifferentiated state of a precursor cell to the highly differentiated phenotype of a mature neuron or photoreceptor could conceivably require a cascade of inductive events (each one of them regulating the expression of one, or at most a few, differentiated properties); alternatively, an early inductive event could set in place a "master program" regulating the expression of many, if not most, of the properties that characterize each particular cell type. Recent experimental approaches to these questions have involved (1) tracing the progeny of individual precursor cells labeled with retroviruses, horseradish peroxidase, or fluorescent molecules (Holt et al., 1988; Turner and Cepko, 1987; Wetts and Fraser, 1988); (2) investigating the expression of cell-specific antigens during retinal development *in vivo* (i.e., Knight and Raymond, 1990; Larison and Bremiller, 1990; McLoon and Barnes, 1989; Wikler and Rakic, 1991; among others), and (3) investigating the differentiation of precursor cells in reaggregation or monolayer cultures (Adler and Hatlee, 1989; Adler et al., 1984; Araki, 1984; Reh and Kljavin, 1989; Taylor and Reh, 1990; Watanake and Raff, 1990). While the mechanisms that regulate the emergence of diversity and order during retinal development have not yet been elucidated, these studies have stimulated new research efforts through the formulation of various working hypotheses amenable to experimental verification. Given that the goal of this chapter is to summarize recent studies from our own laboratory, it will not be possible to review in detail the recent literature in this field; however, a brief comparison of our results with those from other laboratories will be presented at the end of this chapter.

LOW-DENSITY RETINAL CELL CULTURES AS AN EXPERIMENTAL SYSTEM TO STUDY CELL DIFFERENTIATION

A series of methods for the generation of retinal cell cultures have been developed in various laboratories (Araki, 1984; Carri and Ebendal, 1983; Combes et al., 1977; Hicks and Courtois, 1988; Johnson et al., 1986; Linser and Moscona, 1979; MacLeish and Townes-Anderson, 1988; Moscona, 1965; Puro, 1983; Schubert and LaCorbiere, 1986; Sparrow et al., 1990; Spoerri et al., 1987; Townes-Anderson et al., 1988). Our own protocols allow the generation of purified, low-density cultures of chick embryo retinal neurons and photoreceptors (for a review, see Adler, 1990). The chick embryo neural retina is devoid of connective tissue and blood vessels, and can be cleanly isolated from other ocular tissues, including the pigment epithelium; therefore, chick retinal cell cultures are free of contamination with nonneural cells, such as fibroblasts or endothelial cells. In addition, glial cells fail to develop under our culture conditions. Also relevant in the context of the studies discussed here is that the retinal cells used in our experiments are isolated at early embryonic stages, when there is very little cell differentiation in this tissue. For example, very few neurons, and no photoreceptors, can be recognized in the chick retina even on embryonic day (ED) 8, the oldest stage used for our experiments. Another important feature of our technique is the use of polyornithine-coated, highly adhesive substrata, to which the dissociated retinal cells attach readily as individual units, free of intercellular contacts. This allows testing their developmental potential in the absence of contact-mediated cell interactions.

That this experimental system is suitable for cell differentiation studies is suggested by the dramatic changes in cellular composition that can be detected in the cultures during the first week *in vitro*. At the beginning of the *in vitro* period the

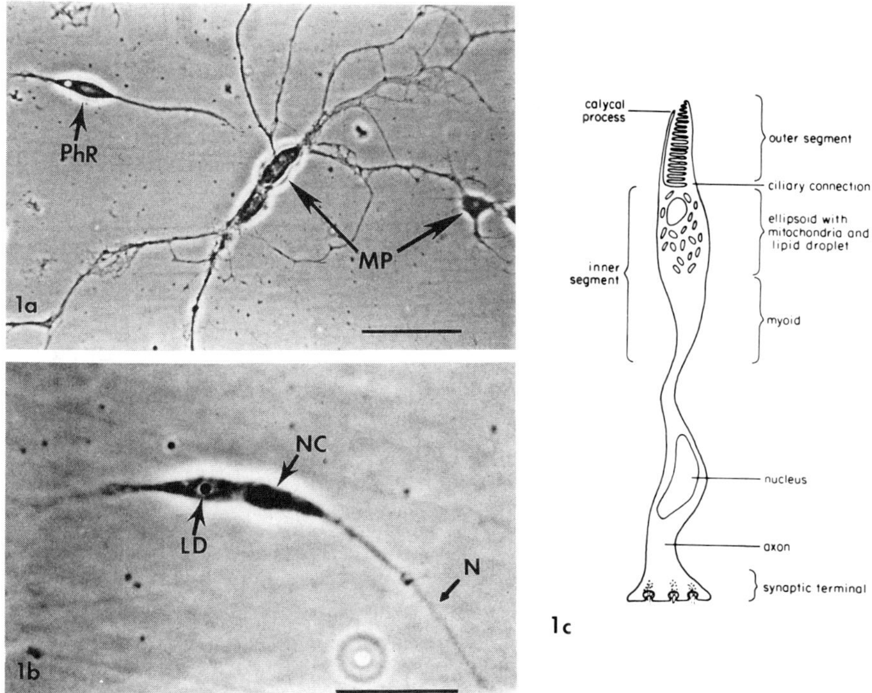

FIGURE 1.1. A. Culture of dissociated cells from 8-day chick embryo retina, grown *in vitro* for 6 days. Multipolar neurons (MP) and a photoreceptor (PhR) are seen. Note that the photoreceptor is free of contacts with other cells. B. Cultured chicken photoreceptor showing an elongated, polarized, compartmentalized organization with a short neurite (N), a nuclear compartment (NC), and an inner segment containing a lipid droplet (LD). Ultrastructural and immunocytochemical studies have also shown the presence of a small outer segment-like process. Magnification bar: 30 μm. C. Diagram of a chick cone photoreceptor as seen *in vivo*. Reprinted with permission of Academic Press from Madreperla SA, Adler R (1989): Opposing microtubule—and actin—dependent forces in the development and maintenance of structural polarity in retinal photoreceptors. *Dev Biol* 131:149–160.

cultures appear as a morphologically homogeneous population of process-free "round" cells; a few days later, however, the same cultures appear very heterogeneous, and it becomes possible to distinguish two main differentiated cell types (see Figure 1.1). Some of these cells have a typical neuronal appearance, with a large cell body and several long neurites. The remaining differentiated cells are the photoreceptors, which under phase contrast microscopy appear as elongated, polarized cells with a single, short neurite and a characteristic lipid droplet (see Figure 1.1). This divergent differentiation of *isolated* precursor cells as either neurons *or* photoreceptors provides an indication of the possible role of intracellular determinants in the control of retinal cell differentiation, thus stimulating the investigation of a putative "photoreceptor master program" of development.

PHOTORECEPTOR DIFFERENTIATION *IN VITRO*, AND THE CONCEPT OF A DEVELOPMENTAL MASTER PROGRAM

The Phenotype Expressed by Isolated Photoreceptors in vitro *Is Very Complex*

The notion that cell differentiation can be regulated by "master genes" controlling the expression of many cell-specific properties has been repeatedly demonstrated in studies with invertebrates, and there is growing evidence that similar mechanisms are also present in higher animals (for a review, see Davidson, 1991). Against this background, the possible existence of a "photoreceptor master program" is suggested by the many similarities between the properties expressed by *isolated* photoreceptor precursors developing *in vitro*, and those present in normal photoreceptors within the intact retina. *In vivo*, cones and rods share a basic plan of organization, but they also have several distinctive features. Mature photoreceptors of both types appear as elongated, compartmentalized, highly polarized cells (see Figure 1.1). Their apical end is the outer segment, which contains stacks of visual pigment-rich membranous disks, as well as a series of second messenger molecules involved in phototransduction. A modified cilium connects the outer segment to the inner segment, where the organelles involved in synthetic activities and energy production are located. The next compartment is the cell body, which is occupied almost exclusively by the nucleus. Finally, photoreceptors have a very short axon, ending in a highly specialized synaptic terminal. Photoreceptors are polarized not only morphologically but also at the molecular level, as illustrated by the accumulation of visual pigments in the outer segment, and of Na^+,K^+-ATPase in the plasma membrane of the inner segment. Features differentiating rods from cones include the shape and morphological organization of their outer segments, which also contain different second messenger molecules and visual pigments. The latter (rhodopsin in rods, one of several cone opsins in cones) are responsible for the particular spectral sensitivity of each type of photoreceptor cells. In many animal species (including chickens) cone inner segments have a conspicuous oil droplet that is not present in the rods.

Evidence gathered during the last few years through the analysis of low-density chick embryo retinal cell cultures has shown that the isolated precursor cells that differentiate as photoreceptors *in vitro* express many cell type-specific properties, which are very different from those acquired by neighboring cells that differentiate as neurons. In cultures grown *in vitro* for 4–7 days (see Figure 1.1), cultured photoreceptors resemble their *in vivo* counterparts in their highly elongated and polarized organization (Adler et al., 1984). Molecular polarity has been recognized in the accumulation of opsin immunoreactive materials in a rudimentary outer segment (Adler, 1986a), and of Na^+,K^+-ATPase in the plasma membrane of the inner segment (Madreperla et al., 1989; see Figure 1.2) As their *in vivo*

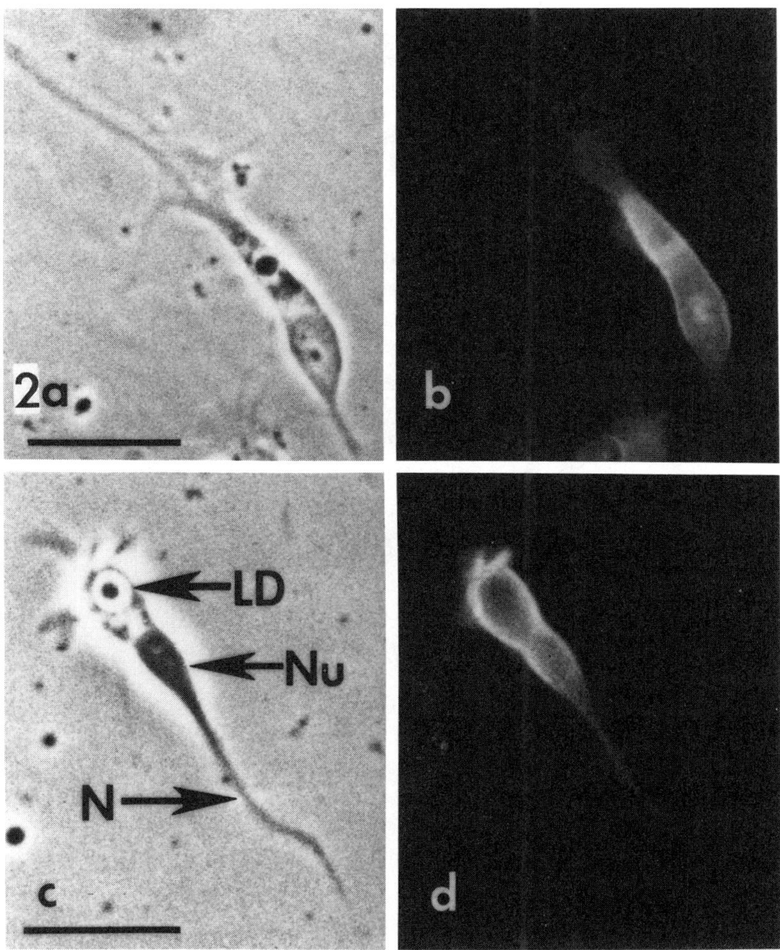

FIGURE 1.2. Na^+,K^+-ATPase and spectrin immunoreactivity in photoreceptors cultured for 6 days. A,B. Cell immunoreacted with an antiserum against ATPase. C,D. Cell immunoreacted with an antiserum against spectrin. A,C. Phase contrast microscopy. B,D. Immunofluorescence. Photoreceptors appear highly polarized, and are subdivided in compartments including the short neurite (N), the cell body, occupied by the nucleus (Nu), and the inner segment with a conspicuous lipid droplet (LD). Both ATPase and spectrin immunoreactive materials are concentrated in the inner segment region as well as over the apical portion of the nuclear compartment. A variable amount of membranous-appearing material is present above the inner segment region of these cells which does not show immunoreactivity for either the ATPase or spectrin. Reprinted with permission of The Rockerfeller University Press from Madreperla SA et al. (1989): Na^+, K^+-adenosine triphosphatase polarity in retinal photoreceptors: A role for cytoskeletal attachments. *J Cell Biol 109:1483–1493*.

counterparts, cultured photoreceptors display a high affinity uptake mechanism for the excitatory neurotransmitter glutamate; on the other hand, aminobutyric acid (GABA) is restricted to nonphotoreceptor neurons both *in vivo* and in the cultures (Politi and Adler, 1986). Biochemical evidence has also suggested that the cultured photoreceptors contain the rate-limiting enzyme in melatonin synthesis, serotonin N-acetyltransferase, which can be regulated by dopamine and by cyclic nucleotides in a manner reminiscent of the *in vivo* situation (Iuvone et al., 1990). The divergent differentiation of neurons and photoreceptors can also be recognized through their susceptibility or resistance to pathological agents. Thus, most of the cultured retinal neurons can be infected by *Herpes simplex* virus and are destroyed by excitotoxic amino acids such as kainic acid; the photoreceptors present in the same cultures are not affected by these agents (Politi and Adler, 1986; Politi et al., 1987). These and related studies have shown that some retinal precursor cells, isolated from the embryo before the onset of overt differentiation, can express many characteristic photoreceptor properties even when developing in the absence of contact-mediated intercellular interactions. Other precursor cells, developing in the same environment and under similar conditions, mature as neurons and express a different set of phenotypic properties.

Intracellular, Cytoskeleton-Dependent Mechanisms Regulate the Development and Maintenance of Photoreceptor Polarity

How do isolated process-free, round-shaped precursor cells give rise to elongated, polarized photoreceptors *in vitro*? Using sequential photomicrographic analysis of identified cultured cells, we were able to show that photoreceptor morphogenesis involves a series of stereotyped changes, which begin after approximately 24 hr in culture with the appearance of an oil droplet in some of the precursor cells (Madreperla and Adler, 1989). This is followed by the emergence of a short neuritelike process at a site opposite to the oil droplet, and by the onset of inner segment elongation. Cultured photoreceptors attain their mature morphological configuration some 24–36 hr after the beginning of their morphogenesis. Immunocytochemical analysis of cells fixed at various stages of their morphological transformation (Madreperla et al., 1989) shows that Na^+,K^+-ATPase immunoreactivity is diffusely and homogeneously distributed in the round photoreceptor precursors present at early culture stages, but becomes polarized toward the inner segment as the photoreceptors elongate (Madreperla et al., 1989). The occurrence of similar and concomitant changes in the immunocytochemical localization of the cytoskeleton-associated protein spectrin suggests that this molecule could play a role in the polarized distribution of Na^+,K^+-ATPase by anchoring it to the photoreceptor inner segment subcortical cytoskeleton, much as has been suggested for asymmetrically distributed plasma membrane components in other cells (i.e., Nelson and Veshnock, 1986). Consistent with this possibility if the observation that, as ATPase and spectrin become localized to the inner segment region during photoreceptor elongation, there are concomitant *increases* in the fraction of immobile ATPase molecules (as detected by fluorescence recovery after photobleaching), and in the resistance of ATPase immunoreactive materials to extraction by detergents (Madreperla et al., 1989). Both results imply that the sequestration of Na^{+},K^{+}-ATPase to the inner segment region of the plasma membrane is develped and maintained thanks to intracellular, cytoskeleton-dependent mechanisms expressed by isolated photoreceptor precursors in an apparently autonomous manner.

Cytoskeleton-dependent intracellular forces also appear to play crucial roles in the development and maintenance of the polarized morphology of isolated photoreceptor cells (Madreperla and Adler, 1989). When "mature" cultured photoreceptors are exposed to microtubule-depolymerizing drugs, such as nocodazole, they quickly lose their elongated, polarized organization, and revert to a configuration resembling that normally found in undifferentiated precursor cells at culture onset (see Figure 1.3). Remarkably, these changes are fully reversible: if the cultures are switched back to nocodazole-free medium, the "collapsed" photoreceptors elongate again, regenerating a pattern of organization identical to the one they had prior to nocodazole treatment. The actin-depolymerizing drug cytochalasin D causes the opposite response, that is, it leads to exaggerated (and also reversible) photoreceptor elongation. These observations suggest that the structural polarity of photoreceptor cells is maintained through an equilibrium between constantly active microtubule-dependent forces, which tend to elongate the cells, and actin-dependent forces, which tend to shorten the cells (Madreperla and Adler, 1989). Experiments with the same cytoskeletal inhibitors suggest that similar forces are also responsible for the transformation of round precursors into elongated photoreceptors during their normal morphogenesis *in vitro*. Two salient features of these mechanisms are, first, their autonomous expression, and second, the apparent "memory mechanism" that allows the cells to regenerate their highly complex pattern of structural organization even after they have been profoundly disorganized through treatments with cytoskeletal inhibitors.

Isolated Cultured Photoreceptors Become Capable of Functional Responses to Light

Photoreceptor responses to light are not limited to phototransduction in the intact organism. In many vertebrates, for example, rod photoreceptors elongate markedly in response to light but contract when returned to darkness, while cones show the opposite responses (for reviews, see Burnside and Dearry, 1986; Dearry and Burnside, 1989). These "photomechanical" behaviors are also affected by neuromodulators such as melatonin (involved in adaptive responses to darkness) and dopamine (which seems to mediate the effects of light).

We have recently observed that isolated photoreceptor cells that differentiate *in vitro* also develop light responsiveness, recognizable through photomechanical behaviors (Stenkamp and Adler, 1990). When retinal cell cultures are maintained in a cycle of 12 hr light/12 hr darkness, most of the photoreceptors elongate during the light phase and contract when exposed to darkness, while a smaller population shows the opposite responses (i.e., elongation in the dark and contraction in

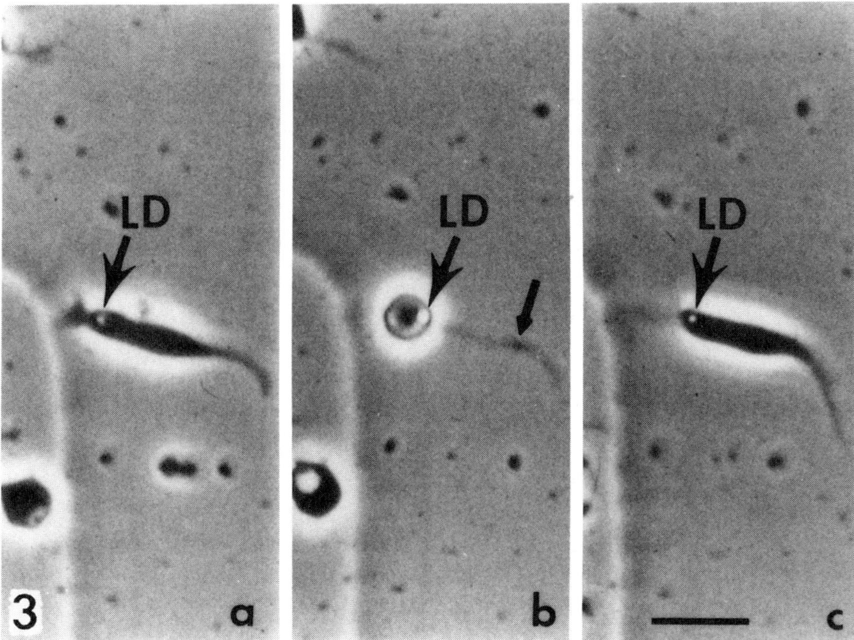

FIGURE 1.3. Reversible effects of nocodazole upon photoreceptor cell length. An elongated photoreceptor is shown before treatment in A. B. Twelve hours after the addition of 2 M nocodazole the cell has lost its elongated configuration, and shows a circular outline. The nucleus has apparently moved to a position more apical than that of the lipid droplet (LD). Some materials appear to remain attached to the substratum in the area previously occupied by the photoreceptor cell body (arrow in (b)). C. The same cell 24 hr after the nocodazole was removed by three 5-min washes with fresh medium; the photoreceptor appeared to have regained an elongated, polarized morphology very similar to that before treatment. Magnification bar: 15 m. Reprinted with permission of Academic Press from Madreperla SA, Adler R (1988): Opposing microtubule—and actin—dependent forces in the development and maintenance of structural polarity in retinal photoreceptors. *Dev Biol* 131:149–160.

response to light). Photoreceptors repeat these photomechanical changes over several light/dark cycles. Pharmacological studies have suggested the involvement of dopamine and melatonin in these responses, in a manner reminiscent of their role *in vivo* (Stenkamp et al., 1990). These observations demonstrate that photoreceptor differentiation *in vitro* includes not only structural and molecular features, but also the development of complex functional activities.

THE "PHOTORECEPTOR DEFAULT HYPOTHESIS"

The final pattern of differentiation expressed by each retinal precursor cell *in vivo* correlates very accurately with its position within one of the retinal layers (see above). At least two different scenarios can in principle be proposed to explain this correlation. One scenario is that each cell becomes specified at the time of terminal mitosis regarding *both* the phenotype that it should express and the retinal layer to which it should migrate. The alternative scenario proposes that cells remain plastic for some time after terminal mitosis, with their differentiated fate being determined by position-dependent inductive signals to which they become exposed as they migrate within the retina. While the distinction between these two possibilities remains unsettled, several experimental results from our laboratory are more consistent with the second scenario, and furthermore suggest that, in the chick embryo, photoreceptor differentiation represents a constitutive or "default" pathway for retinal precursor cells.

Given that photoreceptor cells are among the last cells to be born and differentiate in the chick embryo retina (Grun, 1982), one would predict that the frequency with which precursor cells would differentiate as photoreceptors *in vitro* would be lower in cultures from younger than older embryos. Our experimental results are exactly opposite to this prediction (Adler and Hatlees, 1989). While over 80% of the differentiated cells become neurons and only 15–20% become photoreceptors in cultures from ED8 neural retina, photoreceptor cells represent the predominant differentiated cell type in cultures from younger embryos (approximately 80% in ED6 cultures, and almost 100% in ED5 cultures). One possible explanation for these surprising results would be the selective degeneration of nonphotoreceptor cells in the younger cultures. This mechanism does not appear very likely for two reasons. One is that overall rates of cell survival are not conspicuously different in cultures from different developmental stages. More importantly, selective cell survival could hardly lead to the *premature* appearance of photoreceptor cells in cultures from younger embryos. An alternative hypothesis to explain our results could be that precursor cells have a constitutive or "default" tendency to differentiate as photoreceptors. The "photoreceptor default hypothesis" (Adler and Hatlee, 1989) postulates that: (1) retinal precursor cells remain plastic after terminal mitosis; (2) their fate during normal develpment is determined by inductive signals present in the retinal microenvironment with which they interact during their migration; and (3) photoreceptor differentiation represents a constitutive or "default" pathway. This pathway would be followed by cells that are not exposed to these inductive intraretinal signals either because they do not migrate toward the inner retina (as is the case with precursors that differentiate as photoreceptors *in vivo*), or because they are isolated from the retinal microenvironment prior to their migration (as in the *in vitro* experiments reported above).

While the validity of the various components of this hypothesis still remains to be established, some experimental results are consistent with prediction derived from it. One set of experiments showed that the *in vitro* fate of precursor cells "born" *in vivo* by ED5 varies as a function of the developmental stage at which they are isolated from the retina (Adler and Hatlees, 1989). In these experiments the time of cell birth is determined by injecting [^{3}H]-thymidine into the embryos on ED5, and then repeating the administration of this DNA precursor both *in vivo* and *in vitro* as frequently as necessary in order to ensure that the only cells that will remain autoradiographically unlabeled are those that are already postmitotic by ED5. The key observation in these cultures is that the unlabeled cells give rise predominantly to photoreceptors when isolated from the embryo on ED6, but differentiate as neurons when they are isolated for culture on ED8.

Another prediction derived from this hypothesis that has been experimentally verified is that isolated retinal precursor cells that undergo their terminal mitosis *in vitro*, in the absence of contact-mediated cell interactions, should follow the default pathway and differentiate as photoreceptors. To test this prediction the cells are dissociated from *unlabeled* ED5–ED8 chick embryo retinas, and are grown in ^{3}HT thymidine-containing medium (Repka and Adler, 1989). This allows identifying the cells that divide one or more times *in vitro*, become postmitotic, and differentiate. The frequency of photoreceptors among the differentiated cells born *in vitro* is almost 100% in ED5 cultures, and is higher than 80% even in ED8 cultures. This photoreceptor bias is expressed by the cells born *in vitro* regardless of the phenotype expressed by other cells present in the cultures, such as those born in vivo prior to their isolation for culture, which in ED8 cultures give rise predominantly to neurons.

CONCLUSIONS AND DISCUSSION

The evidence presented above provides strong support for the notion that many aspects of photoreceptor differentiation are regulated by a developmental "master program." The autonomy with which cells can implement this developmental program, however, is far from absolute. For example, outer segment development does not seem to progress in the cultured cells beyond a fairly rudimentary state (Adler, 1986a; Araki, 1984; Spoerri et al., 1987), in agreement with observations suggesting that the development and maintenance of photoreceptor outer segments are regulated by the pigment epithelium, by retinoids, and perhaps also by other factors (Dowling, 1964; Hollyfield and Witkovsky, 1974). Nevertheless, many of the molecular, structural, and even functional attributes characteristic of photoreceptor cells appear to be controlled by a "master program" that some precursor cells acquire early in embryogenesis. While the existence of this master program appears likely, however, we know practically nothing about the cellular and molecular mechanisms that are responsible for it.

Molecular studies of retinal cell differentiation in vertebrates have investigated the expression of cell-specific genes with the aid of probes derived from recent progress in the identification and cloning of important visual genes (Applebury and Hargrave, 1986; Bowes et al., 1990; Brann and Young, 1986; Dryja et al., 1990; Lem et al., 1991; Nathans, 1987; Nathans et al., 1989; Treisman et al., 1988; Zak et al., 1991; among others). On the other hand, several genes involved in the induction of photoreceptor differentiation have been characterized in invertebrates (Rubin, 1989), but not yet in vertebrates. There is reason to believe, however, that vertebrates have many genes homologous to the invertebrate "master genes" that determine the identity of individual cell types, or even of entire body segments (Davidson, 1991). Recent technical developments in molecular biology allow the investigation of gene expression even with the limited amount of material that can be obtained from embryos and cell cultures. Therefore, there is ground for optimism about the prospects for progress in the search for master genes regulating cell differentiation in the vertebrate retina.

When do precursor cells become "determined" to a specific fate? The possible existence of cell lineages involving dividing precursors already committed to a particular differentiated fate is likely to operate in some regions of the nervous system (see Luskin's chapter, this volume). In the retina, however, lineage tracing studies suggest that precursor cells remain multipotential at least until a time very close to terminal mitosis (Holt et al., 1988; Turner and Cepko, 1987; Wetts and Fraser, 1988). Our own *in vitro* data are consistent with this contention, and in fact suggest that precursor cells remain plastic for some time after terminal mitosis (Adler and Hatlee, 1989). The issue is far from settled, because "adult" cell surface markers have been observed in some retinal cells almost immediately after terminal mitosis (Larison and Bremiller, 1990; McLoon and Barnes, 1989). This could imply that those cells are already committed to an adult phenotype, but it is also possible that the expression of adult cell markers by early postmitotic cells could be transient and/or labile (Larison and Bremiller, 1990). The timing of cell differentiation may also have other effects on cell fate, as suggested by studies with cultured rat retinal cells in which the developmental potential of cells born at different stages of embryogenesis appeared to be different (Reh and Kljavin, 1989; Watanabe and Raff, 1990). It remains to be established whether the observed developmental changes reflect intrinsic changes in cell determination, and/or changes in responsiveness to environmental cues. However, stage-dependent changes in the repertory of differentiated pathways available to the cells is an interesting possibility that deserves to be investigated further.

The existence of intraretinal signals capable of changing the fate of one or another retinal cell type has been suggested by many experiments over the years. In the goldfish retina, for example, precursor cells that would normally give rise to rod photoreceptors will give rise to other cell types when the retinal microenvironment is changed as a result of the experimental elimination of retinal neurons (Raymond, 1985; Raymond et al., 1988); these results suggest that the retinal microenvironment plays a crucial role in selecting one of the several differentiated pathways available to the precursor cells. Equivalent results have been obtained in other species (Reh, 1987). There is practically no information, however, about the molecular nature of the putative inductive signals involved in retinal cell determination. This lack of information is certainly not due to a lack of candidate molecules, since the retina is very rich in growth factors and their receptors (Adler, 1986b; Campochiaro et al., 1989; Fassio et al., 1989; Fayein et al., 1990; Hewitt et al., 1990; Hicks et al., 1992; Johnson et al., 1986; Von Bartheld et al., 1991; among others), in retinoids (Chader, 1989), and in biologically active extracellular matrix molecules (Hewitt, 1986). Several of these agents have been shown to regulate the survival and/or function of various retinal cell types *in vivo* and *in vitro*, but none has so far been shown to induce precursor cells to follow a particular differentiated pathway. It is noteworthy, however, that the fibroblast growth factor (FGF) not only prevents photoreceptor degeneration in genetically abnormal rats (Faktorovic et al., 1990) and stimulates survival and opsin immunoreactivity of rat photoreceptors in culture (Hicks and Courtois, 1988), but can also induce the regeneration of a complete retina from the chick pigment epithelium (Park and Hollenberg, 1989). It is not unreasonable to expect that significant progress will soon be made in the investigation of the possible involvement of these, and other still-undiscovered microenvironmental factors, in the induction of the differentiation of specific cell types in the retina.

Acknowledgments. The original work described in this chapter was supported by United States Public Health Service grant EY04859. The author is grateful to Dr. Ann Repka for comments on the manuscript and to Ms. Doris Golembieski for secretarial help.

REFERENCES

Adler R (1986a): Developmental predetermination of the structural and molecular polarization of photoreceptor cells. *Dev Biol* 117:520–527

Adler R (1986b): Trophic interactions in retinal development and in retinal degenerations: *In vivo* and *in vitro* studies. In: *The Retina: A Model for Cell Biology Studies, Part 1*, Adler R, Farber D, eds. Orlando, FL: Academic Press

Adler R (1990): Preparation, enrichment, and growth of purified cultures of neurons and photoreceptors from chick embryos and from normal and mutant mice. In: *Methods in Neurosciences: Vol. 2. Cell Culture*, Conn PM, ed. Orlando, FL: Academic Press

Adler R, Hatlee M (1989): Plasticity and differentiation of embryonic retinal cells after terminal mitosis. *Science* 243:391–393

Adler R, Lindsey JD, Elsner CL (1984): Expression of cone-like properties by chick embryo neural retina cells in glial-free monolayer cultures. *J Cell Biol* 99:1173–1178

Applebury ML, Hargrave PA (1986): Molecular biology of the visual pigments. *Vision Res* 26:1881–1895

Araki M (1984): Immunocytochemical study of photoreceptor cell differentiation in the cultured retina of the chick. *Dev Biol* 103:313–318

Bowes C, Li T, Dancinger M, Baxter LC, Applebury ML, Farber DB (1990): Retinal degeneration in the *rd* mouse is caused by a defect in the beta subunit of rod cGMP phosphodiesterase. *Nature* 347:677–680

Brann MR, Young WS III (1986): Localization and quantitation of opsin and transducin mRNAs in bovine retina by in situ hybridization histochemistry. *FEBS Lett* 200:275–278

Burnside B, Dearry A (1986): Cell motility in the retina. In: *The Retina: A Model for Cell Biology Studies, Part 2*, Adler R, Farber D, eds. Orlando, FL: Academic Press

Campochiaro PA, Sugg R, Grotendorst G, Hjelmeland LM (1989): Retinal pigment epithelial cells produce PDGF-like proteins and secrete them into their media. *Exp Eye Res* 49:217–227

Carri NG, Ebendal T (1983): Organotypic cultures of neural retina: Neurite outgrowth stimulated by brain extracts. *Dev Brain Res* 6:219–229

Chader GJ (1989): Interphotoreceptor retinoid-binding protein (IRBP): A model protein for molecular biological and clinically relevant studies. *Invest Ophthalmol Vis Sci* 30:7–22

Combes PC, Privat A, Pessac B, Calothy G (1977): Differentiation of chick embryo neuroretina cells in monolayer cultures. An ultrast victoral study: 1. Seven day retina. *Cell Tissue Res* 185:159–173

Davidson EH (1991): Spatial mechanisms of gene regulation in metazoan embryos. *Development* 113:1–26

Dearry A, Burnside B (1989): Light-induced dopamine release from teleost retinas acts as a light-adaptive signal to the retinal pigment epithelium. *J Neurochem* 53:870–878

Dowling JE (1964): Nutritional and inherited blindness in the rat. *Exp Eye Res* 3:348–356

Dryja TP, McGee TL, Reichel E, Hahn LB, Cowley GS, Yandell DW, Sandberg MA, Berson EL (1990): A point mutation of the rhodopsin gene in one form of retinitis pigmentosa. *Nature* 343:364–366

Faktorovich EG, Steinberg RH, Yasumura D, Matthes MT, Lavail MM (1990): Photoreceptor degeneration in inherited retinal dystrophy delayed by basic fibroblast growth factor. *Nature* 347:83–86

Fassio JB, Brockman EB, Jumblatt M, Greaton C, Henry JL, Geoghegan TE, Barr CSG, Schultz GS (1989): Transforming growth factor-alpha and its receptor in neural retina. *Invest Ophthalmol Vis Sci* 30:1916–1922

Fayein NA, Courtois Y, Jeanny JC (1990): Ontogeny of basic fibroblast growth factor binding sites in mouse ocular tissues. *Exp Cell Res 188:75–88*

Grun G (1982): The development of the vertebrate retinas: A comparative survey. *Adv Anat Embryol Cell Biol* 78:1–85

Hewitt AT (1986): Extracellular matrix molecules: Their importance in the structure and function of the retina. In: *The Retina: A Model for Cell Biology Studies, Part 2*, Adler R, Farber D, eds. Orlando, FL: Academic Press

Hewitt AT, Lindsey JD, Carbott D, Adler R (1990): Photoreceptor survival-promoting activity in interphotoreceptor matrix preparations: Characterization and partial purification. *Exp Eye Res* 50:79–88

Hicks D, Bugra K, Faucheux B, Jeanny J-C, Laurent M, Malecase F, Mascarelli F, Raulais D, Cohen Y, Courtois Y (1992): Fibroblast growth factors in the retina. *Prog Ret Res*: in press

Hicks D, Courtois Y (1988): Acidic fibroblast growth factor stimulates opsin in retinal photoreceptor cells *in vitro*. *FEBS Lett* 234:475–479

Hollyfield JG, Witkovsky P (1974): Pigmented retinal epithelium involvement in photoreceptor development and function. *J Exp Zool* 189:357–378

Holt CE, Bertsch TW, Ellis HM, Harris WA (1988): Cellular determination in the *Xenopus* retina is independent of lineage and birth date. *Neuron* 1:15–26

Iuvone PM, Avendano G, Butler BJ, Adler R (1990): Cyclic AMP-dependent induction of serotonin N-acetyltransferase activity in photoreceptor-enriched chick retinal cell cultures: Characterization and inhibition by dopamine. *J Neurochem* 55:673–682

Johnson JE, Barde Y-A, Schwab M, Thoenen H (1986): Brain-derived neurotrophic factor supports the survival of cultured rat retinal ganglion cells. *J Neurosci* 6:3031–3038

Knight JK, Raymond PA (1990): Time course of opsin expression in developing rod photoreceptors. *Development* 110:1115–1120

Larison KD, Bremiller R (1990): Early onset of phenotype and cell patterning in the embryonic zebra fish retina. *Development* 109:567–576

Lem J, Applebury ML, Falk JD, Flannery JG, Simon MI (1991): Tissue-specific and developmental regulation of rod opsin chimeric genes in transgenic mice. *Neuron* 6:201–210

Linser P, Moscona AA (1979): Induction of glutamine synthetase in embryonic neural retina: Localization in Muller fibres and dependence on cell interactions. *Proc Natl Acad Sci U S A* 78:6476–6480

MacLeish PR, Townes-Anderson E (1988): Growth and synapse formation among major classes of adult salamander retinal neurons *in vitro*. *Neuron* 1:751–760

Madreperla SA, Adler R (1989): Opposing microtubule- and actin-dependent forces in the development and maintenance of structural polarity in retinal photoreceptors. *Dev Biol* 131:149–160

Madreperla SA, Edidin M, Adler R (1989): Na^+,K^+-adenosine triphosphatase polarity in retinal photoreceptors: A role for cytoskeletal attachments. *J Cell Biol* 109:1483–1493

McLoon SC, Barnes RB (1989): Early differentiation of retinal ganglion cells: An axonal protein expressed by premigratory and migrating retinal ganglion cells. *J Neurosci* 9:1424–1432

Moscona AA (1965): Recombination of dissociated cells and the development of cell aggregates. In: *Cells and Tissues in Culture*, Wilmer ED, ed. New York: Academic Press

Nathans J (1987): Molecular biology of visual pigments. *Ann Rev Neurosci* 10:163–194

Nathans J, Davenport CM, Maumenee IH, Lewis RA, Hejtmancik JF, Litt M, Lovrien E, Walar R, Bachynski B, Zwas F, Klingaman R, Fishman G (1989): Molecular genetics of human blue cone monochromacy. *Science* 245:4920

Nelson WJ, Veshnock PJ (1986): Dynamics of membrane-skeleton (fodrin) organization during development of polarity in Madin-Darby canine kidney epithelial cells. *J Cell Biol* 103:1751–1765

Park CM, Hollenberg MJ (1989): Basic fibroblast growth factor induces retinal regeneration *in vivo*. *Dev Biol* 134:201–205

Politi L, Adler R (1986): Generation of enriched populations of cultured photoreceptor cells. *Invest Ophthalmol Vis Sci* 27:656–665

Politi L, Adler R, Whittum-Hudson JA (1987): Differential sensitivity of cultured retinal neurons and photoreceptors to *Herpes simplex* infection. *Exp Eye Res* 44:923–937

Puro DG (1983): Cholinergic transmission by embryonic retinal neurons in culture: Inhibition by dopamine. *Brain Res* 273:79–86

Raymond PA (1985): The unique origin of rod photoreceptors in the teleostretina. *Trends Neurosci* 8:12–17

Raymond PA, Reifler MJ, Rivlin PK (1988): Regeneration of goldfish retina: Rod precursors are a likely source of regenerated cells. *J Neurobiol* 19:431–463

Reh RA (1987): Cell-specific regulation of neuronal production in the larval frog retina. *J Neurosci* 1:3317–3324

Reh TA, Kljavin IJ (1989): Age of differentiation determines rat retinal germinal cell phenotype: Induction of differentiation by dissociation. *J Neurosci* 9:4179–4189

Repka A, Adler R (1990): Clonal analysis of retinal cell differentiation *in vitro*. *Invest Ophthalmol Vis Sci (Suppl)* 32:923

Rubin GM (1989): Development of the *Drosophila* retina: Inductive events studied at single cell resolution. *Cell* 57:519–520

Schubert D, LaCorbiere M (1986): Role of purpurin in neural retina histogenesis. *Prog Clin Biol Res* 217B:3–16

Sparrow JR, Hicks D, Barnstable CJ (1990): Cell commitment and differentiation in explants of embryonic rat neural retina—comparison with the developmental potential of dissociated retina. *Dev Brain Res* 51:69–84

Spoerri PE, Kelley KC, Allen CB, Ulshafer RJ (1987): Conditioned medium-mediated photoreceptor differentiation in retina from embryonic rd chickens. *Europ J Cell Biol* 44:105–111

Stenkamp D, Adler R (1990): Photomechanical responses of isolated embryonic chick photoreceptors in cell culture. *Invest Ophthalmol Vis Sch (Suppl)* 31:4

Stenkamp D, Adler-Graschinshy E, Adler R (1990): Responses of cultured chick embryo photoreceptor cells to light: A role for neuromodulators. *Soc Neurosci Abstr* 16:406

Taylor M, Reh TA (1990): Induction of differentiation of rat retinal, germinal, neuroepithelial cells by dbcAMP. *J Neurobiol* 21:470–481

Townes-Anderson E, Dacheux RF, Raviola E (1988): Rod photoreceptors dissociated from the adult rabbit retina. *J Neurosci* 8:320–331

Treisman JE, Morabito MA, Barnstable CJ (1988): Opsin expression in the rat retina is developmentally regulated by transcriptional activation. *Mol Cell Biol* 8:1570–1579

Turner DL, Cepko CL (1987): A common progenitor for neurons and glia persists in rat retinas late in development. *Nature* 328:131–136

Vonbartheld CS, Heuer JG, Bothwell M (1991): Expression of nerve growth factor (NGF) receptors in the brain and retina of chick embryos—comparison with cholinergic development. *J Comp Neurol* 310:103–129

Watanabe T, Raff MC (1990): Rod photoreceptors development *in vitro:* Intrinsic properties of proliferating neuroepithelial cells change as development proceeds in the rat retina. *Neuron* 2:461–467

Wetts R, Fraser SE (1988): Multipotent precursors can give rise to all major cell types of the frog retina. *Science* 239:1142–1145

Wikler KC, Rakic P (1991): Relation of an array of early-differentiating cones to the photoreceptor mosaic in the primate retina. *Nature* 351:397–400

Zak DJ, Bennett J, Wang Y, Davenport C, Klaunberg B, Gearhard J, Nathans J (1991): Unusual topography of bovine rhodopsin promoter-*IacZ* fusion gene expression in transgenic mouse retinas. *Neuron* 6:187–199

2

Differentiation of the GABAergic System in the Avian Retina: Control of Glutamic Acid Decarboxylase Expression by GABA

FERNANDO G. DE MELLO, JAN N. HOKOÇ,
ANA L. M. VENTURA, AND PATRICIA F. GARDINO

Due to its histo-anatomical characteristics, the avian retina has been extensively used as a system to approach the basic principles of central nervous system (CNS) function and differentiation. Cells from the embryonic chick retina can be easily dissociated and cultured as dispersed or aggregated cells. Most, if not all, of retina biochemical markers differentiate properly *in vitro*, making this system useful to follow the differentiation of neurochemical properties of the tissue (Akagawa and Barnstable, 1987; Akagawa et al., 1987).

GABA is one of the major neurotransmitters in the CNS including the retina of most species. The synthesis of GABA in the CNS is accomplished mainly by the decarboxylation of glutamic acid catalyzed by the enzyme glutamic acid decarboxylase (GAD; L-glutamate 1-carboxylase, EC 4.1.1.15). Because this enzyme is widely distributed in the brain, it has been routinely used as a marker of GABAergic neuron maturation during CNS development.

In the mature CNS most of the GABA comes from glutamic acid (Morgan, 1985). However, in the embryonic tissue a considerable portion of this compound can be synthesized from putrescine (Seiler and Al-Therib, 1974).

In previous publications we have shown that GABA is present in the chick retina from very early in development when GAD activity is undetectable. A significant amount of GABA in the early stages of retina ontogeny is synthesized from putrescine, a compound that is abundant in the embryonic tissue (De Mello et al., 1976).

IMMUNOREACTIVITY OF GABA AND GAD DURING RETINA DEVELOPMENT

GABA-immunoreactivity is first detected in retinas on embryonic day 6 (ED6) and is clearly detectable in a few cells neighboring the central retina, along a row one-third from the pigmented epithelium (see Figure 2.1A). Coulombre (1955) described retina cells at this age as elongated and forming an epithelium with nuclei arranged in several layers, except for the ganglion cell bodies, some of which begin to assume a round shape and send out axons. Thus, GABA-positive

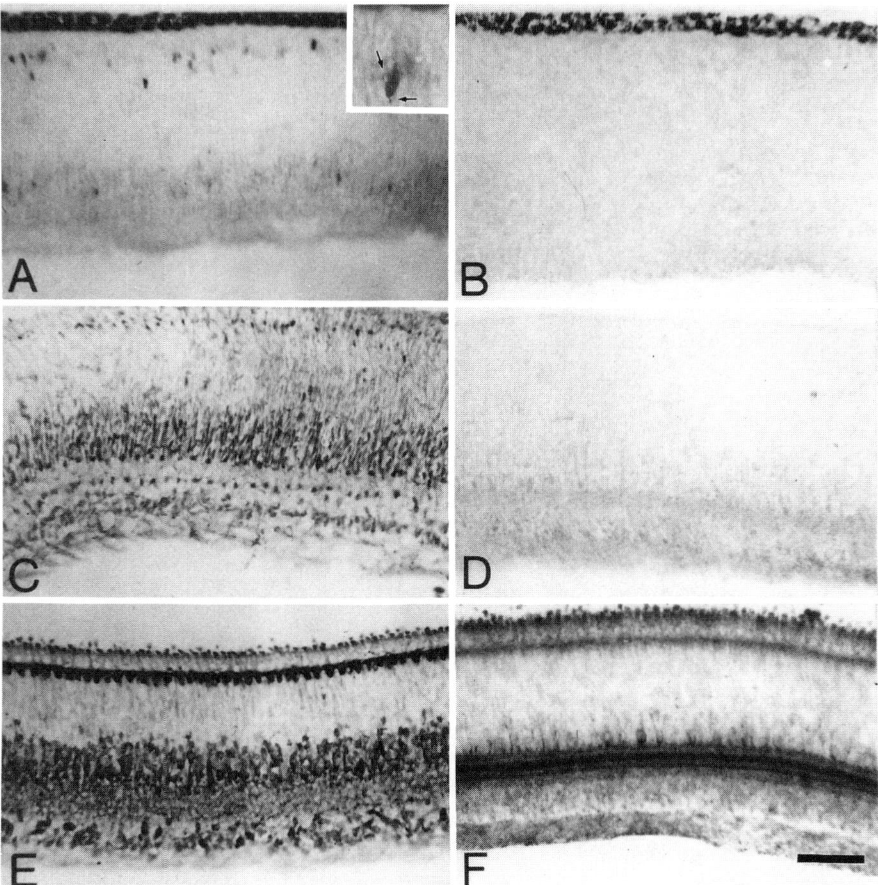

FIGURE 2.1. Light micrographs of 20 μm thick cryostat radial section from embryonic chick retinas at several stages of development reacted for GABA- or GAD-immunoreactivity. The primary reaction was visualized by the peroxidase-anti-peroxidase (PAP) method. A. Retina section through the central region of an ED6 retina. GABA immunoreactivity is detected in neuroblastlike cells in a row close to the pigment epithelium. Insert: High magnification of GABA-immunoreactive presumptive horizontal cells with processes (arrows). B. No immunoreactivity for GAD is detected in ED6 retina. C. Retina section through the central region on ED8. GABA-immunoreactive cell bodies are localized in the inner nuclear layer, including horizontal and amacrine cells, and in the ganglion cell layer. Migrating cells can be seen in the middle of the inner plexiform layer. D. No immunoreactivity to GAD can be detected in ED8 retina. E. Retina section of a ED12 embryo incubated with GABA antiserum showing a profuse number of horizontal cell bodies, with processes extending to only one stratum of the outer plexiform layer. Amacrine cell bodies and perikarya in the ganglion cell layer are also present. F. First detection of GAD-containing cell bodies in ED12 retina. Most of the cells are characterized as amacrine cells with processes extending into three well-defined strata in the inner plexiform layer. Very weak label can be seen in horizontal cell bodies. Bar = 20 μm. (Modified from Hokoç et al., 1990).

cells detected at ED6 can be either migrating neuroblasts or presumptive horizontal cell bodies. At higher magnification, some processes emerging from these cells can be observed (see Figure 2.1A insert), possibly indicating horizontal cells undergoing differentiation. Unexpectedly, no GAD-immunoreactivity is detected at this stage of retina development (see Figure 2.1B).

On ED8 the retina is characterized by the appearance of the inner and outer plexiform layers in the central region (Coulombre, 1955). The morphological aspect of cell bodies in the inner nuclear layer is still elongated, when compared to ganglion and horizontal cell somata.

The immunohistochemistry for GABA shows labeled cell bodies in the neuroblastic layer adjacent to the inner plexiform layer (see Figure 2.1C). A row of cell bodies can also be distinguished along the presumptive location of horizontal cells, close to the prospective outer plexiform site. Near the central region of the retina, some GABA-positive cell somata are localized in the middle of the inner plexiform layer, resembling migrating cells. A few cells have processes extending toward the inner plexiform layer. These cell bodies can be displaced amacrine cells in late migratory process, as suggested by Layer and Vollmer (1982).

As for ED6, GAD-labeled cell bodies or processes are not detected at any level in any region of ED8 retinas (see Figure 2.1D).

Sections from 12-day-old embryo (ED12) retinas already display the characteristic layered structure, except for the outer segments of the photoreceptors, which are not completely differentiated yet. At this stage, horizontal cell bodies at the outer portion of the inner nuclear layer have their processes extending along the outer plexiform layer (Coulombre, 1955). Thick processes from amacrine cells appear and extend themselves into the inner plexiform layer. By this time synaptogenesis begins in this layer (Coulombre, 1955).

GABA-immunoreactivity is detectable in ED12, in the same types of cells as in the adult retina (Figure 2.1E). Strongly reactive cell bodies of horizontal cells, with processes lying along the outer plexiform layer, seem to be more numerous than in the mature retina. Although weakly labeled, amacrine and ganglion cell somata are also present, resembling the adult pattern. A three-strata pattern can be detected in the inner plexiform layer, which disappears in older retinas (see Figure 2.1E).

At the ED12 stage, GAD-containing amacrine cells are present along the inner portion of the inner nuclear layer. GAD-positive cytoplasm is concentrated at their vitreal side and processes of these cells extend radially along three laminae in the inner plexiform layer (se Figure 2.1F). GAD immunoreactivity in the chick retina appears at a later embryonic stage, 3 or 4 days after GABA immunoreactivity is detected.

GAD EXPRESSION IN CULTURED RETINA CELLS IS CONTROLLED BY GABA

The presence of GABA at the early stages of development suggests that this compound may have functions related to the control of embryological parameters

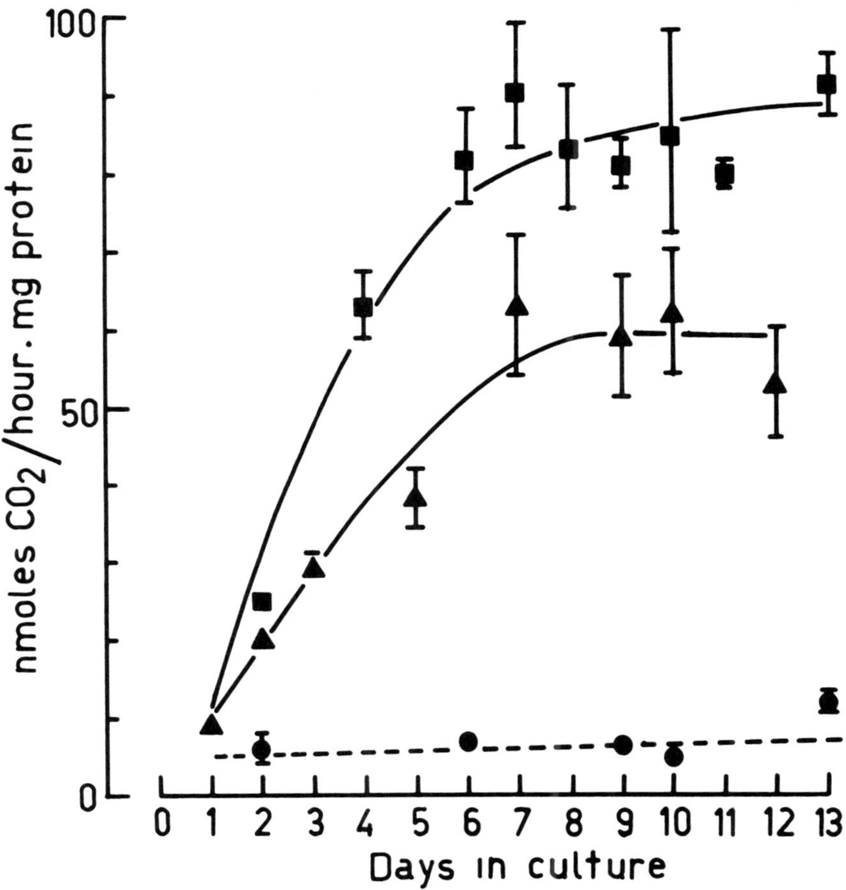

FIGURE 2.2. Effect of GABA upon the expression of GAD activity of cultured retina cells. Twenty-four hours after the onset of the cultures a group of aggregates was fed with medium containing 5 mM GABA, with subsequent changes of medium containing GABA at 24-hr intervals (circles). Control aggregates were fed with medium without GABA at 24- and 48-hr intervals (squares and triangles, respectively). The addition of GABA fully prevented the expression of GAD activity of aggregates. The maximal enzyme activity of aggregates that had their medium changed less frequently was 30% lower than the enzyme activity of aggregates submitted to daily changes of medium. Each point is the mean of 3–5 experiments ± *SD*. For GABA-treated cultures (circles) the values are the mean of 2 independent experiments ± the deviation of individual values from the mean. Reprinted with permission from De Mello et al. (1991): Glutamic acid decarboxylase of embryonic avian retina cells in culture: Regulation by -aminobutyric acid (GABA). Cell Mol Neurobiol 11:485–496.

rather than to its classical neurotransmitter role. Trophic influences of GABA on embryonic CNS neurons have also been suggested by others (Spoerri, 1988).

Dissociated retina cells from 8-day-old embryos can be cultured and under specific conditons they form aggregates that display a histotypic organization similar to that of the retina. Using these cultures, the specific activity of GAD was measured during the course of aggregate differentiation. The enzyme activity is low in the initial phases of the cultures and increases 8-fold up to day 7, if the culture medium is changed every 24 hr (see Figure 2.2, squares). This activity remains high until culture day 13. However, if the culture medium is changed at 48-hr intervals, the enzyme reaches an equilibrium level that is approximately 30% lower than that of aggregates whose medium had been changed daily (see Figure 2.2, triangles). One possible interpretation of these results is that aggregate products are released into the medium and lower the equilibrium level of GAD activity of aggregates whose medium is changed less frequently.

Figure 2.2 (circles) shows that the addition of 5 mM GABA to the medium right after cell seeding and in subsequent changes of medium completely inhibits the expression of GAD activity. The high concentration of GABA used in these experiments was chosen to provide saturating levels of GABA throughout the period studied. However, the EC50 for GABA necessary to inhibit GAD activity is approximately 10 μM and maximal inhibition is obtained with approximately 100 μM GABA concentration (De Mello, 1984).

The fact that GABA is present in the avian retina from early stages of differentiation, when the levels of its synthesizing enzyme cannot be detected (De Mello et al., 1976), suggests that the release of GABA in the medium might regulate GAD activity of GABAergic neurons. The immunoreactivity of aggregate sections to GABA antiserum reveals that indeed aggregates are enriched in GABA-containing cells (see Figure 2.3). Random sections through aggregates show abundant GABA-containing cell bodies that are detected through the whole section, preferentially along the periphery of the aggregates. Although GABA labeling does not allow the precise definition of the retinal GABAergic cell type in aggregates, by analogy with the *in vivo* retinal tissue (see Figure 2.1), it is likely that GABA immunoreactivity of aggregates may reflect the presence of horizontal and amacrine cells, and possibly a type of GABA-containing cell that is also observed in the ganglion cell layer of the intact tissue (Hokoç et al., 1990). The data suggest the possibility that GABA released in the medium may limit the extent of GAD expression. Support for this hypothesis comes from the fact that aggregates do release [^{3}H]-GABA taken up by cells. Figure 2.4 shows the result of an experiment in which aggregates (E8C8) were first incubated with [^{3}H]-GABA at 5×10^{-7} M for 2 hr and then perfused as described previously (Do Nascimento and De Mello, 1985). A steady and continuous efflux of [^{3}H]-GABA equivalent to approximately 2 fmoles GABA/min is reached after 5 min perfusion of aggregates. Stimulation of the aggregates with 1 mM glutamate for 4 min results in a dramatic, 17-fold increase in the level of [^{3}H]-GABA release, which returns to the basal efflux rate 6 min after stopping glutamate infusion. The characteristics of GABA release by cultured retina cells have been reported previously (De Mello et al., 1988).

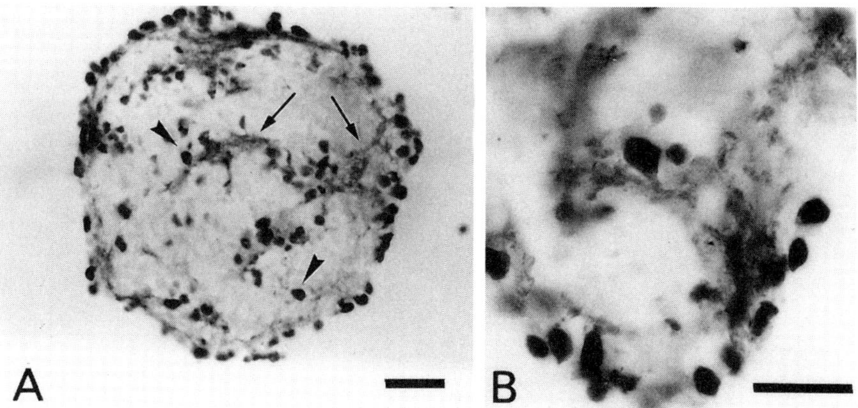

FIGURE 2.3. A. GABA-immunoreactivity of aggregate sections from chick retina cells cultured for 6 days with medium change every other day. Abundant GABA-containing cell bodies (arrowhead) and few plexus (arrow) are shown throughout the whole section. B. High magnification of GABA-containing cells. Bar = 50 μm (A) 20 μm (B).

The kinetics of GAD inhibition and recovery of aggregates exposed to GABA suggests that changes in the levels of GAD activity may reflect changes in the number of GAD molecules found in GABAergic neurons (De Mello, 1984). In an attempt to further explore this hypothesis we carried out a series of histochemical experiments in which polyclonal antibodies against GAD, obtained by Oertel and colleagues (1981), were used to detect GAD immunoreactivity of aggregates. This antibody has been shown to react with GAD molecules of different species (Brandon, 1985), including the avian enzyme (see Figure 2.1F and 2.5C). GAD immunoreactivity of control untreated aggregates shows an intense labeling for GAD observed mostly in patches of possible neurite regions (see Figure 2.5A). At higher magnification, the label is clearly visible in the cytoplasm and more intense in the region of the emergence of the main neurite (see Figure 2.5B). No immunoreactivity is observed in the nuclei. For comparison, Figure 2.5C shows a retina section of a posthatched chicken, where the characteristic distribution of GAD immunoreactivity can be observed. As described above, immunoreactivity is observed in cell bodies of horizontal and amacrine cells in the inner nuclear layer, with labeled cytoplasm surrounding the nuclei. The inner plexiform layer is also strongly immunoreactive, with three distinct labeled bands. The outer plexiform layer is weakly immunoreactive.

Aggregates obtained from the same pool of the control group were exposed to 0.3 mM GABA for 48 hr, starting on the second day after plating, and then processed for GAD immunocytochemistry. Such aggregates are completely devoid of GAD immunoreactivity (see Figure 2.5D). Thus the loss of enzyme activity observed with GABA treatment is followed by loss of GAD immunoreactivity.

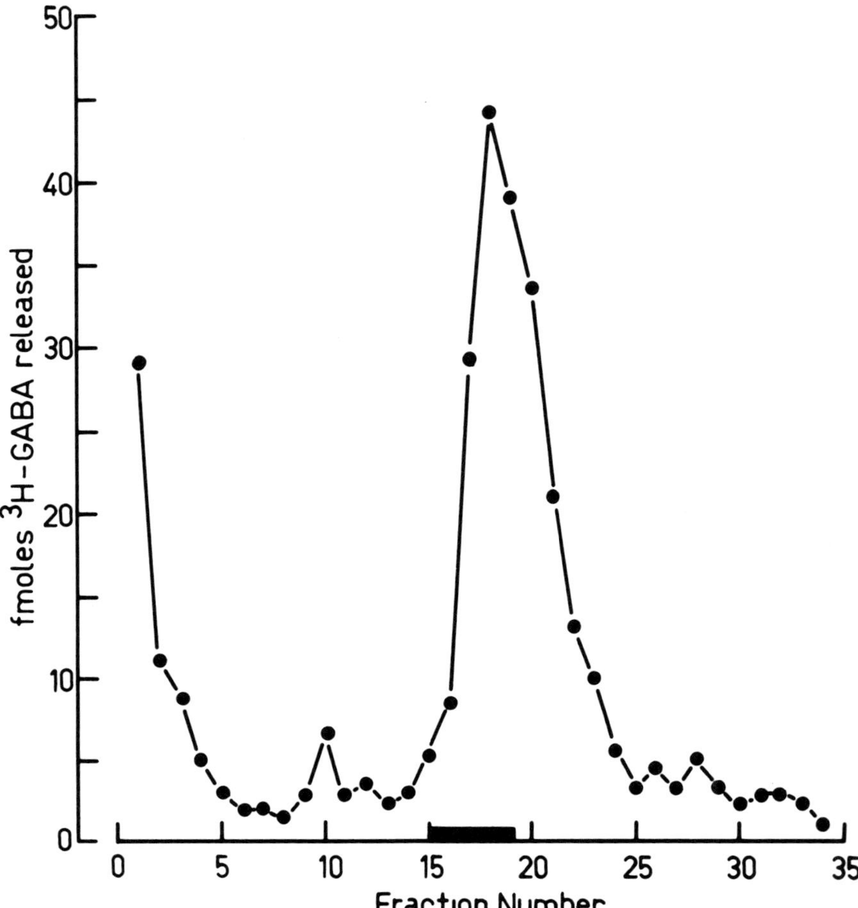

FIGURE 2.4. [^{3}H]-GABA release by retina cell aggregate culture. Aggregates were used 8 days after onset of cultures with medium changes every other day. Then [^{3}H]-GABA (85 Ci/mM) was added to a sample of aggregates to a final concentration of 5×10^{-8}M. The aggregates were further incubated for 2 hr at 37°C under an atmosphere of 5% CO_2/95% air. The aggregates were then washed 5 times with 5 ml Hanks saline to remove nonincorporated GABA and superfused with the same solution as described elsewhere (Do Nascimento & De Mello, 1985). The dark bar in abscissa indicates the exposure time of aggregates to 1 mM L-glutamate. Reprinted with permission from De Mello et al. (1991): Glutamic acid decarboxylase of embryonic avian retina cells in culture: Regulation by -aminobutyric acid (GABA). Cell Mol Neurobiol 11:485–496.

GABA at a concentration as high as 10 mM does not interfere directly with GAD activity of retina homogenates (De Mello, 1984). This observation indicates that the GABA effect requires cells to be intact in order to respond either to GABA metabolization or to the decodification of signals carried by GABA, possibly via membrane receptors. In fact, aggregates cotreated for 48 hr with GABA (0.3 mM) and picrotoxin (0.05 mM), an antagonist of GABA receptor at the level of chloride channel, display intense GAD immunoreactivity (see Figure 2.5E), indicating that picrotoxin prevents the loss of GAD immunoreactivity induced by GABA.

Our data reveal that GAD activity is expressed in chick embryo retina cells cultured under conditions that favor the formation of aggregates. The expression of the enzyme is almost completely prevented if aggregates are incubated in culture medium containing GABA.

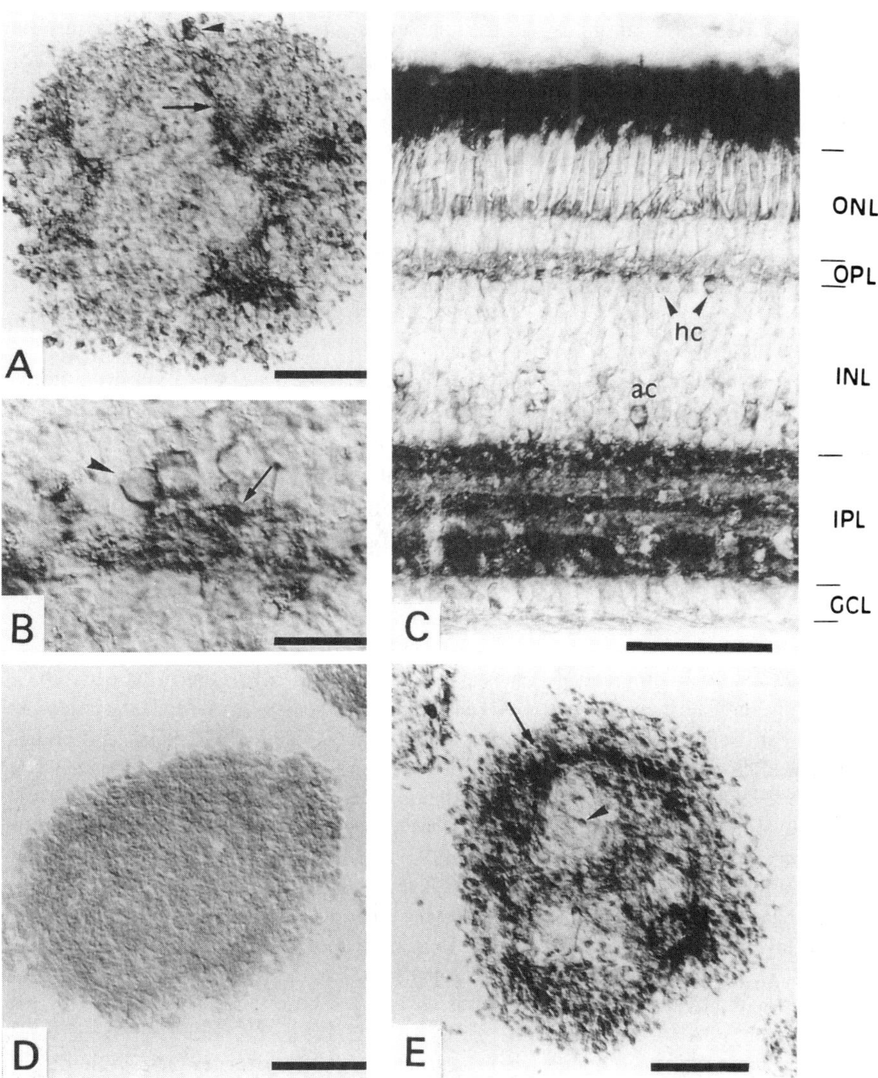

See caption FIGURE 2.5 on next page.

The immunoreactivity of aggregates to GAD antibody shows that, in association with decreased enzyme activity, aggregates exposed to GABA for 48 hr or more do not display immunoreactivity for GAD, indicating that either GAD molecules are lost from GABAergic neurons with GABA treatment or that GABA induces modifications in GAD molecules that change antigenic sites of the enzyme responsible for recognizing the antibody. This should imply that the modified sites should be important to confer catalytic activity to the enzyme. In control cultures, GAD is localized in cell bodies and also in sites that mimic neurite regions. Another possible explanation for our observations would be that GABA could hinder the antigenic sites recognizable by GAD antibody. However, if we prepare control untreated aggregates in the presence of 5 mM GABA during the histochemical procedure, the aggregates do show intense immunoreactivity for GAD. This indicates that GABA does not interfere with the interaction of the antibody with enzyme antigenic sites.

The GABA-mediated GAD inhibition is a reversible phenomenon as observed by measuring enzyme activity (De Mello, 1984). Thus the GABA effect seems not to reflect a toxic effect of this compound over GABAergic neurons. In addition, other retina neurotransmitters do not interfere with GAD activity, nor does GABA affect other decarboxylases of the retina (De Mello, 1984). Thus, the effect of GABA in controlling the level of GAD molecules of sensitive neurons seems to be selective and very specific.

A few possibilities could explain the changes in the level of GAD immunoreactivity of GABA-treated aggregates. Control of the expression of GAD genes, posttranscriptional control of GAD synthesis, degradation of GAD by regular cell metabolism, or, as mentioned above, metabolic modifications of GAD antigenic sites could also change the activity of the enzyme. The recent availability of genetic clones for GAD, described by several groups, may help to approach this

FIGURE 2.5. A. Micrograph of a cryostat section (20 μm thick) of control untreated chick retina cell aggregate incubated with GAD antiserum. Cell bodies (arrowhead) and processes (arrow) can be noticed. Bars = 50 μm. B. High magnification of labeled cell bodies (arrowhead) and processes (arrow) of control untreated aggregates. Note immunoreactivity in cell bodies and in presumptive neurite region where cell processes seem to arborize. Bar = 20 μm. C. Radial section (20 μm thick) through a mature chick retina stained for GAD-immunoreactivity. Labeled horizontal (hc) and amacrine cells (ac) are found in the outer and inner portion of the inner nuclear layer (INL), respectively. Three intensively labeled bands of processes can be detected in the inner plexiform layer (IPL). ONL, outer nuclear layer; OPL, outer plexiform layer; INL, inner nuclear layer; IPL, inner plexiform layer; GCL, ganglion cell layer. D. Aggregate exposed to 0.3 mM GABA for 2 days and incubated with GAD antiserum. No label is observed. E. Aggregate cotreated for 2 days with 0.3 mM GABA and 0.05 mM picrotoxin and incubated with GAD antiserum. GAD-immunoreactive cell bodies (arrowhead) and patches of neurites (arrow) are distinguished, indicating that picrotoxin prevents the loss of GAD immunoreactivity induced by GABA. Bar = 50 μm for D and E.

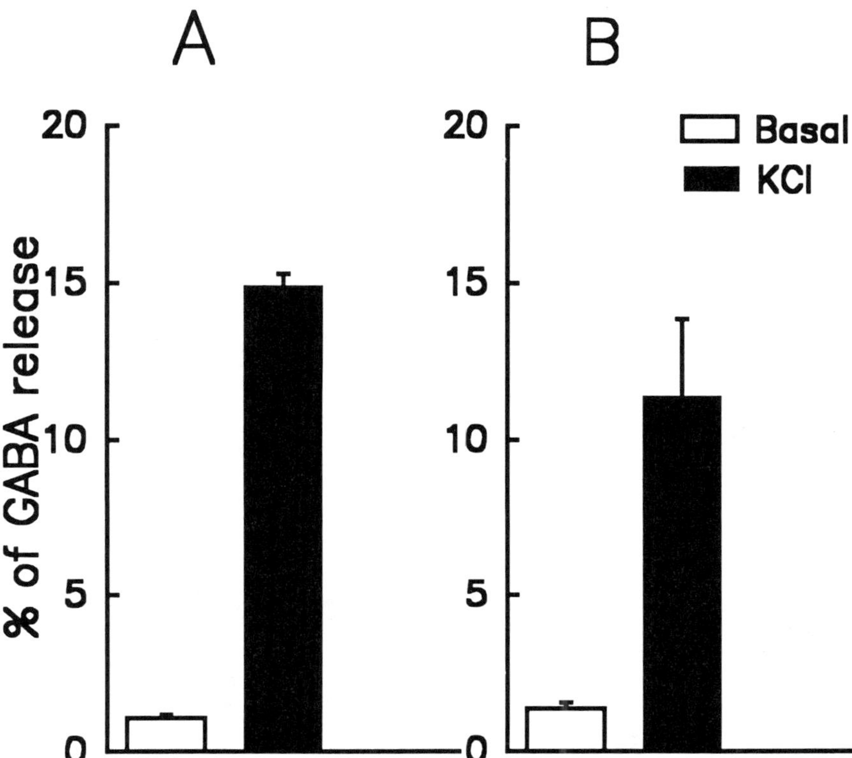

FIGURE 2.6. Putrescine-derived GABA in cultured retina neurons: potassium-stimulated release. Cells were incubated in BME medium with 1% FCS containing 20 μCi/ml of [^{3}H]-putrescine for 4 hr. Then the cells were washed 8 times with 2 ml of BME to remove nonincorporated putrescine. Superfusion proceeded by the addition of 0.5 ml of modified Hanks solution (Do Nascimento and De Mello, 1985) at 12-min intervals, with addition of the desired KCl concentration (empty columns equal to 5 mM KCl, dark columns equal to 80 mM KCl). A. Superfusion was conducted in the presence of 3mM $CaCl_2$ and 1 mM $MgCl_2$. B. Superfusion was conducted in the absence of $CaCl_2$ plus 10 mM $MgCl_2$. [^{3}H]-GABA was separated from [^{3}H]-putrescine by thin layer chromatography as dansyl derivatives (De Mello et al., 1976). Note that high potassium stimulates the release of GABA derivated from putrescine by approximately 5- to 10-fold in a calcium-independent manner, which is the characteristic of GABA release in the avian retina (Do Nascimento and De Mello, 1985).

problem more directly, assaying for the expression of GAD mRNAs of GABA-treated cells (Bond et al., 1988; Vernier et al., 1988).

As mentioned above, GABA can be detected in chick embryo retinas as early as the ED6 of incubation (De Mello et al., 1976) in neuroblastlike cells (see Figure 2.1A). The source of GABA at this stage is likely to be putrescine, a compound that is quite abundant in undifferentiated tissue (De Mello et al., 1976). Since GAD activity and immunoreactivity can be eliminated from cells that are exposed

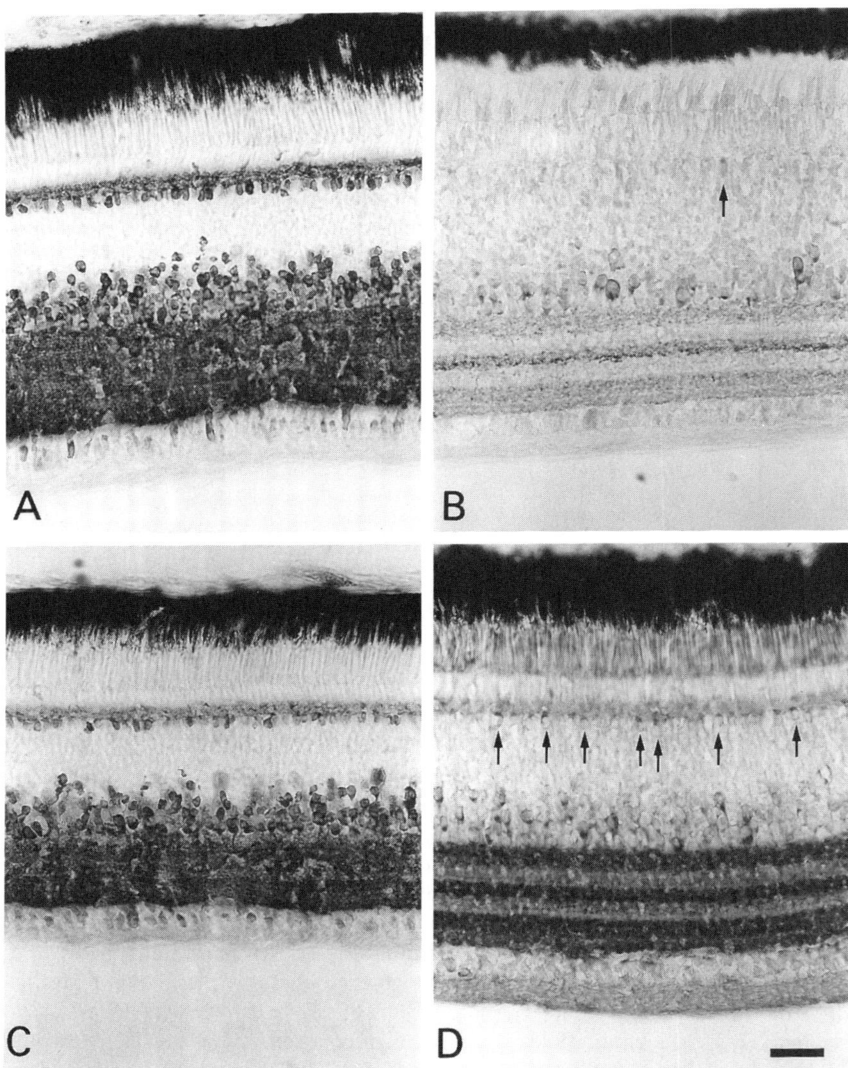

FIGURE 2.7. GABA and GAD immunoreactivity of 20 μm thick cryostat radial retinal sections from posthatched chicken under different light adaptation conditions. A. GABA immunoreactivity of a retina from a chicken maintained under constant light for 24 hr (2000 lux). B. GAD immunoreactivity of the retina from the same light-adapted animal. C. Retina section reacted for GABA immunohistochemistry from a chicken kept in constant darkness for 24 hr. The retina was dissected and manipulated under dim red light. D. Retina from the same dark-adapted chicken reacted for GAD immunohistochemistry. Dark arrows indicate GAD-containing horizontal cell bodies. Bar = 30 μm.

to GABA at the early stages of aggregate differentiation, one attractive suggestion is that GABA synthesized from putrescine in the early stages of retina development may serve as a modulator of GAD expression during the course of GABAergic neuron differentiation. In support of this hypothesis is the fact that cultured retina neurons release GABA at early stages of development (see Figure 2.2) by mechanisms that depend on cell exposure to excitatory amino acids (De Mello et al., 1988; Do Nascimento and De Mello, 1985). Furthermore, GABA synthesized from putrescine can be released in the extracellular space (see Figure 2.6), which reinforces the possibility that GABA from this source may have embryological function.

Although the observations reported above refer to cells of embryological origin, the data in Figure 2.7 reveal that GAD immunoreactivity of horizontal cells in mature avian retina changes significantly as a function of light adaptation. Retinas from posthatched chicks maintained under constant light (2000 lux) for 24 hr show a prominent content of GABA in horizontal and amacrine cell bodies as well as neurites of the interplexiform layer (see Figure 2.7A). In contrast, GAD immunoreactivity is low in the retinas of light-adapted chicks (see Figure 2.7B). A significant decrease in the immunoreactivity for GABA is seen mainly in horizontal cells of retinas from posthatched chicks kept in constant darkness for 24 hr (see Figure 2.7C). In the dark-adapted retina immunoreactivity is intense in several horizontal cell bodies (see Figure 2.7D, arrows), as well as in amacrine cells and in the inner plexiform region. The data suggest that the change in GAD immunoreactivity of the retina is inversely correlated with the amount of GABA present in the tissue, as evidenced by immunocytochemistry. Thus, the mechanism of GAD control by GABA that is quite evident in cells of embryonic origin, seems to be conserved in retinas of posthatched young chickens.

Acknowledgments. We are grateful to Dr. Wenthold for his generous gift of antiserum to Gamma-aminobutyric acid (#2) and GABA-BSA conjugate. Antiserum to GAD, as well as preimmune serum, was provided through the Laboratory of Clinical Science, National Institute of Mental Health. We thank Dr. M.C.F. de Mello for critically reading and correcting this manuscript, and Ms. Maír Medeiros and Ms. Neila de Almeida Soares for their technical assistance. This work was supported by grants from Financiadora de Estudos e Projetos, Conselho Nacional de Desenvolvimento Cientifico e Tecnológico, Fundaçáo de Amparo à Pesquisa do Estado do Rio de Janeiro, and Conselho de Ensino Para Graduados to Fernando G. de Mello, and grants from CNPq, FAPERJ, and FINEP to Jan N. Hokoç.

REFERENCES

Akagawa K, Barnstable CJ (1987): Identification and characterization of cell types accumulating GABA in rat retina cultures using cell type specific monoclonal antibody. *Brain Res* 408:154–162

Akagawa K, Hicks D, Barnstable CJ (1987): Histotypic organization and cell differentiation in rat retinal reaggregate cultures. *Brain Res* 437:298–308

Bond RW, Jansen KR, Gottlieb DI (1988): Pattern of expression of glutamic acid decarboxylase mRNA in the developing rat brain. *Proc Natl Acad Sci U S A* 85:3231–3234

Brandon C (1985): Retinal GABA neurons: Localization in vertebrate species using an antiserum to rabbit brain glutamate decarboxylase. *Brain Res* 344:286–295

Coulombre AJ (1955): Correlations of structural and biochemical changes in the developing retina of the chick. *Am J Anat* 96:153–189

De Mello FG (1984): GABA-medited control of glutamate decarboxylase (GAD) in cell aggregate culture of chick embryo retina. *Dev Brain Res* 14:7–13

De Mello FG, Bachrach U, Nirenberg M (1976): Ornithine and glutamic acid decarboxylase activities in developing chick retina. *J Neurochem* 27:847–851

De Mello, FG, Hokoç JN, Ventura ALM, Gardino PF (1991): Glutamic acid decarboxylase of embryonic avian retina cells in culture: Regulation by -aminobutyric acid *(GABA)*. *Cell Mol Neurobiol* 11:485–496

De Mello MCF, Klein WL, De Mello FG (1988): L-glutamate evoked release of GABA from cultured avian retina cells does not require glutamate receptor activation. *Brain Res* 443:166–172

Do Nascimento JL, De Mello FG (1985): Induced release of gamma-aminobutyric acid by a carrier-mediated, high-affinity uptake of L-glutamate in cultured chick retina cells. *J Neurochem* 45:1820–1827

Hokoç JN, Ventura ALM, Gardino PF, De Mello FG (1990): Developmental immunoreactivity for GABA and GAD in the avian retina: Possible alternative pathway for GABA synthesis. *Brain Res* 532:197–202

Layer PG, Vollmer G (1982): Lucifer yellow stains displaced amacrine cells of the chicken retina during embryonic development. *Neurosci Lett* 31:979–984

Morgan WW. (1985): GABA: A potential neurotransmitter in the retina. In: *Retinal Transmitters and Modulators: Models for the Brain*, Morgan WW, ed. Boca Raton, FL: CRC Press

Oertel WH, Schmechel DE, Tappaz ML, Kopin IJ (1981): Production of a specific antiserum to rat brain glutamic acid decarboxylase by injection of an antigen-antibody complex. *Neuroscience* 6:2689–2700

Seiler N, Al-Therib MJ (1974): Putrescine catabolism in mammalian brain. *Biochem J* 144:29–35

Spoerri PE (1988): Neurotrophic effects of GABA in culture of embryonic chick brain and retina. *Synapse* 2:11–22

Vernier P, Julien JF, Rataboul P, Fourrier O, Feuerstein C, Mallet J (1988): Similar time changes in striatal levels of glutamic acid decarboxylase and proenkephalin mRNA following dopaminergic deafferentation in rat. *J Neurochem* 51:1375–1380

3

Role of Acetylated Gangliosides on Neuronal Migration and Axonal Outgrowth

ROSALIA MENDEZ-OTERO, BURKHARD SCHLOSSHAUER, AND MARTHA CONSTANTINE-PATON

INTRODUCTION

The establishment of form and pattern within the nevous system is dependent on cellular interactions that are initiated early in development. These interactions regulate cell proliferation and differentiation, cell migration, axonal growth and guidance, target recognition, and synapse formation. Over the past decade the analysis of these sequential developmental steps has advanced from the descriptive to the molecular level. In large part, progress has resulted from a better appreciation of the way in which developing neurons respond to their environment. Considerable attention has been focused on proteins that are involved in these responses. Thus, the amino acid sequences and functional domains of many cell surface adhesive and repulsive proteins are now relatively well understood.

Although the importance of cell surface carbohydrates in mediating many of the responses of developing neurons has also been emphasized in many current studies, in comparison to the proteins, our knowledge of the critical structural features and of the mechanisms of carbohydrate function is still rudimentary. The initial suggestions that carbohydrate structures might mediate cell-cell interactions were based on the expression of complex oligosaccharides, gangliosides, and cell surface glycosyltransferases on developing neural cells (Barbera 1975; Barbera et al., 1973; Marchase, 1977; Roth et al., 1971). With the availability of monoclonal antibodies (Mabs), the complex expression patterns of oligosaccharides in developing brain has become more readily apparent. Many are restricted to subsets of neurons during particular stages of differentiation. Recently, several carbohydrate-binding proteins with specificity for neural cell surface oligosaccharides have been detected in vertebrate nervous tissue (Begovac and Shur, 1990). However, most of the studies of the mechanisms through which cell surface carbohydrates work have progressed slowly because sufficient quantities of particular species are generally unavailable and their cellular regulation is poorly defined. It is difficult to generate oligosaccharide structure synthetically. Moreover, the cellular synthetic or degradative pathways for these molecules involve complex biochemical cascades in which many of the endogenous enzymes have not yet been isolated or even identified.

The gangliosides constitute a major group of cell surface carbohydrate molecules that are particularly enigmatic. These small glycolipids derive their name because of their prevalence on the surface of nervous system cells (ganglia). The sialic acid residues account for a significant proportion of the charge on neuronal and glial cell cell surfaces and most studies have failed to localize gangliosides to any location other than the outer leaflet of the cell membrane. Indeed, many of the proposed cellular functions of gangliosides as receptors or a modulators of protein receptors are consistent with this outer leaflet location. However, *in vitro* studies have demonstrated that gangliosides can alter levels of kinase activity and levels of intracellular free Ca^{++}(Kim et al., 1986). The involvement of sphingosine, the nonsugar "tail" of all gangliosides, in a host of lipid signaled interactions is now well established. These latter studies imply a cytoplasmic site of ganglioside mediated action.

In other words, there are major outstanding questions as to the actual cellular role played by these small glycolipids. These questions, compounded by the technical difficulties associated with studying carbohydrates in general, have led many developmental biologists to abandon the field of glycolipid involvement in cell-cell interactions in favor of the more easily studied proteins and glycoproteins. However, the older biochemical and tissue culture data, as well as a host of recent studies utilizing antibody species, testify to the significance of these molecules in neuronal differentiation and axon growth. In this review we will describe some of the older studies on ganglioside function in neuronal development and some of the newer work from a variety of different laboratories. We also will outline our own work and that of others on developmentally prevalent gangliosides O-acetylated at the 9 position of their sialic acid residues. A number of independent studies now directly implicate this species in the process of cell adhesion and/or growth cone extension. Our hope is to motivate more intensive mechanistic studies of cell surface glycolipids in general and gangliosides in particular in order to promote an understanding of the interactions controlling brain development that is broader and more realistic than is now possible from the intensive focus on proteins and glycoproteins alone.

GANGLIOSIDE-MEDIATED ADHESION IN THE DEVELOPING RETINOTECTAL SYSTEM

The topographic projection of retinal ganglion cell axons onto the tectal surface represents one of the best-studied neural systems with which to examine the formation of specific connections in the nervous system. Molecules involved in establishing visual system topography are thought to be expressed in a graded pattern across the population of retinal cells so that they bias the selectivity of optic tract axons for different subregions of their central terminal fields (Bonhoeffer and Gierer, 1984; Sperry, 1963). Recent studies by Bonhoeffer and others suggest that topography along the anterior-posterior axis in chick, fish, and mouse, involves tectal proteins that inhibit the migration of incoming temporal retinal axons

(Godement and Bonhoeffer, 1989; Vielmetter and Stuermer, 1989; Walter et al., 1987). The work in chicks has resulted in the identification of a glycoprotein that is apparently responsible for the *in vivo* inhibition of axon invasion because it locally stimulates growth cone collapse (Baier and Bonhoeffer, 1992). Other studies have identified several other proteins which, because of a differential topographic distribution in the eye, are believed to be associated with the establishment of retinotopic central projections. However, no specific functional role in axon growth or guidance has, as yet, been described for any of these proteins (McLoon, 1991; Rabacchi et al., 1990; Trisler and Collins, 1987; Trisler et al., 1981.

Biochemical support for the existence of carbohydrate gradients in the retinotectal system was originally obtained by measuring the adhesion of dissociated dorsal or ventral retinal cells to topographically appropriate tectal regions (Barbera, 1975; Barbera et al., 1973; Marchase 1977). Such experiments found that cells from dorsal retina adhere preferentially to ventral tectum and cells from ventral retina adhere preferentially to dorsal tectum. This adhesive selectivity mimicked the normal pattern of innervation of the tectum by the retina. Marchase (1977) established that treatment of dorsal retinal cells or dorsal tectum with the enzyme hexosaminidase blocked the preferential adhesion of the retinal cells to their matching tectal halves, whereas similar protease treatment of ventral retinal cells or ventral tectum did not alter adhesion. This enzyme cleaves N-acetylhexosamines from the ends of glycoproteins or glycolipids. Marchase proposed the existence of two opposing gradients of complementary molecules on the cell surfaces of the retina and the tectum. The first gradient is that of a glycoprotein or glycolipid carrying a carbohydrate chain ending in N-acetylgalactosamine. The concentration of this molecule is highest at the dorsal ends of both retina and tectum. The second gradient consists of a molecule capable of recognizing the N-acetylgalactosamine. This was hypothesized to be glycosyltransferase, an enzyme capable of forming a lock-and-key complex with the N-acetylgalactosamine. Thus, the dorsal portion of the retina would be specifically recognized by the ventral portion of the tectum and vice versa. Complete identification of these molecules was not possible. Support for this model was obtained with the demonstration that the binding of GM2 ganglioside to retina and tectum exhibits a ventral-to-dorsal gradient. One enzyme, UDP-galactose GM2 galactosyltransferase, which recognizes GM2 and converts it to GM1 by the addition of a terminal galactose residue, was found to be 30% more concentrated in ventral retina than dorsal retina (Marchase, 1977).

Functional evidence that gangliosides may be involved in the specification of retinotectal projections has been produced by independent experiments that revealed selective adhesion of retinal ganglion cells to immobilized gangliosides, including GM2 (Blackburn et al., 1986). These adhesive interactions appear to be specific in that they are not detected between neural retinal cells and other charged lipids (e.g., sulphatides, phospholipids) or between gangliosides and other cell types, such as hepatocytes. In addition, of several gangliosides tested, GM2, GD3, and GD1a supported a greater strength and extent of adhesion than GT1b,

GM1, and GD1b, a result that suggested the presence of a receptor on retinal cells that can distinguish between gangliosides (Blackburn et al., 1986). In further support, immunocytochemical studies have revealed selective expression patterns of several gangliosides in the retinotectal system. The monoclonal antibody 18B8 detects a number of developmentally regulated ganglioside species in chicken retina and brain. The major ganglioside species recognized by antibody 18B8 in retina, GT3, is associated with the cell bodies of most neurons in the retina in early development but becomes progressively restricted to synaptic layers during later development (Grunwald et al., 1985). Also, monoclonal antibody 8A2, that recognizes a 9-O-acetyl group on a set of gangliosides, reveals a complex staining pattern in the chick retina during development (Drazba et al., 1991). Neither 18B8 nor 8A2, however, shows a topographically varied staining pattern from dorsal to ventral retina.

We have been using an immunological approach to identify molecules that show unequal distribution in the developing rat retina. Using embryonic rat retina as immunogen, a Mab, named Jones, was generated that identifies gangliosides expressed on the surface of retinal cells. The expression of these gangliosides is first detected in the central region of embryonic day (ED) 12–13 retinas (Constantine-Paton et al., 1986). By ED 17–18 the antigen was clearly distributed in a gradient and antibody binding was high in the dorsal retina, decreasing gradually in more ventral regions (see Figure 3.1). The dorsal-to-ventral gradient was still detectable on postnatal day (PD) zero and was present though less pronounced on PD3. Upon high performance thin layer chromatography (HPTLC) analysis the epitope recognized by the Jones Mab was carried on two major bands (Blum and Barnstable, 1987; Schlosshauer et al., 1988). The removal of Jones reactivity by base treatment and the consequent production of GD3 showed that one of the bands is a modified form of GD3. The ability of the modification to protect the molecule from periodate oxidation and the reconstitution of Jones immunoreactivity that migrates to the same position as the authentic molecule all indicate that the antigen recognized is a 9-O-acetyl GD3. The slower migrating immunoreactive band had HPTLC mobility between GD3 and GD1a. Mild base treatment also abolished Jones immunoreactivity of this band and changed its mobility to the region where GQ1c migrates, suggesting that the second band is probably a 9-O-acetyl-GQ1c (Multani et al., 1988). The 9-O-acetyl gangliosides recognized by Jones Mab are present in retinas of all mammalian species studies so far (see Figure 3.2), but are not present in extracts of frog retinas (Schlosshauer et al., 1986).

At least two other Mabs have been described (Mab D1.1 and RB13-2) that recognize the same ganglioside (9-O-acetyl-GD3) in nervous tissue as Jones Mab, and, in addition, two other minor gangliosides (Levine et al., 1984; Reinardt-Maelicke et al., 1990). Ganglioside 9-O-acetyl-GD3 is also expressed on the surface of human malignant melanomas and other tumors of neuroectodermal origin and also on several cell lines (Blum and Barnstable, 1987; Bonafede et al., 1989; Cheresh et al., 1984).

The pattern of expression of the 9-O-acetylated form of the GD3 ganglioside is independent of that of the nonacetylated GD3 ganglioside recognized with Mab

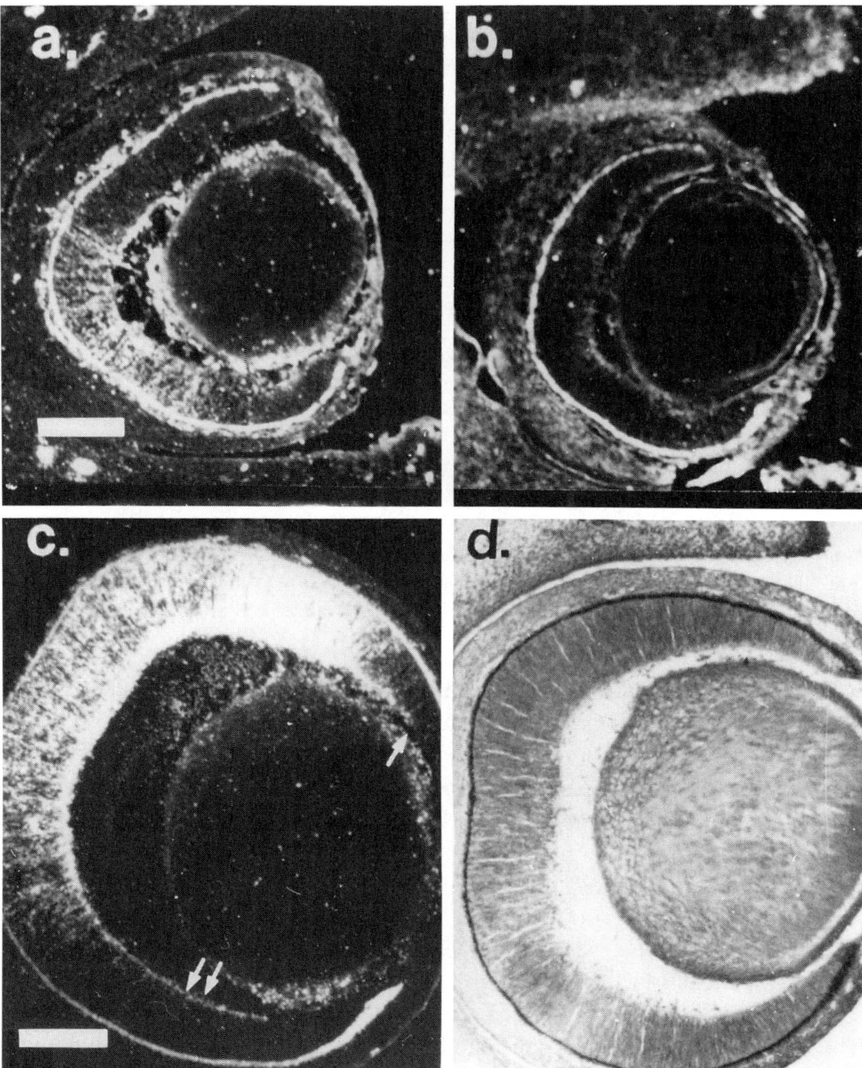

FIGURE 3.1. Immunocytochemical distribution of 9-O-acetyl gangliosides on developing rat retinas. A. Darkfield photomicrograph of immunogold-stained parasagittal section of ED13 retina. Labeling is restricted to the central retina. B. An adjacent control section reacted with the second antibody alone. C. Darkfield photomicrograph of a ED17 retina stained with Jones Mab and reacted with immunogold. Notice the dorsoventral gradient of antigen. Arrows point to the optic fiber layer. D. Brightfield photomicrograph of an adjacent section stained with toluidin blue. Scale bar = 150 μm.

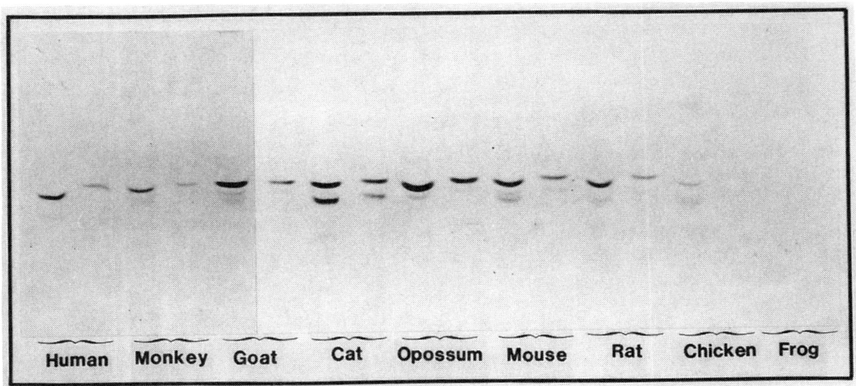

FIGURE 3.2. 9-O-acetyl gangliosides in retinas of different species. Enriched gangliosides from adult retinas were separated by thin layer chromatography with a solvent system chloroform/methanol/0.2% $CaCl_2/H_2O$ (50:45:10). Thereafter, Jones binding gangliosides were identified in an overlay assay with a secondary peroxidase-labeled antimouse antibody. The approximate amount applied for each lane was: 13 μg sialic acid (left-hand lane) and 6 μg sialic acid (right-hand lane). All samples were run on the same chromatogram.

R24. In retina and tectum, the Jones antigen is distributed on neurons and glia in a dorso-ventral gradient (Constantine-Paton et al., 1986), whereas GD3 staining appears to be uniformly distributed, thus suggesting that the selectivity of expression of the acetylated form is dependent on the spatial restriction of specific biosynthetic or degradation enzymes (Blum and Barnstable, 1987; Sparrow and Barnstable, 1988).

CORRELATION OF GANGLIOSIDE EXPRESSION WITH CELL AND PROCESS MOTILITY

The 9-O-acetyl gangliosides are found in a number of other developing neural tissues. More extensive mapping of the distribution of these molecules in the developing rat nervous system has shown that they are found in very discrete groups of cells or processes and that its expression correlates well with migration of cell bodies or cell processes (Mendez-Otero and Constantine-Paton, 1990; Mendez-Otero et al., 1988). In all regions studied, during embryonic and early postnatal life, the labeling is very intense in the ventricular zone and shows a radial array in the adjacent intermediate and marginal zones. The expression of these gangliosides correlates particularly with times of cell migration in the retina, superior colliculus, cerebellum, and telencephalon.

The cerebellum is a region particularly adequate to study cell migration. Cerebellar differentiation involves three distinct patterns of cell migration (Altman, 1969; Altman and Bayer, 1985a, 1985b, 1985c). Prenatally, Purkinje cells,

Golgi cells, and deep nuclear neurons originate in the ventricular zone and migrate into the body of the cerebellar anlage. This is a radial migration from the ependymal region toward the pia. At ED18 the positions of these dorsally migrating cells correspond closely to the radial pattern of Jones staining extending from the ependymal border upward through the anlage. The primary tangential migration of cells destined for the external germinal zone, or the external granular layer (EGL), is also underway in the cerebellum at this age. This migration can be observed as a thin layer of densely stained cells in the subpial region in the caudal half of the anlage, and it colocalizes with a subpial strip of Jones labeling (Mendez-Otero et al., 1989).

In the postnatal period, the EGL gives rise to several types of neurons but predominantly to the granule cells which migrate inward from the pial surface into the future granular layer. The postnatal migration of granule cells has been shown to follow the processes of Bergman glia (Rakic, 1971) and to occur in different subdivisions of the cerebellar cortex at different ages (Altman, 1969). During the first postnatal days, Jones immunoreactivity exhibits a stereotyped but discontinuous pattern in the cerebellar cortex, which appears to correspond to differences in the maturity of different regions. In the most rostral regions, Jones binding is detected throughout the folium, with a radially oriented pattern extending from the EGL and through the presumptive molecular layer. Caudal to the fissura prima, the presumptive cortical region is completely undstained (see Figure 3.3) By the end of the first postnatal week, Jones staining reveals labeling uniformly distributed throughout the cerebellar cortex. The distribution correlates with the presence of migrating neurons, which suggests that the function of the Jones antigen in the retina and other regions of the embryonic central nervous system is in the regulation of early neuroephitelial cell migration (Mendez-Otero et al., 1988).

The radially oriented pattern of Jones binding in the cerebellar folia suggested that the gangliosides might be expressed by the radial glia which were serving as substrates for the migrating granule cells. Staining of adjacent sections with RAT-401 (generous gift of S. Hockfield) revealed radial glia present in all folia (see Figure 3.3D, 3.3E), but only the most anterior and posterior folia showed Jones staining at this stage (see Figure 3.3A, 3.3B, and 3.3C). One possibility is that only the radial glial supporting migration expressed the gangliosides. We have used an *in vitro* system to further investigate this hypothesis. Identification of cell types expressing the Jones antigens in dissociated cerebellar cell cultures has shown that granule cells in isolation and in association with glia expresss the antigens, but, in the absence of neurons, glial cells are Jones-negative. Moreover, most glial cells with an elongated morphology are also Jones-positive and only a small proportion of the stellate astroglia express these gangliosides (Mendez-Otero and Constantine-Paton, 1990). Hatten and colleagues have shown that only glial cells with an elongated (radial) morphology support extensive neuronal migration *in vitro* (Edmondson and Hatten, 1987; Hatten et al., 1984). The absence of Jones staining in glial cell cultures in the absence of neurons and the different expression in the two morphological types of glia led us to suggest that 9-O-acetyl gangliosides could be associated with the functional state of the glia and

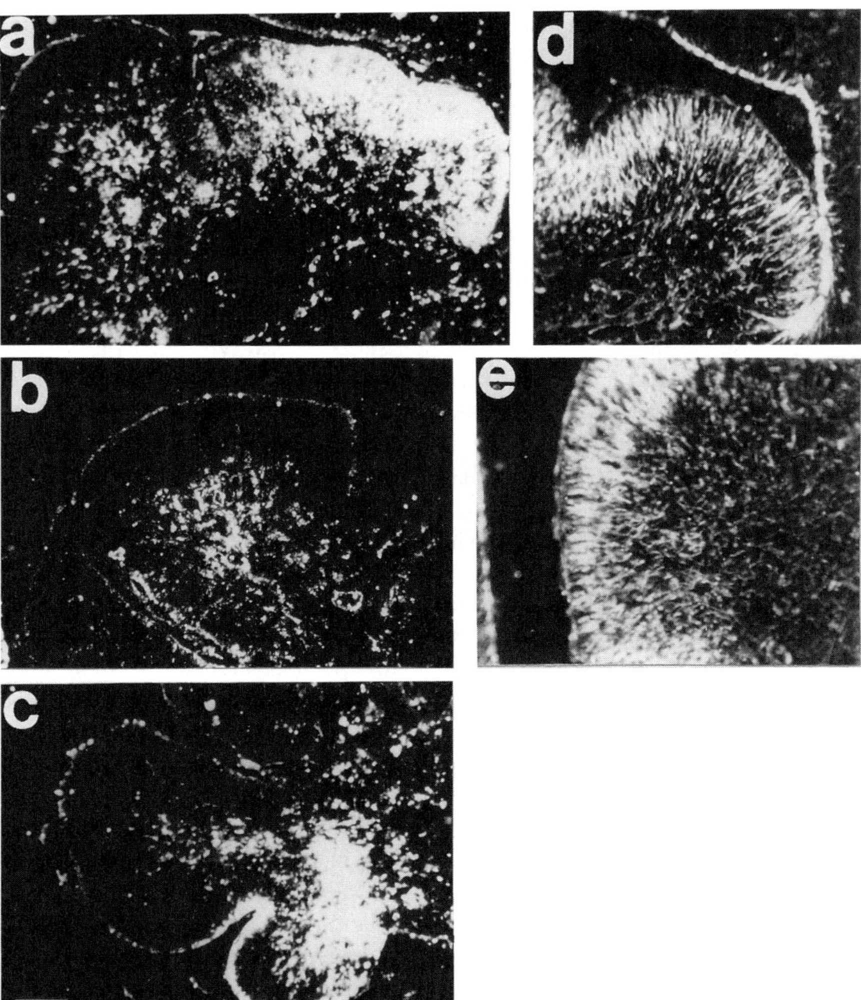

FIGURE 3.3. Expression of 9-O-acetyl gangliosides on early postnatal cerebellum. A–C. Darkfield immunogold staining of Jones binding in parasagittal sections through the cerebellum of a postnatal day zero rat. Observe Jones staining throughout the molecular layer in the most anterior folium (A), and the external granular layer in the most posterior folium (C). The intermediatee folia (B) do not stain with Jones at this age. D and E, Immunofluorescent staining of adjacent sections of the anterior folium and an intermediate folium using RAT 401 (kindly supplied by S. Hockfield) as the primary antibody. Jones binding is similar to RAT 401 binding in the anterior folium but is absent from the intermediate folium where RAT 401 demonstrates the presence of radial glia indicating that the 9-O-acetyl gangliosides are expressed by a subset of radial glia. Scale bar = 150 μm.

modulate motility either directly or by modifying the efficacy of some other component of the system.

In numerous regions of the developing rat nervous system Jones antigenicity is colocalized with differentiating axon tracts and appears to be roughly related in amount and timing of reactivity to periods when the largest numbers of axons would be growing. This occurs, for example, in the optic pathway, the central and peripheral roots of the trigeminal and dorsal root ganglia, fiber tracts entering the cerebellum, dorsal fiber tracts of the spinal cord, and the olfactory nerve. Furthermore, *in vitro*, Jones immunoreactivity is present not only in cell bodies but also in neurites (Mendez-Otero and Constantine-Paton, 1990; Mendez-Otero et al., 1989). In spite of such an intriguing correlation of 9-O-acetyl gangliosides expression with cell and process motility, little is known about the specific function of these molecules during development. Results from a series of experiments discussed below suggest a direct involvement of gangliosides, more specifically 9-O-acetylated gangliosides, on cell migration and axonal outgrowth.

ROLE OF GANGLIOSIDES IN CELL MIGRATION AND PROCESS OUTGROWTH

In recent experiments Varki and colleagues (1991) demonstrated that the selective destruction of 9-O-acetyl groups on sialic acids caused significant abnormalities in murine embryogenesis. In these experiments, they have developed transgenic mice expressing the 9-O-acetyl-sialic acid-specific esterase of influenza C virus in a tissue-specific and developmentally regulated fashion. Under physiological conditions (37°C and neutral pH) this protein functions only as a highly specific sialate: 9-O-acetylesterase. The promoter used in these experiments causes the expression of the enzyme, selectively, in the adrenal gland and retina. In these tissues, the ganglioside 9-O-acetyl-GD3 was specifically destroyed, resulting in significant abnormalities, while all other tissues examined appeared normal. In the abnormal adrenal gland, the peripheral zona glomerulosa and the central medullary region appeared relatively normal. However, the zona fasciculata of the cortex appeared somewhat disorganized, with loss of the orderly columns seen in the normal gland. The zona reticulosa (normally immediately adjacent to the medulla) was the most abnormal and was almost completely replaced by fatty tissue. The retina also showed severe gross disorganization with loss of many normal layers. The appearance suggested degeneration of the photoreceptor layer and fusion and/or failure of formation of some of the other layers (Varki et al., 1991). Although it is possible that 0-acetyl groups had been removed from other glycoconjugate, it is tempting to speculate that this phenotype was caused by disruption of the normal pattern of cell migration due to the specific removal of 9-O-acetyl groups from GD3.

Abundant work has been done describing the effects of exogenous ganglioside treatment on the rates of neuritogenesis both *in vitro* and *in vivo* (for reviews, see Bahr et al., 1989; Sable, 1988). Also, antibodies to gangliosides have been shown to inhibit goldfish optic nerve regeneration (Sloan et al., 1989; Sparrow et al., 1984) and nerve growth factor–induced neurite outgrowth from dorsal root ganglion (Schwartz and Spirman, 1982). However, these experiments did not address the functional significance of individual gangliosides and the mechanisms underlying these effects.

Recently, we have been studying the effect of Jones Mab on axonal elongation using dorsal root ganglia explanted on laminin. Using a time-lapse video-enhanced imaging system, it was possible to demonstrate that the advance of the growth cone on laminin is reversibly halted in the presence of the Jones Mab. The onset of effects is rapid and signaled by an immediate cessation of elongation, a loss of lamellipodia, and an emergence of a prominent filopodial morphology. A retrieval of axoplasm begins in distal filopodia, and the material accumulates at the base of the growth cone. After several minutes of washing with the culture medium, the growth cone starts growing again. Mab A2B5 (Eisenbarth et al., 1979), which also recognizes gangliosides expressed on these growth cones, does not induce any change on the growth rate (Mendez-Otero, Friedman, and Constantine-Paton, unpublished results). Similar results were described on regenerating retinal axons from goldfish explants with monoclonal antibody 8A2 which recognizes 9-O-acetyl gangliosides in chicks (Sloan et al., 1989).

Although gangliosides have been suggested as candidate receptors for fibronectin (Kleinman et al., 1979; Yamada et al., 1983), it is becoming apparent that the role of gangliosides in integrin-based adhesion systems is that of accessory molecules to facilitate binding or to modulate receptor function rather than to function themselves as specific receptors (Stallcup, 1988; Stallcup et al., 1989). According to this hypothesis, gangliosides, being strongly anionic, may play a role in the electrostatic requirements for optimal cell-substratum interactions. In this respect, studies with a rat neural cell line have shown that the D1.1 ganglioside, an 0-acetylated derivative of GD3, is involved in cellular adhesion to fibronectin (Stallcup, 1988). In addition, it was also demonstrated that D1.1-positive cells derived from embryonic brain and from postnatal cerebellum have biochemically identifiable cell surface fibronectin receptors that mediate their attachment to surfaces coated with fibronectin. The distribution of the D1.1 antigen has been shown to overlap with that of fibronectin and the fibronectin receptor. The adhesion of postnatal rat cerebellar cells to fibronectin is inhibited by antibodies to the fibronectin receptor, by Arg-Gly-Asp peptides that block the recognition of fibronectin by its receptor, and also by antibodies to D1.1 itself, demonstrating the involvement of both types of cell surface components in the adhesion process. The D1.1 ganglioside may therefore be required to enhance the attachment to fibronectin of cells expressing the fibronectin receptor (Stallcup, 1988; Stallcup et al., 1989).

The sialic acid residues of gangliosides are known to be efficient chelators of divalent cations. One possibility, therefore, is that gangliosides may serve to modulate the availability of divalent cations to a number of different calcium- or magnesium-dependent adhesion systems (Cheresh, 1987; Cheresh et al., 1986; Cheresh et al., 1987). Alternatively, they may also regulate the amount of free calcium on the external side of the plasma membrane, affecting the influx of calcium in the growth cone and consequently influencing a multitude of Ca^{++}-dependent processes in that structure. Thus, a protein-based adhesion system could be broadly distributed, but parts of the pathway could be favored by subsets of moving cells or processes because of the presence of the ganglioside, which might facilitate an adhesion/migration cycle or otherwise optimize the cell-substrate bond for maximal probability of advancement.

In view of all the potential roles gangliosides can play and the very strong correlation of the expression of 9-O-acetylated gangliosides with cell migration and process outgrowth, it is likely that they are actively involved in the complex process that results in the stereotyped cytoarchitecture of different nervous system regions.

Acknowledgments. These studies were supported in part by Conselho Nacional de Desenvolvimento Científico e Tecnológico, Fundação de Amparo a Pesquisa do Estado do Rio de Janeiro, Financiadora de Estudos e Projetos, and Conselho de Pesquisas e Ensino Para Graduados grants to Rosalia Mendez-Otero and American Paralysis Association grant to Martha Constantine-Paton.

REFERENCES

Altman J (1969): Autoradiographic and histological studies of postnatal neurogenesis: 3. Dating the time of production and onset of differentiation of cerebellar microneurons in rats. *J Comp Neurol* 136:269–294

Altman J, Bayer SA (1985a): Embryonic development of the rat cerebellum: 1. Delineation of the cerebellar primordium and early cell movements. *J Comp Neurol* 231:1–26

Altman J, Bayer SA (1985b): Embryonic development of the rat cerebellum: 2. Translocation and regional distribution of the deep neurons. *J Comp Neurol* 231:27–41

Altman J, Bayer SA (1985c): Embryonic development of the rat cerebellum: 3. Regional differences in the time of origin, migration, and settling of Purkinje cells. *J Comp Neurol* 231:42–65

Bahr M, Vanselow J, Thano S (1989): Ability of adult rat ganglion cells to regrow axons *in vitro* can be influenced by fibroblast growth factor and gangliosides. *Neurosci Lett* 96:197–201

Baier H, Bonhoeffer F (1991): Axon guidance in vitro by a target-derived cell membrane component. In: *The Nerve Growth Cone*, Letourneau PC, Kater SB, Macagno ER, eds. New York: Raven Press

Barbera AJ (1975): Adhesive recognition between developing retinal cells and the optic-tecta of the chick embryo. *Dev Biol* 46:167–191

Barbera AJ, Marchase RB, Roth S (1973): Adhesive recognition and retino-tectal specificity. *Proc Natl Acad Sci U S A* 70:2482–2486

Begovac PC, Shur BD (1990): Cell surface galactosyltransferase mediates the initiation of neurite outgrowth from PC12 cells on laminin. *J Cell Biol* 110:461–470

Blackburn CC, Swank-Hill P, Schnaar RL (1986): Gangliosides support neural retina cell adhesion. *J Biol Chem* 261:2873–2881

Blum AS, Barnstable CJ (1987): O-acetylation of a cell-surface carbohydrate creates discrete molecular patterns during neural development. *Proc Natl Acad Sci U S A* 84:8716–8720

Bonafede DM, Missias AC, Constantine-Paton M (1989): Expression and synthesis of ganglioside 9-O-acetyl GD3 in mouse glioma subclones. *Soc Neurosci Abstr* 15:567

Bonhoeffer F, Gierer A (1984): How do retinal axons find their targets on the tectum? *Trends Neurosci* 7:378–381

Cheresh DA (1987): Ganglioside involvement in tumor cell-substratum interactions. In: *Development and Recognition of the Transformed Cell* Greene MI, Hamaoka T, eds. New York: Plenum Press

Cheresh DA, Piersbacher MD, Herzig MA, Mujoo J (1986): Disialogangliosides GD2 and GD3 are involved in the attachment of human melanoma and neuroblastoma cells to extracellular matrix proteins. *J Cell Biol* 102:688–696

Cheresh DA, Pytela R, Pierschbacher MD, Klier FG, Ruoslahti E, Reisfeld RA (1987): An Arg-Gly-Asp-directed receptor on the surface of human melanoma cells exist in a divalent cation-dependent functional complex with the disialoganglioside GD2. *J Cell Biol* 105:1163–1172

Cheresh DA, Varki AP, Varki NW, Stallcup WB, Levine J, Reisfeld RA (1984): A monoclonal antibody recognizes an 0-acetylated sialic acid in a human melanoma-associated ganglioside. *J Biol Chem* 259:7453–7459

Constantine-Paton M, Blum AS, Mendez-Otero R, Barnstable CJ (1986): A cell surface molecule distributed in a dorso-ventral gradient in the perinatal rat retina. *Nature* 324:459–462

Drazba J, Pierce M, Lemmon V (1991): Studies of the developing chick retina using monoclonal antibody 8A2 that recognizes a novel set of gangliosides. *Dev Biol* 145:154–163

Edmondson JC, Hatten ME (1987): Glial-guided granule neuron migration *in vitro*: A high-resolution time-lapse video microscopic study. *J Neurosci* 7:1928–1934

Eisenbarth GS, Walsh FS, Nirenberg M (1979): Monoclonal antibody to a plasma membrane antigen of neurons. *Proc Natl Acad Sci U S A* 76:4913–4917

Godement P, Bonhoeffer F (1989): Cross-species recognition of tectal cues by retinal fibers in vitro. *Development* 106:313–320

Grunwald GB, Fredman P, Magnani JL, Triesler D, Ginsburg V, Nirenberg M (1985): Monoclonal antibody 18B8 detects gangliosides associated with neuronal differentiation and synapse formation. *Proc Natl Acad Sci U S A* 82:4008–4012

Hatten ME, Liem RK, Mason CA (1984): Two forms of cerebellar glial cells interact differently with nervous system. *J Cell Biol* 90:622–630

Kim JYH, Goldering JR, DeLorenzo RJ, Yu RK (1986): Gangliosides inhibit phospholipid-sensitive Ca^{2+} -dependent kinase phosphorylation of rat myelin basic proteins. *J Neurosci Res* 15:159–166

Kleinman HK, Martin GR, Fishman PH (1979): Ganglioside inhibition of fibronectin-mediated cell adhesion to collgen. *Proc Natl Acad Sci U S A* 76:3367–3371

Levine JM, Beasly L, Stallcup WB (1984): The D.1.1 antigen: A cell surface marker for germinal cells of the central nervous system. *J Neurosci* 4:820–831

Marchase RB (1977): Biochemical investigation of retinotectal adhesive specificity. *J Cell Biol* 75:237–257

McLoon SC (1991): A monoclonal antibody that distinguishes between temporal and nasal retinal axons. *J Neurosci* 11:1470–1477

Mendez-Otero R, Constantine-Paton M (1990): Granule cell induction of 9-O-acetyl gangliosides on cerebellar glia in microcultures. *Dev Biol* 138:400–409

Mendez-Otero R, Schlosshauer B, Barnstable CJ, Constantine-Paton M (1988): A developmentally regulated antigen associated with neural cell and process migration. *J Neurosci* 8:564–579

Multani P, Bonafede DM, Yu RK, Constantine-Paton M (1988): Biochemical characterization of Jones immunoreactive gangliosides in rat. *Soc Neurosc Abstr* 14:1016

Rabacchi SA, Neve RL, Drager UC (1990): A positional marker for the dorsal retina is homologous to the high-affinity laminin receptor. *Development* 109:521–531

Rakic P (1971): Neuron-glia relationship during granule cell migration in developing cerebellar cortex. A Golgi and electronmicroscopic study in *Macacus rhesus J Comp Neurol* 141:283–312

Reinhardt-Maelicke S, Cleeves V, Kindler-Rohrborn A, Rajewsky MF (1990): Differential recognition of a set of O-acetylated gangliosides by monoclonal antibodies RB13-2, D1.1, and Jones during rat brain development. *Dev Brain Res* 51:279–282

Roth S, McGuire EJ, Roseman S (1971): Evidence for cell-surface glycosyltransferases: Their potential role in cell recognition. *J Cell Biol* 51:536–547

Sabel B (1988): Anatomic mechanisms whereby ganglioside treatment induces brain repair. In: *Pharmacological approaches to the Treatment of Brain and Spinal Cord Injury*, Stein DG, Sabel B, eds. New York: Plenum Press

Schlosshauer B, Blum AS, Mendez-Otero R, Barnstable CJ, Constantine-Paton M (1988): Developmental regulation of ganglioside antigens recognized by the Jones antibody. *J Neurosci* 8:580–592

Schlosshauer B, Mendez-Otero R, Constantine-Paton M (1986): Developmental regulation of JONES gangliosides in the mammalian nervous sytem. *Soc Neurosci Abstr* 12:317

Schwartz M, Spirman N (1982): Sprouting from chick embryo dorsal root ganglia induced by nerve growth factor is specifically inhibited by affinity purified antiganglioside antibodies. *Proc Natl Acad Sci U S A* 79:6080–6083

Sloan SF, Koening E, Lemmon V (1989): Monoclonal antibody 8A2 triggers evacuation of growth cone contents and disto-proximal bulk redistribution of axoplasm in growing axons. *Soc Neurosci Abstr* 15:1027

Sparrow JR, Barnstable CJ (1988): A gradient molecule in rat retina: Expression of 9-O-acetyl GD3 in relation to cell type, developmental age, and GD3 ganglioside. *J Neurosci Res* 21:398–409

Sparrow JR, McGuiness C, Schwartz M, Grafstein B (1984): Antibodies to gangliosides inhibit goldfish optic nerve regeneration in vivo. *J Neurosci Res* 12:233–243

Sperry RW (1963): Chemoaffinity in the orderly growth of nerve fiber patterns and connections. *Proc Natl Acad Sci U S A* 5:703–710

Stallcup WB (1988): Involvement of gangliosides and glycoprotein fibronectin receptors in cellular adhesion to fibronectin. *Exp Cell Res* 177:90–102

Stallcup WB, Pytela R, Ruoslahti E (1989): A neuroectoderm-associated ganglioside participates in fibronectin receptor mediated adhesion of germinal cells to fibronectin. *Dev Biol* 132:212–229

Trisler D, Collins F (1987): Corresponding spatial gradients of TOP molecules in the developing retina and optic tectum. *Science* 237:1208–1209

Trisler D, Schneider MD, Nirenberg M (1981): A topographic gradient of molecules in retina can be used to identify neuron position. *Proc Natl Acad Sci U S A* 78:2145–2149

Varki A, Hooshmand F, Diaz S, Varki NM, Hedrick SM (1991): Developmental abnormalities in transgenic mice expressing a sialic acid-specific 9-O-acetylesterase. *Cell* 65:65–74

Vielmetter J, Stuermer CAO (1989): Goldfish retinal axons respond to position-specific properties of tectal cell membranes *in vitro*. *Neuron* 2:1331–1339
Walter JS, Henke-Fahle S, Bonhoeffer F (1987): Avoidance of posterior tectal membranes by temporal retina axons. *Development* 101:909–914
Yamada KD, Critchley D, Fishman P, Mossi J (1983): Exogenous gangliosides enhance the interaction of fibronectin with ganglioside-deficient cells. *Exp Cell Res* 143:295–302

4

The Activity-Dependent Mechanism in the Development of the Refined Retinotopic Map

HOLLIS T. CLINE

A central problem in the development of the central nervous system (CNS) concerns the formation of specific or appropriate synaptic connections. The developing visual system of amphibia is an optimal system in which to study this problem. The frog visual system is characterized by a highly refined topographic projection from each retina to the contralateral optic tectum, the central visual processing area in these animals. We know that at least two discreet, but interacting mechanisms operate during the development of the retinotectal projection: (1) an activity-independent mechanism, based on differential distributions of cell surface molecules on retinal axon terminals and the tectal cells, can account for a crude topographic organization of retinal ganglion cell axons within the tectal neuropil; and (2) an activity-dependent mechanism, which is thought to recognize correlated patterns of afferent activity, refines the projection to yield a point-to-point specificity in the position of the retinal axons within the tectal neuropil. The activity-dependent mechanism is the subject of this review.

Many of the experiments investigating the mechanisms underlying the development of the retinotectal system have used the three-eyed tadpole (Law and Constantine-Paton, 1981). In these animals an extra eye primordium is surgically implanted into the embryo prior to the time when retinal axons have exited from the eye. The tadpole grows up with three functional retinas, each of which sees a different part of visual space. The extra eye projects axons into the CNS, where they innervate one or both of the optic tectal lobes. The retinal axons from the supernumerary eye must share the available target space with the eye that normally projects to that tectal lobe. It does so by dividing the tectal neuropil into rostrocaudally oriented striped termination zones which are alternately innervated by the normal eye or the supernumerary eye. It is thought that the same rules that govern the formation of the topographic projection in the normal tadpoles also operate to form the stripe pattern in the three-eyed tadpoles. Namely, the activity-independent mechanism spreads the inputs from both eyes evenly across the tectum in a crudely topographic distribution, and the activity-dependent mechanism segregates the axons from one each eye.

The activity-dependent mechanism is so called because it requires normal patterns of action potential activity in the retinal ganglion cell (RGC) axons. This was shown by several labs working with lower vertebrates and mammals. In the three-eyed tadpole, Reh and Constantine-Paton (1985) applied tetrodotoxin in the slow release plastic Elvax to the optic nerves to silence the activity in the RGC axons. This treatment caused a desegregation of the stripe pattern in the optic tectum. At about the same time several reports demonstrated that neighboring RGCs of the same response type exhibit similar patterns of action potential activity, while nonneighboring RGCs exhibit disparate patterns of activity (Mastronarde, 1990). These contributions led many investigators to hypothesize that the formation of topographic projections does indeed involve two discreet mechanisms and that one of them is activity-dependent in that it can be blocked by TTX. Furthermore, the hypothesis suggested that the correlated patterns of action potentials in axons arising from neighboring RGCs might convey the neighbor relations of the RGC somata to the axon arbors within the tectal neuropil.

A more specific version of this correlated activity hypothesis has received support from a series of experiments performed over the last 5 years in several lower vertebrates. This version of the hypothesis suggested that the NMDA type of glutamate receptor might serve to recognize the correlated patterns of input activity and as such, would be required for the development and maintenance of retinal topography. The NMDA receptor might be suited for this job because both membrane depolarization and binding of the transmitter glutamate are required in order for the associated channel to conduct ions. Glutamate is now thought to be the neurotransmitter of the RGCs and both NMDA and non-NMDA type receptors are present in the tectal neuropil where the retinal afferents form synaptic contacts with the postsynaptic tectal neurons (McDonald et al., 1989).

An initial set of experiments tested whether NMDA receptor activity was necessary for the maintenance of the stripe pattern in three-eyed tadpoles and young postmetamorphic froglets (Cline et al., 1987). Competitive (2-aminophosphonovaleric acid, [APV]) or noncompetitive (MK801) NMDA receptor antagonists were applied to the optic tectum of three-eyed animals, using the slow release plastic Elvax. After a period of 4–8 wk, the integrity of the stripe pattern was tested by applying HRP to the supernumerary optic nerve to label its projection in the tectum. The stripes desegregate after constant exposure to NMDA receptor antagonists. Drug exposure for shorter periods resulted in a partial desegregation, indicating that desegregation is a gradual process. Furthermore, the stripe desegregation was reversible: APV-Elvax was implanted over the tecta of three-eyed tadpoles for 4–6 wk. Some of the animals were tested for stripe desegregation as mentioned above. In others the APV-Elvax was removed from the tectum and the stripe pattern was assayed after another 2 wk. In these animals the stripe pattern appeared normal. The reversibility of the APV-induced desegregation serves as a reminder that the retinotectal synapses are labile and that the axon arbors are capable of considerable mobility within the tectal neuropil over relatively short periods of time. The remarkable mobility of the RGC axon terminals within the tectal neuropil and the repercussions of this mobility on the development and maintanence of retinotectal maps has been reviewed elsewhere (Cline, 1991).

Similar experiments performed in two-eyed tadpoles indicated that treatments with NMDA receptor antagonists also disrupted retinal topography in normal tadpoles (Cline and Constantine-Paton, 1989). In these experiments the fidelity of the projection was tested by injecting a focal region of the optic tectum with HRP. The HRP was taken up into the RGC axons that project to that tectal region and was transported back to the RGC somata. In control or sham-operated tadpoles the HRP-labeled RGCs are confined to a small portion of the retina, whereas in the APV- or MK801-treated animals the HRP-labeled somata spread over an unusually large portion of the retina. Importantly, APV-treated arbors are not larger in either two-eyed or three-eyed tadpoles. We interpret these results to mean that RGCs that would normally project their axons to a different tectal location now project at least a part of their arbor to the tectal region labeled by the HRP injection, thus decreasing the point-to-point specificity of the topographic map.

Treatment of the optic tecta with APV does not interfere with action potential activity in the retinal afferents (Cline et al., 1987) and does not decrease the amplitude of evoked potentials recording from the tectal neuropil following optic nerve stimulation (Debski et al., 1989). Other studies have shown that optic nerve shock is not the optimal method for activating the NMDA type receptor, presumably because the nerve shock will activate axons of similar diameter and threshold, without regard to whether the fibers originate from neighboring RGCs. However, the NMDA receptor is likely to be activated preferentially by activity in RGCs that converge on the same tectal cell dendrite. Therefore, it remains a possibility that synaptic transmission in response to visual input may be somewhat decremented by NMDA receptor antagonists, as it is in the mammalian visual system and in the hippocampus.

The experiments in the two- and three-eyed frogs demonstrate that NMDA receptor activity is necessary to maintain the correct retinotectal topography during the growth of the visual system. The retinal arbors are capable of great mobility within the tectal neuropil and the active maintenance of topography must be accomplished by a mechanism that will maintain the arbors within the "correct tectal locale" and limit their lateral migration. The "correct tectal locale" is defined by the location where the activity of one retinal input is most highly correlated with the activity patterns of the other inputs in that part of the tectum. Consequently, the specific part of the tectum that is "topographically correct" changes during development. The NMDA receptor operates to recognize afferent coactivity and send an intracellular signal, notably a rise in intracellular calcium (Cline and Tsien, 1991) that such coactivity has occurred. The rise in intracellular calcium may then trigger events that temporarily stabilize the coactive synapses in that location. The stabilization must be temporary because shortly, with additional retinal and tectal cell proliferation, that particular tectal location will no longer be topographically correct for those afferents and they will migrate to a new site, based on requisite coactivity with neighboring inputs.

These studies suggested that the NMDA receptor may be required for the development and maintanence of retinotopic maps. This conclusion has received support from work by Schmidt (1990) on the regenerating retinotectal projection and by Udin and her co-workers (Scherer and Udin, 1989) on the development of binocular inputs. Schmidt has also demonstrated that another type of neuronal plasticity, long-term potentiation (LTP) of synaptic transmission, can be elicited in the optic tectum and that it is sensitive to APV treatment. The APV-sensitivity of the LTP recorded in the optic tectum supports the idea that the cellular mechanisms underlying plasticity in developing systems may be comparable to those underlying plasticity in mature neural systems (Cline et al., 1987).

Further information about the possible role of the NMDA receptor in shaping retinotectal connections was obtained from experiments in which the tecta of three-eyed tadpoles or postmetamorphic froglets were treated with the receptor agonist NMDA. This treatment resulted in a dramatic change in the structure of the axonal arbors. Specifically, the number of branchtips was severely reduced. This observation may be particularly relevant to retinotectal connectivity since electron microscopic studies by Lai-Hsing Yen have shown that the majority of synaptic contacts are made by the fine terminal branchtips of the retinal axon arbors (Yen and Constantine-Paton, 1988, 1991). It is precisely these branchtips that are lost with NMDA treatment.

However, comparable treatments of two-eyed animals did not result in a comparable reduction in the number of branchtips. What differences between the normal and the striped tecta could account for the different responses to NMDA treatment? One major difference is in the axon arbors themselves. Arbors drawn from sham-operated but drug-free striped tecta were compared to those drawn from sham-operated normal tecta. Arbors from striped tecta cover the same tangential area as those from normal tecta, but arbors from striped tecta have about twice the number of branchtips as the arbors from normal tecta (Cline and Constantine-Paton, 1990). Therefore, arbors from striped tecta are more densely arborized and the density of the arborization can be reduced by NMDA treatment. In contrast, the arbors from normal tecta are less densely arborized and are not subject to loss of arbor density by NMDA treatment.

Several differences between the striped tecta and the normal tecta might account for the different arbor morphologies seen and for the different responses to NMDA treatment. Striped tecta have twice the number of RGCs projecting to a tectal neuropil which is increased in size by only 40% (Constantine-Paton and Ferrari-Eastman, 1987; Constantine-Paton and Norden, 1986). This suggests that in the striped tecta, the same number of RGC arbors are forced to converge on a relatively smaller target area. Therefore arbors that in the normal tecta would terminate in neighboring regions of the tectal neuropil now terminate in the same region of neuropil. If one accepts the hypothesis that arbors terminate in neighboring regions of the neuropil in normal tecta because they exhibit different patterns of electrical activity, then the increased convergence in the striped tecta means that any one site in the striped neuropil now has inputs from coactive afferents that would normally project to that site in addition to inputs from less-coactive afferents, which would normally project to a neighboring site. If one furhter accepts that NMDA receptor activity is an indication of the degree of afferent coactivity at a tectal site, then the increased convergence in the striped tectum would mean that NMDA receptor activity would be decreased relative to the total amount of postsynaptic activity in the striped tecta. We would like to suggest that it is this decrease in the degree of NMDA receptor activity that leads to the increase in branch elaboration within the axon arbors in striped tecta.

Thus these results indicate that NMDA receptor activity is involved in two seemingly discreet events in the development of the retinotectal projection: the organization of the positions of the axon arbors within the target space and the regulation of the addition and retraction of individual branches within the axon arbor. These results suggest that the mechanism underlying these events may not be as unrelated as previously thought.

Our conclusions and interpretations of the experiments discussed above are based on the use of glutamate as the RGC transmitter and on the presence of glutamate receptors (NMDA and non-NMDA type) in the optic tectum. The evidence for the use of glutamate as the RGC transmitter has been reviewed elsewhere (Debski et al., 1990). Using receptor autoradiography, we have shown that NMDA and non-NMDA subtypes of glutamate receptors are located in the tectal neuropil (McDonald et al., 1989), where both pre- and postsynaptic elements are located. Several studies indicate that the glutamate receptors are on the postsynaptic tectal cells (Cline and Tsien, 1991; Debski et al., 1989). However, a possible location of glutamate receptors on retinal afferents has not been ruled out. Indeed, chronic exposure of the optic tectum to elevated concentrations of NMDA results in a beaded appearence of the retinal arbors (Cline and Constantine-Paton, 1990), reminiscent of glutamate-induced excito-toxic responses seen in other neurons. This observation suggested either that some type of excitotory amino acid receptor exists on the afferents or that the damage might be a secondary response following damage to the tectal cells. To address the question of the relative contributions of retinal afferents and tectal cell dendrites to the total glutamate binding seen in the tectal neuropil, we performed receptor binding of the tectal neuropil following unilateral enucleation with various postenucleation survival times. Previous studies have shown that within the first 3–5 days following enucleation the retinal afferents begin to degenerate (Ostberg and Norden, 1979). However, the tectal cell dendrites do not respond to the trauma until 10–14 days after enucleation. Therefore, a rapid loss of glutamate receptor binding in the enucleated tectum would indicate a presynaptic component to the total binding, whereas a delayed loss in binding would indicate that the majority of binding is postsynaptic.

Enucleation resulted in a 25% loss of NMDA and Quisqualate (Quis) binding sites only after 7–14 days of survival time (Cline et al., 1991). In animals with shorter postenucleation survival times, no decrease in binding was seen (see Figure 4.1). Comparable losses of NMDA and Quis sites were seen. These data indicated that these glutamate binding sites are located exclusively on the postsynaptic tectal cell dendrites. This finding is consistent with the observations that glutamate does not cause a rise in intracellular calcium in retinal axon growth cones (H.T. Cline, unpublished observations) and that retinal axons do not retract when exposed to glutamate (Constantine-Paton and Prusky, 1991). Furthermore,

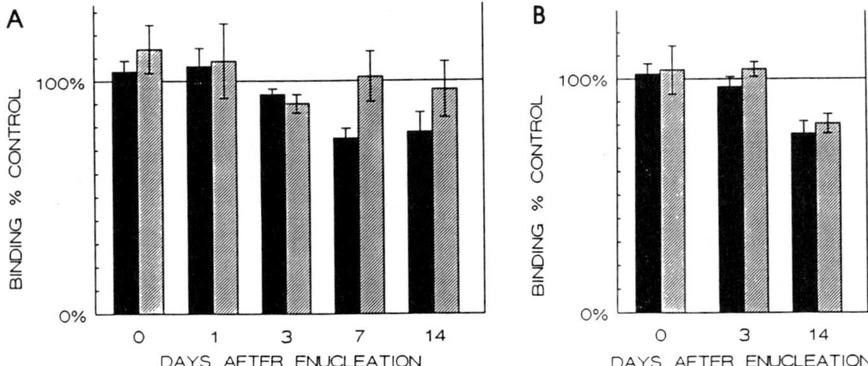

FIGURE 4.1. Enucleation results in a loss of NMDA and Quis binding sites in the tectal neuropil with 7–14 days survival time. A. NMDA receptor binding, expressed as percent of binding in the intact tectum. B. Quisqualate binding, as percent of binding in the intact tectum. Layer 9 (black) is the retinorecipient neuropil. Layer 6 (gray) is the cell body layer. These cells send their dendrites into layer 9.

since only 25% of the total binding in the tectal neuropil was effected by the enucleation, the data suggest that the majority of the glutamate binding sites may exist at synapses with presynaptic elements other than the retina. As these data might suggest, the majority of tectal neurons are immunoreactive with antibodies against glutamate (H.T. Cline, unpublished observations).

The biological basis of the developmental plasticity of the frog retinotectal system is the dramatically asymetrical patterns of cell proliferation in the retina and tectum (Gaze et al., 1979), which is compensated for by shifting retinotectal synaptic connections in order to maintain retinal topography (Reh and Constantine-Paton, 1984). However, the retinotectal projection continues to shift in young postmetamorphic frogs (postmets) and in mature adults as the tectal lobes enlarge up to 5 times the postmet diameter, largely due to intersticial glial proliferation, and the elaboration of both afferent and tectal cell processes. In addition, the plasticity of the regenerated projection in adult frogs is well known (Sperry, 1956). NMDA receptor density may reflect the relative “plasticity” of the tectum at various stages.

We determined the glutamate binding in several brain regions in adult and postmetamorphic frogs and figured the ratio of NMDA binding to Quis binding for postmets and adults as a possible indicator of the relative degree of plasticity of the brain region (Cline et al., 1991). In most regions (i.e., telencephalon, torus, optic nerve, layer 6 of optic tectum) the NMDA/Quis binding ratio is about 0.1 in both postmets and adults. However, in the superficial neuropil layer of the optic tectum, where NMDA receptor-mediated plasticity has been shown to occur, NMDA binding is enriched relative to Quis binding. In postmets the NMDA/Quis binding ratio is 0.256, about 4 times the ratio in other regions. In the adult the tectal neuropil continues to exhibit the highest NMDA/Quis ratio of 0.18. It is interesting that the lower rim cortex of the nucleus isthmi also exhibits an NMDA/Quis ratio of 0.18. The deep layers of the tectum, which receive auditory input, display an elevated NMDA/Quis ratio of 0.18 in postmets, and this value returns to about 0.1 in adults.

In conclusion, considerable evidence from a variety of studies using several means of investigation indicates a prominent role for the NMDA receptor in the maintenance of retinotectal topographic projections during development and even after metamorphosis.

REFERENCES

Cline HT (1991): Activity-dependent plasticity in the visual systems of frogs and fish. *Trends Neurosci* 14:104–111

Cline HT, Constantine-Paton M (1989): NMDA receptor antagonists disrupt the retinotectal topographic map. *Neuron* 3:413–426

Cline HT, Constantine-Paton M (1990): NMDA receptor agonist and antagonists after retinal ganglion cell arbor structure in developing frog retinotectal projection. *J Neurosci* 10:1197–1216

Cline HT, Debski E, Constantine-Paton M (1987): NMDA receptor antagonist desegregates eye-specific stripes. *Proc Natl Acad Sci U S A* 84:4342–4345

Cline HT, McDonald J, Constantine-Paton M (1991): Glutamate receptor binding in the frog brain. *Proc Soc Neurosci* 17:317

Cline HT, Tsien RW (1991): Glutamate-induced increases in intracellular Ca2+ in cultured frog tectal cells mediated by direct activation of NMDA receptor channels. *Neuron* 6:256–267

Constantine-Paton M, Ferrari-Eastman P (1987): Pre- and postsynaptic correlates of interocular competition and segregation in the frog. *J Comp Neurol* 255:178–195

Constantine-Paton M, Norden JJ (1986): Development of order in the visual system. In: *Cell and Developmental Biology of the Eye*, Hilfer SR, Sheffield JB, eds. New York: Springer-Verlag

Constantine-Paton M, Prusky G (1991): Responses of cultured frog retinal axons and dissociated tectal cells to glutamatergic and cholinergic agents. *Proc Soc Neurosci* 17:1134

Debski EA, Cline HT, Constantine-Paton M (1989): Chronic application of NMDA or APV affects the NMDA sensitivity of the evoked tectal response in *Rana pipiens*. *Proc Soc Neurosci* 15:495

Debski EA, Cline HT, Constantine-Paton M (1990): Activity-dependent fine-tuning and the NMDA receptor. *J Neurobiol* 21:18–32

Gaze RM, Keating MJ, Ostberg A, Chung S-H (1979): The relationship between retinal and tectal growth in larval *Xenopus*: Implications for the development of retino-tectal projection. *J Embryol Exp Morph* 53:103–143

Law MI, Constantine-Paton M (1981): Anatomy and physiology of experimentally produced striped tecta. *J Neurosci* 1:741–759

Mastronarde DN (1989): Correlated firing of retinal ganglion cells. *Trends Neurosci* 12:75–80

McDonald JM, Cline HT, Constantine-Paton M, Maragos WE, Johnston MV, and Young AB (1989): Quantitative autoradiographic localization of NMDA, quisqualate and PCP receptors in the frog brain. *Brain Res* 482:155–158

Ostberg A, Norden J (1979): Ultrastructural study of degeneration and regeneration in the amphibian system. *Brain Res* 168:441–455

Reh TA, Constantine-Paton M (1984): Retinal ganglion cells change their projection sites during larval development of *Rana pipiens*. *J Neurosci*. 4:442–457

Reh TA, Constantine-Paton M (1985): Eye-specific segregation requires neural activity in three-eyed *Rana pipiens*. *J Neurosci* 5:1132–1143

Scherer WS, Udin SB (1989): N-methyl-D-aspartate antagonists prevent interaction of binocular maps in *Xenopus* tectum. *J Neurosci* 9:3837–3943

Schmidt JT (1990): Long-term potentiation and activity-dependent retinotopic sharpening in the regenerating retinotectal projection of goldfish: Common sensitive period and sensitivity to NMDA blockers. *J Neurosci* 10:233–246

Sperry RW (1956): The eye and the brain. *Sci Am* 194:48–52

Yen L-H, Constantine-Paton M (1988): EM analysis of single retinal ganglion cell terminals in developing Rana pipiens. *Proc Soc Neurosci* 14:674

Yen L-H, Constantine-Paton M (1991): EM analysis of single retinal ganglion cell terminals in NMDA-treated tecta of Rana pipiens. *Proc Soc Neurosci* 17:1134

5

Mechanisms of Dendritic Tree Development in Mammalian Retinal Ganglion Cells

ARY S. RAMOA AND EDNA N. YAMASAKI

In the adult mammalian retina, several classes of ganglion cells have been described based on morphological parameters such as dendritic field size, soma size, and the pattern of dendritic branching (Boycott and Wässle, 1974; Kolb et al., 1981; Rodieck, 1983; Wässle and Boycott, 1991). Furthermore, some of the morphologically defined classes also display distinctive electrophysiological properties (Fukuda et al., 1984; Saito, 1983). In the retina of the cat, one of the best studied models, alpha and beta classes have been examined in detail. Alpha cells display large somata and dendritic trees, while beta cells have medium-sized somata and the smallest dendritic trees of any class (Boycott and Wässle, 1974). In addition, the dendritic trees of beta cells are more highly branched than those of alpha cells. In the retina of other mammals, ganglion cells have been described that share with alpha cells some of their distinctive features. Type I cells in the rat's retina, for example, resemble alpha cells in that they display large dendritic trees and large somata. Other ganglion cells found in the rat's retina include type II cells, which have small-to-medium size somata and small, highly branched dendritic trees studded with spines (Perry, 1979), and type III cells, which are characterized by large, sparsely branched dendritic trees and the smallest somata among retinal ganglion cells (RGCs).

This brief description of a relatively small number of RGC types is sufficient to convey the idea of a remarkable diversity of form and function in the adult mammalian retina, thus raising the issue of how the mature state is achieved. Several questions can be asked concerning the mechanisms that govern RGC dendritic development and the establishment of intraretinal connections. When do RGCs that resemble the adult classes first appear during development? How do their dendritic trees achieve the adult form? What role do intraretinal factors and the target nuclei play in the establishment of the adult form and function? To answer some of these questions, in recent years we have examined the dendritic morphology and the electrophysiological properties of ganglion cells in the developing cat and rat retina. It is the purpose of this review to summarize and discuss these findings. Moreover, we will discuss preliminary results that suggest that electrical activity at the intraretinal, but not target, site plays an important role in dendritic growth and remodeling of ganglion cells.

The following discussion will be divided into three sections. First, we will summarize descriptive studies of the morphological development of retinal ganglion cells in both rats and cats. Second, we will discuss some of the electrophysiological properties of growing RGCs. Evidence will be provided that functionally mature N-methyl-D-aspartate (NMDA) and acetylcholine receptors (AChR) are present very early in the membrane of RGCs, even before synaptogenesis starts. These findings are relevant in the context of the working hypothesis that intraretinal activity modulates dendritic growth. Third, we will discuss the results of experiments in which we have attempted to alter electrical activity by pharmacologically manipulating the retina and one of its primary targets, the lateral geniculate nucleus.

DENDRITIC REMODELING OF RETINAL GANGLION CELLS DURING DEVELOPMENT

Retinal ganglion cells resembling the adult types are present very early in development, as early as embryonic day 50 (ED50) in the cat (Maslim et al., 1986; Ramoa et al., 1987, Ramoa, Campbell and Shatz 1988) and postnatal day 3 (PD3) in the rat (Maslim et al., 1986; Perry and Walker, 1980). These results were obtained using a variety of techniques, including the Golgi method, extracellular injections of horseradish peroxidase, and intracellular injections of Lucifer yellow in living or fixed tissue. Intracellular dye injection has the advantage of reliably filling the axons and the dendrites, including growth cones, filopodia, and fine dendritic branchlets, of developing neurons. This technique has allowed the demonstration that developing ganglion cells are not just miniature replicas of the mature neurons, but express several transient morphological features (Dann et al., 1987, 1988; Ramoa et al., 1987, Ramoa, Campbell and Shatz 1988). The dendrites of ganglion cells in the fetal and early postnatal cat retina have an excessive number of spines and branches, and the axons give rise to delicate branches and long collaterals which disappear later in development (Ramoa et al., 1988). The number of transient dendritic filaments increases, reaching a peak by the end of the first week of postnatal life, just before eye opening, then falls dramatically to reach adult levels by the end of the first postnatal month. These results indicate that extensive remodeling of the dendritic trees occurs before the final pattern of synaptic connections is established, raising the possibility that a transient neuronal network is present during retinal development. Such a network should not be considered merely a simplified, immature version of that present in the adult, since it could conceivably subserve specific functional roles in the developing retina. For example, transient connections may contribute to the generation of the synchronous bursts of action potentials that have been found in the developing mammalian retina (Meister et al., 1991). It is unfortunate, therefore, that little is known about the types of neuronal contacts, if any, that are made by the transient processes. Although mature synapses are not expected to be present during early development, developing ganglion cells could conceivably interact through gap juctions or relatively simple chemical synaptic specializations. More observations at the ultrastructural level are therefore needed to shed light into the functional significance of transient processes.

The possibility that transient processes play an important role in visual system development has raised the question of whether the findings in cat alpha and beta cells can be generalized to other ganglion cell classes and other species. In nonmammalian species, RGC development seems to be achieved by a process of continuous growth, and a transient exuberance of dendritic and axonal processes has not been found (Bloomfield and Hitchcock, 1991; Sakaguchi et al., 1984). In cats, a differential magnitude of remodeling is present among different classes: alpha cells lose more dendritic branches and spines than beta cells (Ramoa et al., 1988). In rats, the magnitude of the loss also is class-dependent. Type I rat RGCs, which resemble the cat alpha cells, undergo extensive loss of spines and dendritic branches with age, as illustrated in the photomicrographs shown in Figure 5.1. The upper cell was obtained from a rat at PD12 (Figure 5.1A) and the other from a mature animal at PD30 (Figure 5.1B). Both cells had very large cell bodies and large dendritic trees when compared with neighboring neurons at the same ages. Note at PD 12 the large number of dendritic branches and spines, which contrasts with the smooth appearance of the mature neuron. Generally, a scarcity of spines characterizes adult type I cells, indicating that most of the short processes in their immature counterparts are transient. Note, however, that a few spines remained in the mature type I cell.

Other cell types in the adult rat retina display considerably more complex dendritic trees. To determine whether remodeling also occurs during development of these cells, we made drawings of ganglion cells ranging in age from ED20–PD 36 filled with Lucifer Yellow. Careful drawings are crucial to provide an accurate determination of the number of spines and branchpoints at each age and have allowed us to show that even ganglion cells that retain into adulthood a highly branched dendritic tree studded with spines, such as rat type II cells, can undergo extensive dendritic remodeling (Yamasaki et al., 1989; however, see Wong, 1990). The results are illustrated in the line drawings of Figures 5.2 and 5.3. On ED20 (Figure 5.2A) and PD zero (Figure 5.2B), ganglion cells are small and relatively simple, their dendritic trees bearing only a small number of dendritic branches and spines. Just a few days later, on PD3 (Figure 5.2C), the range of morphological diversity has increased substantially and some RGCs that resemble the adult types can already be recognized. The morphological complexity of RGCs continues to increase at a fast pace within the next few days and the most exuberant dendritic trees are observed by PD5–12. After this age, transient processes are gradually lost, so that around PD30 the mature state has been achieved. This is illustrated in Figure 5.3, which shows examples of type I, II, and III cells found on PD5 (A,C, and E, respectively) and PD30 (B,D, and F). Type I cells exhibit a large dendritic field and a very large cell body when compared to their neighbors, while type II RGCs reveal a typically small and highly branched dendritic tree and a small cell body. The other cells were classified as type III due to their small cell bodies and wide, sparsely branched dendritic fields. Notwithstanding the extreme range of morphological diversity exhibited by the PD5 cells in Figure 5.3, it is clear that within each type a major loss of dendritic processes has taken place by PD30. The loss is class-specific in the sense that its magnitude is distinct for each type. Note that type III neurons undergo relatively little remodeling, possibly losing only spines during development (compare Figure 5.3E and 5.3F), and despite the complex appearance of mature type II neurons, they also seem to lose many dendritic processes (compare Figure 5.3C and 5.3D). Quantitative evidence for these conclusions will be provided elsewhere (Yamasaki and Ramoa, 1992).

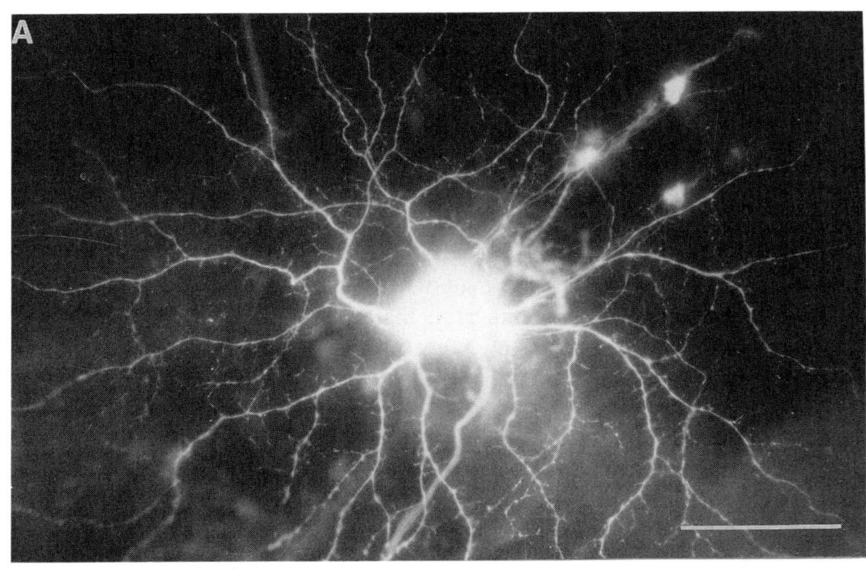

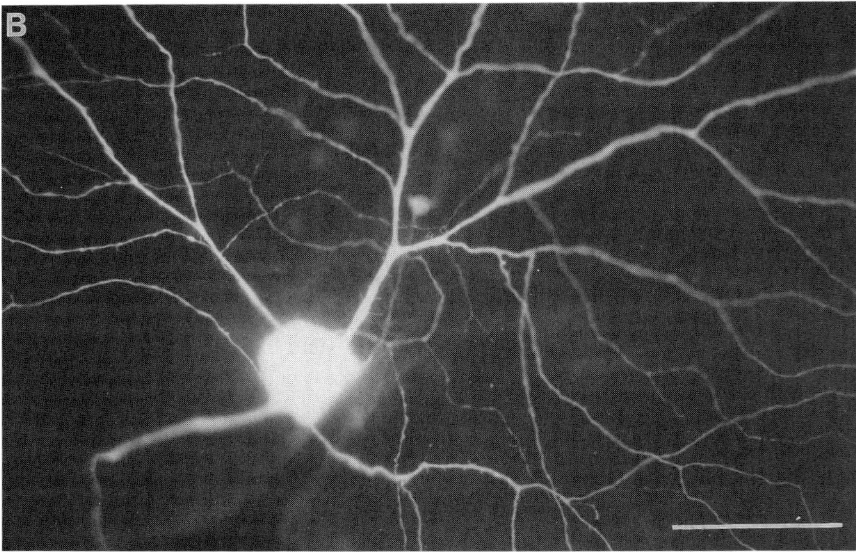

In conclusion, these results show that regressive phenomena, which are common during development of other neuronal systems (Jacobson, 1991), occur in the retina. The easy accessibility of the retina for experimental manipulations, the fact that it is one of the best studied systems in the brain, the almost planar organization of the ganglion cell dendritic trees, and the possibility that some of these findings can be generalized to more complex systems, make the developing retina an extremely valuable model in the study of mechanisms that contribute to dendritic tree maturation and the establishment of neuronal connections.

ELECTROPHYSIOLOGICAL PROPERTIES OF GANGLION CELLS DURING DEVELOPMENT

To further advance our understanding of the mechanisms that play a role in the establishment of the mature dendritic morphology and pattern of connections of ganglion cells, it may be useful to examine intraretinal developmental events that occur concurrent with the changes in morphology described above. Dendritic remodeling overlaps in time with synapse formation at the inner plexiform layer (Maslim and Stone, 1986). During maturation of a synaptic substrate for interneuronal communication, the electrophysiological properties of RGCs also undergo major changes (Rusoff and Dubin, 1977). Unfortunately, although a good

---

FIGURE 5.1. Two type I cells at PD 12 (A) and PD30 (B). Note the exuberant dendritic morphology displayed by the younger cell, and the smooth appearance of the dendritic tree in the mature neuron. Generally, relatively few spines were present in the dendritic trees of adult type I RGCs, suggesting that most spines present in developing type I neurons are transient processes lost after PD12 in the rat. Scale bar = 50 μm. *Methods*: Rats were studied between embryonic day 20 and postnatal day 36. The early ages were selected to reveal RGC morphology just a few days after the first retinal cell axons arrive in their target nuclei (Bunt et al., 1983). The oldest ages were selected to reveal the mature dendritic tree morphology. At every postnatal age studied, we unambiguously identified RGCs with injections of an undiluted suspension of rhodamine-labeled latex microspheres (Lumafluor, Bardonia, NY) made into the target nuclei. The details of the procedures involved in the preparation of the retina, intracellular labeling, and tissue processing can be found elsewhere (Ramoa et al., 1988a) and only a brief description will be provided here. After injection of rhodamine-labeled microspheres the animals were returned to the colony and studied after 24 hr and up to several weeks later, to allow for retrograde transport of the label. Retinas were gently dissected under deep anesthesia and maintained alive in a chamber mounted on the stage of a microscope equipped for epifluorescence. To reveal the fine dendritic morphology of labeled ganglion cells, Lucifer yellow (20% LY-Ch, Sigma, in 0.1 M LiCl) was injected intracellularly into selected cells using a microelectrode made of borosilicate (final resistance of about 80 MOhms). The time needed to completely fill the dendritic tree varied from a few seconds in fetal cells up to several minutes in adults. Next, the retinas were fixed and mounted onto gelatin coated slides and processed for later analysis and photography.

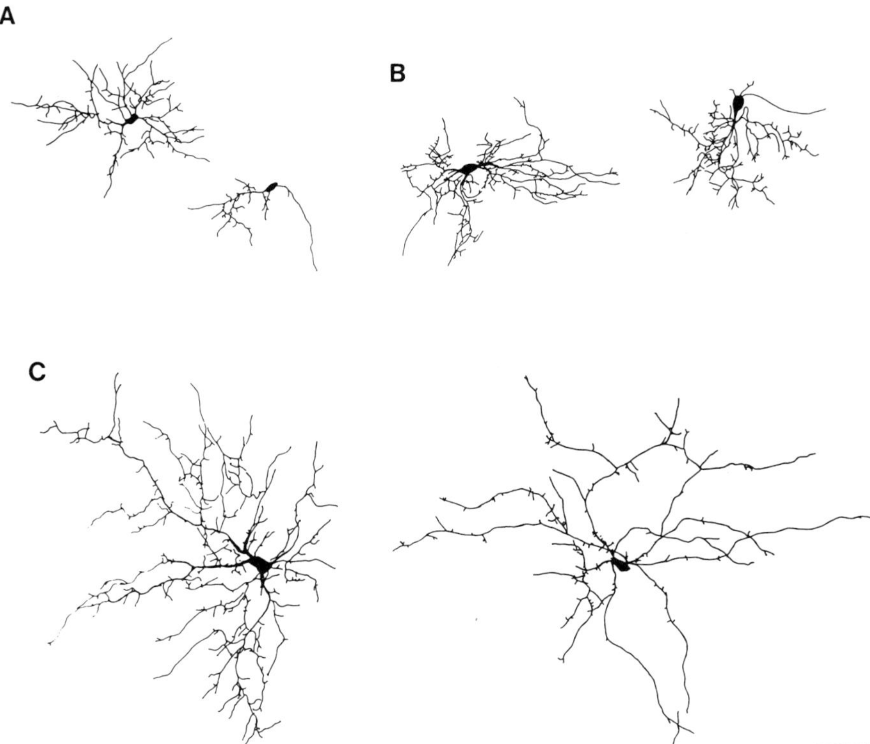

FIGURE 5.2. Camera lucida drawings of selected sets of rat RGCs injected with Lucifer yellow in fetal retina at ED20 (A), and after birth at PD zero (B) and PD3 (C). Note that morphological diversity is an attribute of the RGC population as early as ED20. Note also the striking increase in cell size and dendritic field complexity that occurs within the first few days after birth. Scale bar = 50 μm.

deal is known about the morphology of developing RGCs, very little is known concerning the concurrent changes in their membrane properties. To address this issue, we have examined the electrophysiological properties of identified ganglion cells that were enzymatically dissociated from retinas at different stages of development ranging from PD1–PD30, and studied within a few hours of plating (Ramoa et al., 1988b). The high-resolution patch-clamp technique was used to record spontaneous action potentials and single channel currents evoked by the application of glutamate, NMDA, kainate, and the selective agonist of the nicotinic acetylcholine receptor (AChR), (+) anatoxin (Spivak et al., 1980).

Action potential (AP) activity is known to be present in the intact retina of fetal rats' RGCs (Galli and Maffei, 1988), and we were therefore not surprised to find that ganglion cells fire APs in vitro as early as PD1. These were blocked by application of tetrodotoxin (1 μM) to the bathing medium, indicating they are generated by the activation of voltage-dependent sodium channels. Of greater

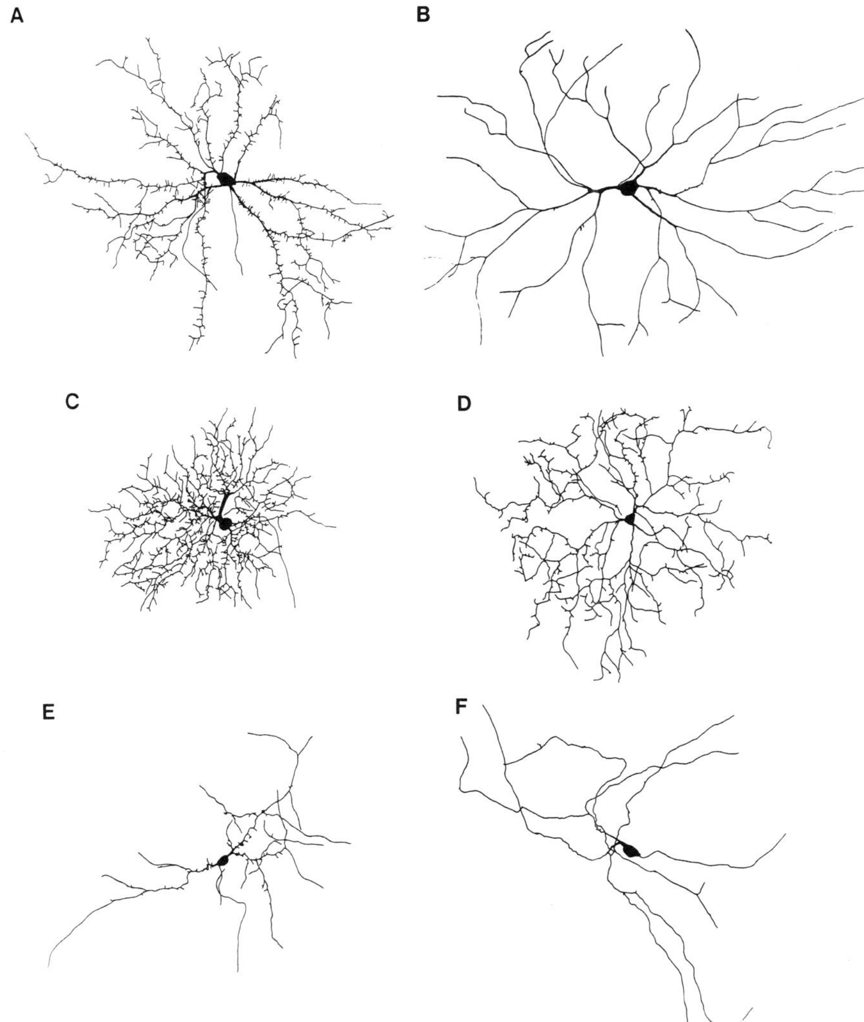

FIGURE 5.3. Camera lucida drawings of Lucifer yellow–fillled ganglion cells from rat retina studied at PD5 and PD30. Sets of 2 cells (one for each age) were selected according to morphological classification into types I (A-PD5; B-PD30), II (C-PD5; D-PD30), and III (E-PD5; F-PD30), to show that transient processes are common to distinct cell types. Although type I RGCs lose most of their spines, the examples indicate that dendritic remodeling also occurs in type II and, possibly to a lesser extent, in type III neurons. Scale bar = 50 μm.

relevance may be the finding that only a relatively small proportion of cells from neonatal retina (less than 20% of the total population) fired spontaneous APs while most cells obtained from animals a few days older (PD10 onward) were spontaneously active. Cells that did not fire APs seemed healthy in other respects:

they responded to chemical agents (see below) and were observed to grow neurites on the cultured substrate. It is important to note that development of AP activity occurs at a time when major changes are taking place in the retina: large numbers of cells are being deleted, dendritic trees are growing at a very fast pace, and synapses are being formed in the inner plexiform layer. Correlation in time between different events raises the possibility, but obviously does not indicate, that a causal relationship exists. The main value of comparing the time course of appearance and disappearance of such distinct phenomena is to originate testable ideas about mechanisms of growth. For example, since pruning of dendritic trees occurs at a time when most RGCs already fire APs, it is tempting to ask whether dendritic remodeling requires AP activity. Consistent with a role of AP activity in the developing retina, it has been shown that blockade of electrical activity by the application of tetrodotoxin (TTX), promotes the death of RGCs in culture (Lipton, 1986). Unfortunately, the role of electrical activity in dendritic growth and remodeling still is not clear. Intraocular application of TTX in cats altered only to a small extent the morphological maturation of RGCs (Wong et al., 1991), while an increased morphological complexity was found in the retina of hamsters injected with the same agent (Thompson, personal communications; Wingate and Thompson, 1991). More selective manipulation of electrical activity in the retina, using blockers and agonists of specific neurotransmitter systems, may help elucidate the role of activity in the developing retina (see next section).

To learn what role neurotransmitter systems play in the developing retina, it is necessary to examine when postsynaptic receptors are present in the membrane of ganglion cells and what their functional properties are. Patch-clamp recordings of singel-channel ion currents revealed that identified RGCs responded to glutamate, N-methyl-D-aspartate (NMDA), quisqualate, and anatoxin (Ramoa et al., 1988b). Responses were present as early as postnatal day 1, and were observed even in cells that did not fire APs. Examples of single-channel ion currents activated by NMDA and anatoxin are shown in Figure 5.4. The conductance values and the mean burst duration of the single-channel currents elicited by NMDA (about 40 pS, and 10 msec) and anatoxin (42 pS and 21 msec) remained unaltered during development. The properties of the response to NMDA, glutamate, and quisqualate (the latter two are not shown here) are similar to those observed in the hippocampus (Jahr and Stevens, 1987). In addition, application of the competitive NMDA antagonist amino-phophono-valerate (APV) reduced the frequency of NMDA-induced openings, and application of the noncompetitive antagonists phencyclidine and MK-801 reduced both the frequency and duration of single ion channel openings, providing evidence that these are NMDA activated channels (Huettner and Bean, 1988; Ramoa and Albuquerque, 1988).

The finding that chemosensitive receptors are present at an early age, even before synaptogenesis, raises the possibility that they play a morphogenetic role in RGC development. Neurotransmitter modulation of neurite growth, for example, which has been shown for neurons maintained in culture (Lankford et al., 1988; Lipton et al., 1988; for a review, see Lipton and Kater, 1989) could underlie dendritic remodeling in the intact retina. This is a question that we address in the following section.

REGULATION OF GANGLION CELL DEVELOPMENT BY NEUROTRANSMITTERS

Several neurotransmitters display effects on neurite outgrowth in vitro (Lipton and Kater, 1989). We will focus here on the possible role of the NMDA subtype of glutamate receptor, since it has often been implicated in synaptic plasticity in the hippocampus and visual system (for a review, see Rauschecker, 1991). Moreover, glutamate is a neurotransmitter known to activate RGCs (Aizemann et al., 1988) and to affect RGC neurite outgrowth *in vitro* (Lipton and Kater, 1988). Indeed, activation of NMDA receptors with the subsequent entry of calcium ions through the activated ion channel have been proposed as key events that lead to increased synaptic strength. We have therefore asked whether NMDA receptors present in immature neurons very early in development modulate neurite growth *in vivo*. To test this possibility, NMDA was applied intraocularly at PD3, a time when the dendritic trees grow and acquire spines and branches at a very fast pace.

The effects of a single intraocular injection of NMDA on the dendritic trees were examined at PD5. Examples of the results obtained for each of the three major cell types are shown in Figure 5.5. Simple inspection of their dendritic trees and a comparison with the normal PD5 cells shown in Figure 5.3 indicates that treated cells are highly abnormal for this age, resembling neurons found in younger animals. They have very small dendritic trees that bear fewer dendritic branches and spines than normal PD5 cells. Results are not yet available from retinas treated with the antagonist of the NMDA receptor, but APV applied to the LGN of developing ferrets has also been shown to affect dendritic tree growth; it induces the appearance of more complex dendritic trees than found in normal LGN cells of equivalent age. (Rocha et al., 1991). Thus, although additional experiments will have to be performed, their preliminary findings raise the possibility that activation of glutamate receptors plays a critical role in the modeling of dendritic trees of neurons in the mammalian visual system.

ROLE OF TARGET NUCLEI ON DENDRITIC TREE DEVELOPMENT

The theory that dendritic growth depends on neuronal interactions with target nuclei was proposed at the beginning of the century by Ramon y Cajal (1909/1911) and had received ample experimental support in both the peripheral and central nervous system (for a review, see Jacobson, 1991). The retina may be distinctive, however, in that evidence from the experiments discussed so far suggests that local interactions play a more important role. Indeed, we will provide additional experimental evidence that the role of target nuclei in the development of RGC dendritic trees is substantially less relevant than expected given previous results from other systems.

Included among the new experimental evidence to support this claim are results showing that developing rat RGCs that make transient projections to anomalous targets, such as the inferior colliculus, cannot be distinguished based on dendritic

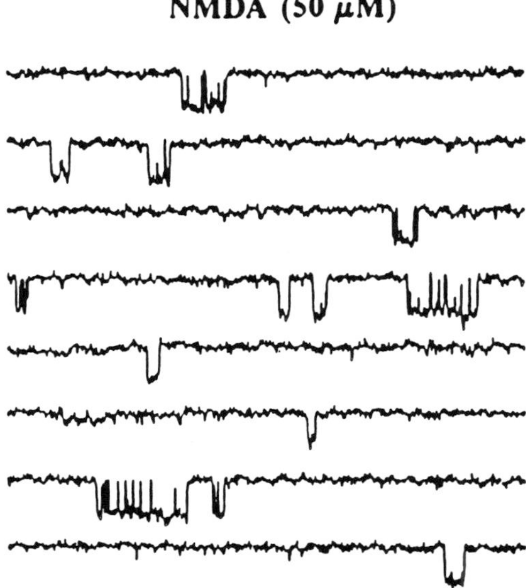
NMDA (50 μM)

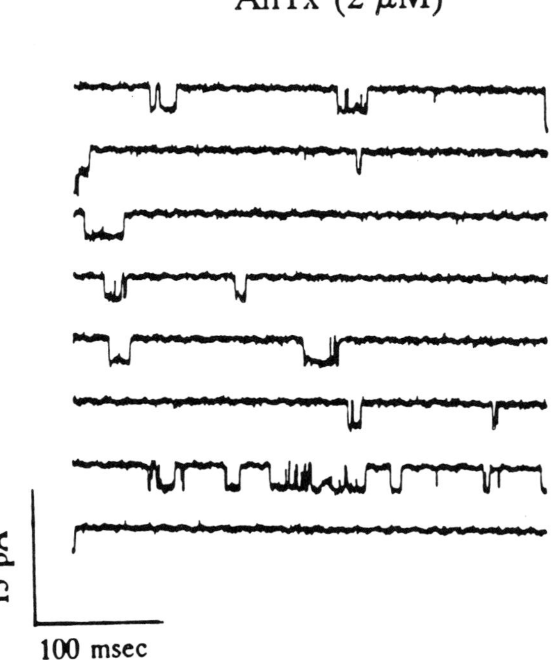
AnTx (2 μM)
15 pA
100 msec

tree forms from other ganglion cells within the same morphological class that project to normal targets such as the superior colliculus (Yamasaki and Ramoa, 1990). An additional type of anomalous retinal projection has been studied yielding similar conclusions. The topographically incorrect projection to the superior colliculus (O'Leary et al., 1986) was found to be composed mainly of type III neurons which, nevertheless, develop a normal dendritic field before being eliminated from the retina (Yamasaki and Ramoa, 1990). A few anomalous cells that survive into adulthood even undergo normal dendritic remodeling.

In another line of experiments, we have attempted to dissociate functionally the cat's retina from its main target, the LGN. Application of TTX into the brain of fetal animals was started at ED43 and continued for 2 weeks. In these animals, examined at ED57, abnormally expanded retinal axonal arbors were seen in the LGN, and eye-specific layers did not form (Sretavan et al., 1988). Nevertheless, normal dendritic trees formed in the retina (Ramoa and Shatz, unpublished observations), suggesting that dendritic growth and remodeling of RGCs does not require interactions with target nuclei. Thus, the role of the target in retinal development may be trophic rather than instructive, modulating RGC survival rate but not directly affecting RGC form.

In conclusion, recent descriptive studies of the developing retina have provided answers to questions posed at the beginning of this review, thus revealing some of the mechanisms that may contribute to the establishment of the remarkable diversity of neuronal form and function in the adult mammalian retina. We now know when cells that resemble the adult RGC types are first present in the

---

FIGURE 5.4. Single-channel ion current recordings were made from outside-out patches of membrane excised from identified RGCs, using the high-resolution patch-clamp technique developed by Hamill et al. (1981). The examples indicate that N-methyl-D-aspartate (NMDA) and anatoxin (AnTx), a selective nicotinic agonist, can activate channels present in immature neurons. Responses were already present by PD1, the earliest age studied. Typical currents are shown on a continuous time scale. *Methods*: All recordings were obtained from the membrane of RGCs identified by retrograde label with rhodamine-labeled latex microspheres. At least 1 day after the injection of rhodamine microspheres into the superior colliculus of rat pups, retinas were dissected from the deeply anesthetized animals. Retinal ganglion cells were obtained according to procedures described elsewhere (Lipton et al., 1987) and plated on cultures of superior colliculus (SC) neurons. The latter procedure seems to induce a quick attachment of RGCs to the culture dish and a higher survival rate. RGCs were healthier when growing on SC neurons than on polylisine-coated dishes or on culture dishes coated with glial cells obtained from the cerebral cortex. This observation may explain the relative ease we had in eliciting NMDA and cholinergic responses in this preparation. Single channel current recordings were made using the high-resolution patch-clamp technique developed by Hamill et al. (1981) using an LM-EPC7 patch-clamp system (List Electronic). Response to NMDA, glutamate, and agonists of the nicotinic acetylcholine receptor were recorded from outside-out patches of membrane excised from the retinal ganglion cells. Information on solutions used and the procedures followed in these experiments can be obtained elsewhere (Ramoa et al., 1990).

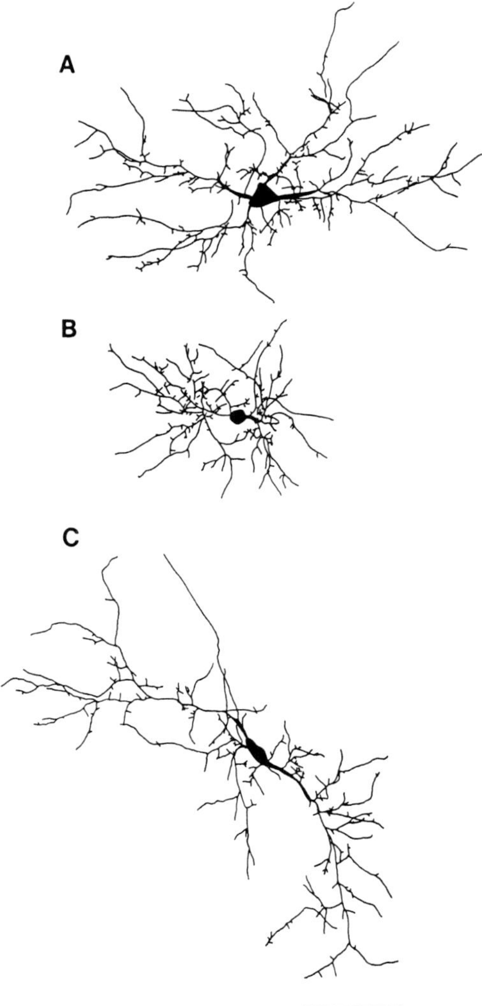

FIGURE 5.5. Camera lucida drawings of Lucifer yellow–filled RGCs observed in the retina of PD5 rats that had been treated with a single intraocular injection of NMDA at PD3. Note that treated RGCs can still be recognized that resemble the types I, II, and III (shown in A,B, and C, respectively) present in normal, untreated animals (compare the drawings with those in Figure 5.3A, 5.3C, and 5.3E). However, the dendritic field size and complexity of the cells resembling types I and II were substantially reduced by the pharmacological treatment. Scale bar = 50 μm (note different size of scale).

developing retina, and it has been established that both growth and remodeling contribute to the maturation of RGC dendritic trees in several species of mammals. Moreover, the finding that RGCs fire action potentials and respond to chemical agents very early in development supports the notion that intraretinal electrical activity plays a role in their development. Not unexpectedly, several issues have also remained unresolved and new questions have been raised by these experiments. Additional studies are needed to establish how intraretinal electrical activity may contribute to dendritic tree development, and the role of target nuclei needs to be further explored. Experimental manipulations in which chemical agents are applied to the developing retina, both *in vitro* and in the intact animal, may help shed some light on some of these issues.

REFERENCES

Aizenman E, Frosch MP, Lipton SA (1988): Responses mediated by excitatory amino acid receptors in solitary retinal ganglion cells from rat. *J Physiol (Lond)* 396:75–91

Bloomfield SA, Hitchcock PF (1991): Dendritic arbors of large-field ganglion cells show scaled growth during expansion of the goldfish retina: A study of morphometric and electrotonic properties. *J Neurosci* 11:910–917

Boycott BB, Wassle H (1974): The morphological types of ganglion cells of the domestic cat's retina. *J Physiol (Lond.)* 240:397–420

Bunt SM, Lund RD, Land PW (1983): Prenatal development of the optic projection in albino and hooded rats. *Dev Brain Res* 6:149–168

Dann JF, Buhl EH, Peichl L (1987): Dendritic maturation in cat retinal ganglion cells: A Lucifer Yellow study. *Neurosci Lett* 80:21–26

Dann JF, Buhl EH, Peichl L (1988): Postnatal dendritic maturation of alpha and beta ganglion cells in cat retina. *J Neurosci* 8:1485–1499

Dubin MW, Stark LA, Archer SM (1986): A role for action-potential activity in the development of neuronal connections in the kitten retinogeniculate pathway. *J Neurosci* 6:1021-1036

Fukuda Y, Hsiao C, Watanabe M, Ito H (1984): Morophological correlates of physiologically identified Y-,X-, and W-cells in the cat retina. *J Neurophysiol* 52:999–1013

Galli L, Maffei L (1988): Spontaneous impulse activity of rat retinal ganglion cells in prenatal life. *Science* 242:90–91

Hamilll OP, Marty A, Neher E, Sakmann B, Sigworth FJ (1981): Improved patch-clamp techniques for high resolution current recording from cells and cell-free membrane patches. *Pflugers Arch* 391:85–100

Huettner JE, Bean BP (1988): Block of N-methyl-D-aspartate-activated current by the anticonvulsant MK-801: Selective binding to open channels. *Proc Natl Acad Sci U S A* 85:1307–1311

Jacobson M (1991): *Developmental Neurobiology*. New York: Plenum Press

Jahr CE, Stevens CF (1987): Glutamate activates multiple single channel conductances in hippocampal neurons. *Nature* 325:522–525

Kolb HR, Nelson R, Mariani A (1981): Amacrine cells, bipolar cells and ganglion cells of the cat retina: A Golgi study. *Vision Res* 21:1081–1114

Lankford KL, DeMello FG, Klein WL (1988): D1 dopamine receptors inhibit growth cone motility in cultured retina neurons: Evidence that neurotransmitters act as morphogenic growth regulators in the developing central nervous system. *Proc Natl Acad Sci U S A* 85:2839–2843

Lipton SA (1986): Blockade of electrical activity promotes the death of mammalian retinal ganglion cells in culture. *Proc Natl Acad Sci U S A* 83:9774–9778

Lipton SA, Aizenman E, Loring RH (1987): Neural nicotinic acetylcholine responses in solitary mammalian retinal ganglion cells. *Pflugers Arch* 410:37–43

Lipton SA, Frosch MR, Phillips MD, Tauck DL, Aizenman E (1988): Nicotinic antagonists enhance process outgrowth by rat retinal ganglion cells in culture. *Science* 239:1293–1296

Lipton SA, Kater SB (1989): Neurotransmitter regulation of neuronal outgrowth, plasticity and survival. *Trends Neurosci* 12:265–269

Maslim J, Stone J (1986): Synaptogenesis in the retina of the cat. *Brain Res* 373:35–48

Maslim J, Webster M, Stone J (1986): Stages in the structural differentiation of retinal ganglion cells. *J Comp Neurol* 254:383–402

Meister M, Wong ROL, Baylor DA, Shatz CJ (1991): Synchronous bursts of action potentials in ganglion cells of the developing mammalian retina. *Science* 252:939–943

O'Leary DDM, Fawcett JW, Cowan WW (1986): Topographic targeting errors in the retinocollicular projection and their elimination by selective ganglion cell death. *J Neurosci* 6:3692–3705

Perry VH (1979): The ganglion cell layer of the retina of the rat: A Golgi study. *Proc R Soc Lond (Biol)* 204:363–375

Perry VH, Walker M (1980): Morphology of cells in the ganglion cell layer during development of the rat retina. *Proc R Soc London (Biol)* 208:433–445

Ramoa AS, Albuquerque EX (1988): Phencyclidine and some of its analogues have distinct effects on NMDA receptors of rat hippocampal neurons. *FEBS Lett* 235:156–162

Ramoa AS, Alkondon M, Aracava Y, Irons J, Lunt GG, Deshpande SS, Wonnacott S, Aronstam RS, Albuquerque EX (1990): The anticonvulsant MK-801 interacts with peripheral and central nicotinic acetylcholine receptor ion channels. *J Pharmacol Exp Therap* 254:71–82

Ramoa AS, Campbell G, Shatz CJ (1987): Transient morphological features of identified ganglion cells in living fetal and neonatal retina. *Science* 237:522–525

Ramoa AS, Campbell G, Shatz CJ (1988a): Dendritic growth and remodelling of cat retinal ganglion cells during fetal and postnatal development. *J Neurosci* 8:4239–4261

Ramoa AS, Deshpande SS, Albuquerque EX (1988b): Development of chemosensitivity and action potential activity in retinal ganglion cells of the rat. *Soc Neurosci Abstr* 14:460

Ramon Y Cajal S (1909/1911): *Histologie du systeme nerveux de l'homme et des vertebres*, 2 vols., Azoulay L, trans. Reprint Madrid: Instituto Ramon Y Cajal.

Rauschecker JP (1991): Mechanisms of visual plasticity: Hebb synapses, NMDA receptors, and beyond. *Physiol Rev* 71:587–615

Rocha M, Ramoa A, Hahm J, Sur M (1991): NMDA antagonist infusion during ON/OFF sublaminar segregation alters dendritic morphology of cells in ferret LGN. *Soc Neurosci Abstr* 17:1136

Rodieck RW, Brening RK (1983): Retinal ganglion cells: Properties, types, genera, pathways and trans-species comparisons. *Brain Behav Evol* 23:121–164

Rusoff AC, Dubin MW (1977): Development of receptive field properties of retinal ganglion cells in kittens. *J Neurophysiol* 40:1188–1198

Saito HA (1983): Morphology of physiologically identified X-,Y-and W-type retinal ganglion cells of the cat. *J Comp Neurol* 221:279–288

Sakaguchi DS, Murphey RK, Hunt RK, Tompkins R (1984): The development of retinal ganglion cells in a tetraploid strain of *Xenopus laevis*: A morphological study utilizing intracellular dye injections. *J Comp Neurol* 224:231–251

Spivak CE, Witkop B, Albuquerque EX (1980): Anatoxin-a: A novel, potent agonist at the nicotinic receptor. *Mol Pharmacol* 18:384–394

Sretavan DW, Shatz CJ, Styker MP (1988): Modification of retinal ganglion cell morphology by prenatal infusion of tetrodotoxin. *Nature* 336:468–471

Wässle H, Boycott BB (1991): Functional architecture of the mammalian retina. *Physiol Rev* 71:447–480

Wingate RJT, Thompson ID (1991): Postnatal dendritic modification in an identified population of hamster retinal ganglion cells. *Soc Neurosci Abstr* 17:559

Wong ROL (1990): Differential growth and remodeling of ganglion cell dendrites in the postnatal rabbit retina. *J Comp Neurol* 294:109–132

Wong ROL, Herrman K, Shatz CJ (1991): Remodeling of retinal ganglion cell dendrites in the absence of action potential activity. *J Neurobiol* 22:685–697

Yamasaki EN, Ramoa AS (1990): Dendritic development of abnormally projecting rat retinal ganglion cells. *Soc Neurosci Abstr* 16:334

Yamasaki EN, Ramoa AS (1992): *Dendritic remodeling of rat retinal ganglion cells during development*. Submitted for publication

Yamasaki EN, Rocha MS, Ramoa AS (1989): Morphology of rat retinal ganglion cells during fetal and postnatal development. *Soc Neurosci Abstr* 15:455

Acknowledgments: This work was supported by Conselho Nacional de Pesquisa, Financiadora de Estudos e Projetos, and Financiadora de Estudos e Projetos/University of Maryland at Baltimore Molecular Pharmacology Training Program. We thank Dr. Edson X. Albuquerque for his generous support, and Dr. Cheryl A. White for her helpful criticisms of the manuscript.

6

Dendritic Competition: A Principle of Retinal Development

RAFAEL LINDEN

The functions of the mature nervous system are determined by the organization of neural circuits. The development of neural connectivity has been the focus of intense research, and a strong body of coherent knowledge has accumulated over the past decades (Easter et al., 1988; Jacobson, 1991; Purves and Lichtman, 1985).

Nonetheless, neural development is still poorly understood in terms of its intimate cellular and molecular mechanisms. In many instances, the classical descriptive approach has been replaced by rules or principles explaining in rather general terms the events of neurogenesis. The present chapter reviews one such principle—dendritic competition— and discusses the limited progress obtained thus far in the search both for its consequences and for the underlying mechanisms.

The idea that developing dendrites are involved in competitive interactions emerged from experimental studies of the mammalian retina (Linden and Perry, 1982; Perry and Linden, 1982), and the implications of dendritic competition became particularly apparent for the same model. Interactions of growing dendrites may, for example, constrain the shape and extent of mature dendritic arbors, and the retinal coverage by the various types of retinal ganglion cells (see Wassle and Boycott, 1991, for a review of retinal architecture). A distinct view is that the retinal ganglion cells compete for survival within the retina, and dendritic geometry is a vehicle for the competitive interactions rather than their final effect. The present review will focus on this latter view, without prejudice to the former.

Accordingly, the principle of dendritic (or dendrodendritic) competition may be generally stated as follows: During development of the nervous system, immature neurons compete for locally derived trophic support available to their growing dendrites, and cells that fail in the competition degenerate during the period of naturally occurring neuron death. Both the original work that led to this principle and the subsequent evidence gathered in various systems suggested that the required trophic support is provided by the afferent supply of the immature neurons. It is possible, however, that the dendrites of devloping cells compete for glial-derived trophic factors.

It is also debatable to what extent competitive-like interactions and other events involved in the overall shaping of the dendritic tree are necessarily linked to the control of neuronal survival. Still, the available evidence is consistent with a consolidated rule as stated above, and it has served as a useful working hypothesis for the past 10 years.

COMPETITIVE INTERACTIONS AND TROPHIC SUPPORT OF DEVELOPING NEURONS

The morphology of adult neurons is attained through a series of developmental steps (Cowan, 1979) that may be summarized, albeit rather sketchily, as follows: Upon withdrawal from the mitotic cycle each cell migrates to its definitive position in an area or nucleus of the nervous system, and projects an axon toward its targets. The dendritic tree grows and contacts afferent neurons. Synaptic connections are formed on both ends of the neuron, thereby determining the role the cell plays within a given neural circuit. Cell interactions around the time of synaptogenesis underlie the selection of neurons that are to survive, among a large number of cells subject to natural degeneration. Naturally occurring neuronal death has been extensively documented in various areas of the developing nervous system, including the population of retinal ganglion cells (see Linden, 1987, for a review).

Experimental studies in various systems and species have shown convincingly that the axonal targets of immature nerve cells are required for neuronal survival during critical stages of embryogenesis. Four major lines of evidence support this general rule (reviewed in Linden, 1988; Oppenheim, 1981, 1991. First, the removal of target cells (e.g., muscle) prior to innervation produces massive cell loss among the target-deprived neurons (e.g., spinal motoneurons) during the period of naturally occurring neuronal death. Second, the enlargement of a target field reduces cell death below control levels during the same period. Third, the removal of one among several immature neuronal populations innervating a common target reduces cell death in the other populations. Fourth, the number of neurons that survive naturally occurring cell death in a given population is quantitatively (sometimes linearly) related to the size of the target.

Such data support the notion that the developing neurons compete for limited target space during neurogenesis. This idea, intuitive as it may seem prima facie, has nonetheless been subject to thoughtful criticism (Guillery, 1988). The major caveat was essentially that severe experimental manipulation may create novel conditions that do not reflect normal developmental mechanisms. These artificial circumstances might particularly vitiate tests of competition between two distinguishable populations of cells (such as the retinal ganglion cells from either eye). Although it is still unclear to what extent neurons that are distinguishable by external criteria such as laterality are necessarily distinct in terms of cell recognition mechanisms, caution should indeed be exerted in accepting the results of single lesion experiments as evidence for competitive interactions.

Here, the term "competition" means that the probability of an element reaching a goal is reduced in the presence of other similar elements. This usage in the context of axonal competition for targets is based on positive results from multiple experiments among the set of evidence described above, and requires the existence of overlap among the presumptive competitors during development. It implies only that there is a limiting supply of, or limited access to, some aspect of the target that is vital for the innervating population. It is *not* meant to imply that individual neurons directly hinder the development of each other, nor that each neuron actively seeks the vital target cells, although neither possibility can be ignored at present.

The competition hypothesis gained strong support from the demonstration that targets do indeed produce specific molecules that support the survival of developing neurons, the paradigm of which is the nerve growth factor (NGF) (Levi-Montalcini, 1987). It is now widely accepted that NGF and other *trophic factors* (Thoenen, 1991) are responsible for the target-dependency inferred from the experiments typified in the preceding paragraphs. Yet, it is uncertain whether neurotrophic factors are synthesized in limiting quantities (Barde, 1988), or whether the actual bounds are imposed by the ability of axons to gain access to the molecules produced by target cells (Blaser et al, 1991; Oppenheim, 1991).

EVIDENCE FOR DENDRITIC COMPETITION IN THE DEVELOPING RETINA

The possibility that afferents might also play a role in the control of neuronal survival during embryogenesis remained largely neglected until the end of the seventies. It was often tacitly assumed that all neurotrophic support for developing neurons is derived from their targets. This is slightly surprising, since it had long been known that deafferented neurons (as well as denervated muscle, for that matter) show marked regressive changes that may lead to degeneration (reviewed by Smith, 1977). Data were available indicating that deafferentation of the isthmo-optic nucleus in the duck led to increased degeneration during the period of naturally occurring neuron death (Sohal, 1976). Furthermore, Cunningham and his associates (1979) reported that the experimental enlargement of the optic pathway in neonatal rats led to increased neuron numbers in hyperinnervated areas of the brain, probably due to reduced rates of cell death.

Against this background, in 1980 we began to test the hypothesis that developing neurons interact competitively by way of their dendrites. We took advantage of the nearly planar arrangement of retinal ganglion cell dendrites within the inner plexiform layer, and of the intermixing of ganglion cells with either crossed or uncrossed axons in certain areas of the rodent retina. In rats, as in other nonprimate mammals, ganglion cells with crossed axons are found throughout the whole retina. Therefore, the ipsilaterally projecting cells that exist mainly, though not exclusively, in the temporal crescent (see below) are located among crossed-projecting cells.

The results of two basic experiments made in the retina of newborn rats led to the principle of dendritic competition (see Figure 6.1). First, the early degeneration of ganglion cells with crossed axons, induced by contralateral optic tract lesions (see

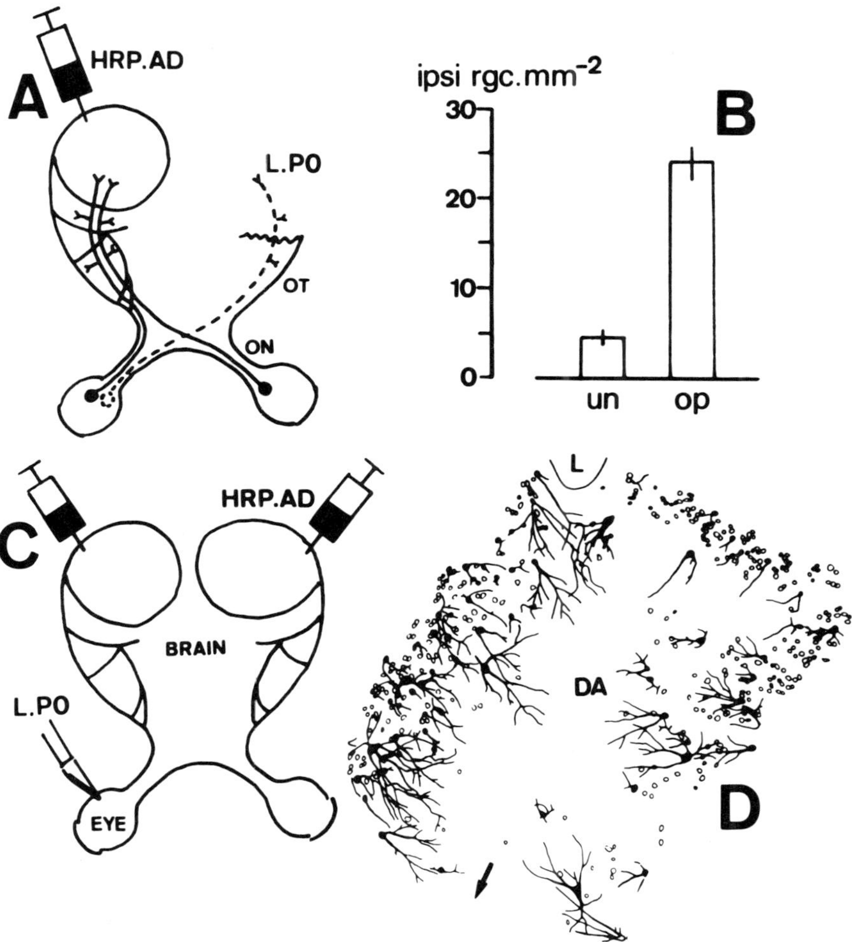

FIGURE 6.1. Evidence for dendritic competition in the developing retina: A. Ganglion cells with uncrossed axons were labeled with horseradish peroxidase (HRP) in adult (AD) rats, following a lesion to the optic tract at birth (L.P0) that causes retrograde degeneration of ganglion cells with crossed axons (interrupted line). OT = optic tract; ON = optic nerve. B. The density of ganglion cells with ipsilateral projections found in nasal parts of the same retina (to the left in A) was increased in operated rats (op) when compared with unoperated controls (un). Data from Linden and Perry (1982). C. Massive injections of HRP in adult rats (HRP.AD) that had received localized retinal lesions at birth (L.P0) labeled dendrites of ganglion cells. D. The cells located at the borders of a depleted area (DA) peripheral to the lesion (L) have the dendrites directed toward the depleted area. The arrow indicates the direction away from the optic nerve head. Mod. from Perry and Linden (1982).

Figure 6.1A), resulted in increased numbers of ganglion cells with uncrossed axons in the same retina (see Figure 6.1B) (Linden and Perry, 1982). Second, depletion of ganglion cells in a restricted area of the retina (see Figure 6.1C) allowed the dendrites of neighboring ganglion cells to orient down the experimentally induced gradient of cell density (see Figure 6.1D), toward the depleted territory (Perry and Linden, 1982).

The interpretation of these findings has been extensively discussed elsewhere (e.g., Linden, 1987), but some key points should be stressed here. In the first experiment (see Figure 6.1A), the availability of target space for the uncrossed retinofugal pathway was not changed as a direct result of the degeneration of cells with crossed projections in the *same* retina. Further, judging from the sizes of the dendritic arbors of immature ganglion cells (Perry, 1979), there is extensive overlap among neighboring dendritic trees in the retina of the rat at a time when natural cell death is occurring (Perry et al., 1983; Potts et al., 1982). A parsimonious explanation for both sets of data was then offered, that the overlapping or abutting dendrites of developing ganglion cells compete for space within the inner plexiform layer, resulting in the degeneration of cells that fail to secure a minimum amount of territory. The increased number of ipsilaterally projecting ganglion cells would thus reflect the survival of part of the neurons that normally degenerate in unoperated rats.

Analogous experiments were made in cats by other investigators, with comparable results. Lesions of the optic tract made in neonatal kittens led to extensive degeneration of ipsilaterally projecting ganglion cells in the temporal part of the retina. As a result, the number of contralaterally projecting ganglion cells in the same part of the retina was increased (Leventhal et al., 1988b). Since the number of contralaterally projecting ganglion cells in the temporal retina of newborn cats is greater than in adults, the results were again interpreted as cell rescue due to diminished dendritic competition. Also in cats, experimental depletion of ganglion cells allows dendrites of neighboring neurons to invade the depleted areas (Eysel et al., 1985; Leventhal et al., 1988a).

In addition, the dendritic trees of retinal ganglion cells in *normal* cats tend to be displaced down the centro-peripheral gradient of cell density, that is, away from the area centralis (Schall and Leventhal, 1987). These data, in particular, lend credence to the idea that the experimentally induced dendritic plasticity represents a capacity intrinsic to normal neurons. It is therefore not an artifact induced by drastic experimental conditions.

On the other hand, Schall and co-workers (1987) also examined the distribution of dendritic arbors of type I (alphalike) ganglion cells in the retina of rats, and did not find a displacement down the density gradient for that cell type. They postulated that the density gradient of ganglion cells in the rat retina, in which the ratio of cell densities between the area centralis and the far periphery is no larger than 5-fold, was too shallow to create biased dendrites. Our recent work, however, suggests that the reason for the difference between the orientations of ganglion cell dendrites in rats and cats may actually be the timing of development of density gradients in the population of retinal ganglion cells.

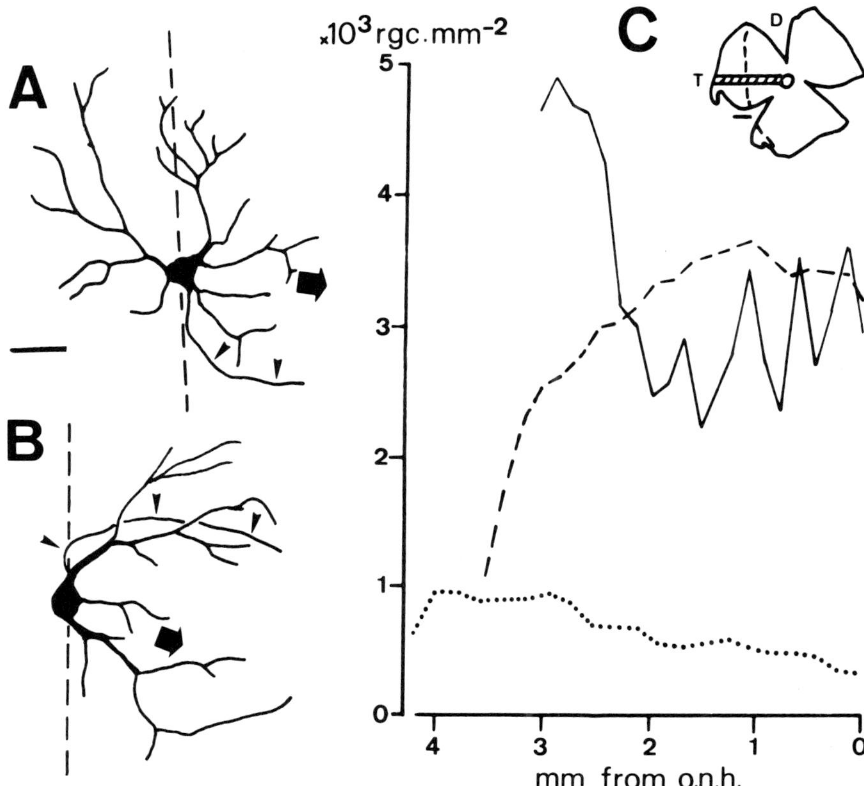

FIGURE 6.2. A, B. Mature ganglion cells located at the nasal border of the temporal crescent (interrupted line) labeled with HRP in an unoperated control (A) and in a rat that had received a contralateral lesion to the optic tract at birth (B, see Figure 6.1A). The thick arrow points to the optic nerve head. Arrowheads indicate the axons. Scale bar = 50 μm. C. Ganglion cell density gradients along a strip of retina (hatched in the inset) from the optic nerve head (o.n.h.) to the periphery through the decussation line scale bar = 1 mm. Interrupted line in the graph is from adult controls, dotted line from rats that received contralateral optic tract lesions at birth (see Figure 6.1A). The solid line is from 5-day-old rats similarly operated at birth. Data from Linden (1992). Notice that the density gradients in operated rats are not steeper than in the unoperated rats.

In normal adult rats, the density of retinal ganglion cells is the highest in areas located close to the optic nerve head, and declines toward the periphery (e.g., Hughes, 1977). On the other hand, optic tract lesions made at birth produce massive retrograde degeneration of ganglion cells with crossed projections (see Figure 6.1A). Since the ipsilaterally projecting ganglion cells are mostly located in a crescent-shaped area located in temporal retina (portrayed as limited by the interrupted line in Figure 6.2, inset), the remaining population of ganglion cells forms a temporonasal gradient of cell density.

In unoperated rats, ganglion cells located along the border of the temporal crescent have dendritic trees symmetrical with respect to the decussation line (see Figure 6.2A), whereas after contralateral optic tract lesions similarly located cells show a significant bias toward nasal retina (see Figure 6.2B; also see Linden and Perry, 1982). Notwithstanding, the decline in cell density across the nasotemporal division toward the temporal retinal periphery of normal rats is much steeper than the opposite gradient found in operated rats (see Figure 6.2C). When the temporonasal gradient of operated rats is examined at postnatal day 5 (P5), that is, during the period of naturally occurring neuronal death (see Figure 6.2C), the immature retina shows a steeper gradient, but it is still not larger than that found in normal adults. In contrast, the distribution of ganglion cells in normal P5 rats is rather homogeneous across the retina (McCall et al., 1987). Thus, a relatively shallow gradient is indeed able to create significant dendritic bias in the retina of the rat, as long as it occurs during the early postnatal period.

Comparing the time course of ganglion cell loss and the development of density gradients in the cat and rat suggests that density gradients create dendritic bias in normal retinas only if occurring simultaneously with ganglion cell death (see Figure 6.3). In the rat retina, a significant gradient appears only after most ganglion cell death has already occurred, especially during an early phase of rapid neuron loss (McCall et al., 1987), while cat retinas reach a 4- to 5-fold centro-peripheral ratio of cell density within the period of rapid ganglion cell loss (Lia et al., 1987). A 4-fold ratio is attained in the rat at P15–20 (McCall et al., 1987), when considerable dendritic growth is still to occur (Perry and Walker, 1980), but even so it is unable to impose a significant dendritic bias.

Therefore, the time course of development, rather than the magnitude, of the cell density gradient appears to be the primary factor in shaping the architecture of the dendritic trees of ganglion cells across the retina. The data are consistent with the idea that ganglion cells are most likely to survive developmental neuron death if their dendrites are preferentially directed toward areas with the lowest cell density.

The local density of neighboring cells not only affects the orientation of dendrites, but also the overall size of the dendritic tree. An early study had detected no change in the size of ganglion cell dendrites located at the borders of a depleted peripheral area of the retina when compared with control cells, notwithstandinng the lateral displacement of the dendritic arbors (Eysel et al., 1985). In contrast, a more recent study of a large sample of cells located within central or peripheral areas in which the density of ganglion cells had been severely depleted showed a significant increase in the sizes of both the cell soma and the dendritic fields of two distinct types (alpha and beta) of ganglion cells (Leventhal et al., 1988a). The dendritic field sizes of other cell types were, however, unaffected in spite of a clear displacement of their dendritic trees down the gradient of cell density (Leventhal et al., 1988a). These results led to the idea that the control of cell soma and overall dendritic field sizes may be independent of the control of the spatial distribution of dendritic fields relative to the cell soma (Leventhal et al., 1988a), although having in common a dependency on local cell densities. It had also been shown that the

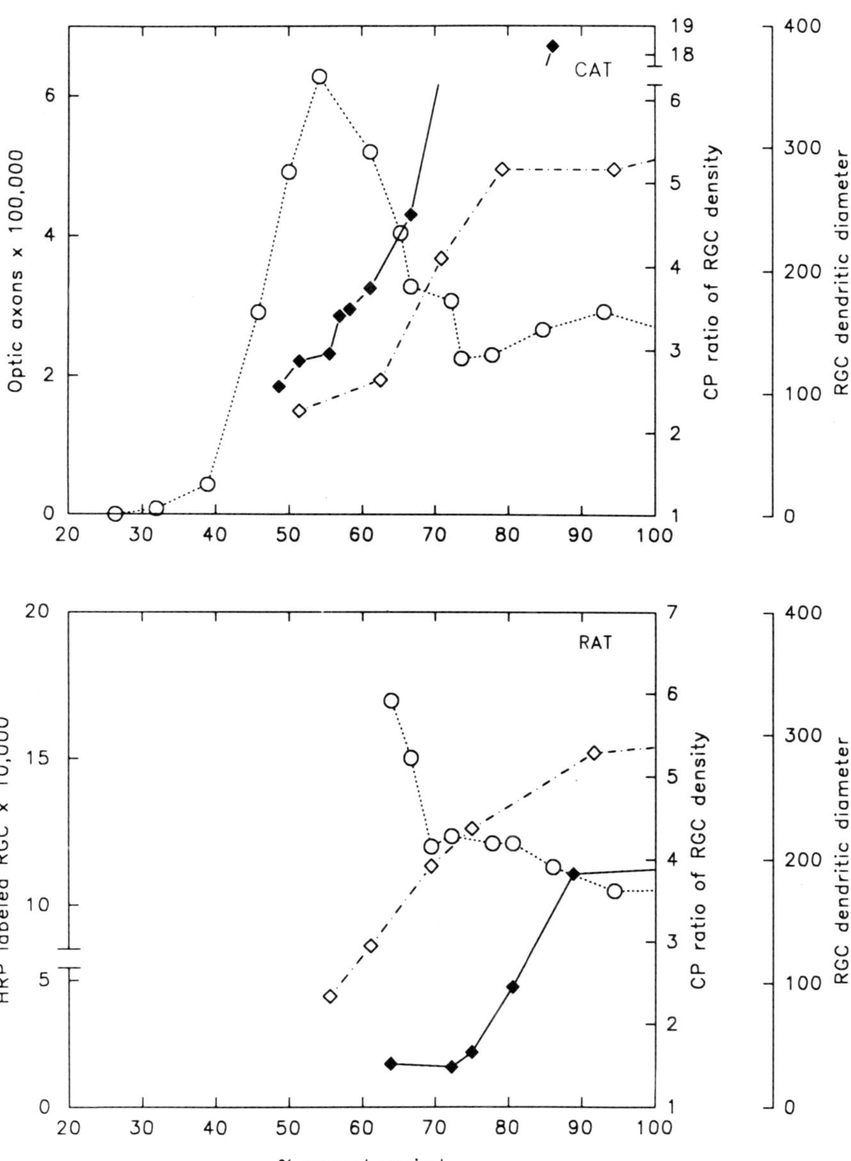

FIGURE 6.3. A. Parameters of retinal development plotted for cat and rat as a function of percentage of the caecal period, i.e., the time from conception to eye opening (Dreher and Robinson, 1988). Circles = number of optic axons or HRP-labeled ganglion cells; open losanges = average diameter of the dendritic tree (in micrometers); closed losanges = ratio of cell densities in central and peripheral retina. Notice the delayed increase in the centroperipheral gradient of cell density in the rat when compared with the cat. Data were taken from the following sources: cat: Williams et al. (1986); Lia et al. (1987); Ramoa et al. (1988); rat: Potts et al. (1982); McCall et al. (1987); Yamasaki (1990).

elongation and overall orienation of the dendritic tree are not exactly correlated with the cell density gradient, but rather appear to reflect the wave of retinal maturation (Schall and Leventhal, 1987). Thus, multiple factors must be involved in shaping altogether the lateral displacement, elongation, and overall size of the mature dendritic tree of each ganglion cell.

AXON ORIENTATION AND AXODENDRITIC POLARITY OF RETINAL GANGLION CELLS

When viewed in retinal flat-mounts, as if looking at the projection of the visual world onto the receptor surface, the axon hillock of ganglion cells in normal adult rats usually points toward the optic nerve head, where axons leave the eye (Maffei and Perry, 1988). The same study showed that the primary dendrites of ganglion cells are preferentially directed away from the axon hillock, resulting in axodendritic polarity on the plane of the retina (Maffei and Perry, 1988). Further studies of the effects of ganglion cell depletion following a localized retinal lesion (see Figure 6.1C, 6.1D) revealed that affected cells not only had their dendrites predominantly directed into the depleted area. The axon hillock of those cells was also directed away from the border of the depleted region, that is, almost at a right angle with respect to the direction of the optic nerve head (Perry and Maffei, 1988).

This rather drastic effect upon *both* dendrites and axons may result from a combination of at least three events: (1) differential growth of the dendrites may favor those located toward the depleted area; (2) differential retraction may stunt the growth of dendrites directed toward neighboring cells; (3) selection of ganglion cells during the period of naturally occurring neuron death may favor cells with the appropriate dendritic orientation.

Perry and Maffei (1988) showed that the cells located at the borders of the depleted area had their dendrites less dispersed around the cell body than control cells, an effect that could result from either (1) or (2) above, or both. Nonetheless, in order to explain the disoriented axon hillocks, one would have to postulate an extensive rotation of the whole cell on the plane of the retina. This possibility cannot be discarded outright, but it seems unlikely given the relatively advanced stage of dendritic development and the intermingling of processes within the inner plexiform layer of newborn rats. It should be noted that the dendritic plasticity observed in cells deprived of neighbors affects *primary* dendrites (Linden, 1992; Perry and Maffei, 1988). Indeed, the biased distribution of primary dendrites is likely to be a major factor in determining the lateral displacement of the whole dendritic tree with respect to the soma. Regression of dendritic appendages and fine, higher order branches has been demonstrated in developing rats (Ramoa and Yamasaki, this volume), similar to other species (Dann et al., 1987; Dunlop, 1990; Ramoa et al., 1987), but there is no evidence for the complete regression of primary dendrites at postnatal stages of development (Nakatani and Linden, 1991; Ramoa and Yamasaki, this volume).

An additional factor contributing to the appearance of disoriented axon hillocks may be found in the experimental design of Perry and Maffei (1988). The localized transection of optic axons at the nerve fiber layer leaves a peripheral wedge of the retina void of ganglion cells. A cell located close to the borders of this region will, however, escape axotomy if the initial portion of its axon is directed away from the wedge and then turns toward the optic disk. Indeed, two-thirds of the ganglion cells located at the border of the depleted region in adult-operated rate had their axon hillocks directed away from the depleted retinal area (see Figure 2 in Perry and Maffei, 1988), although no significant dendritic distortion was observed in these rats.

Thus, the dramatic effect observed by Perry and Maffei (1988) on the distribution of axon orientations and axodendritic polarities after neonatal lesions may conform to the following scenario: First, the surgical procedure would create a depleted area and, simultaneously, kill all neighboring ganglion cells with axons directed through this area. As an immediate result, a population with the axons directed away from the depleted region would be selected. Then, the steep gradient of cell density created at the border of the depleted region would favor the selective survival of ganglion cells with dendrites pointing therein, over neurons with less appropriate dendritic orientations. Since the majority of ganglion cells have their primary dendrites directed away from the axon hillock both in adult (Maffei and Perry, 1988) and in neonatal rats (Nakatani and Linden, 1991; see below), then this whole scenario should favor strongly the survival of neurons with both the axon directed away from *and* the dendrites directed toward the depleted area.

Given the extreme degree of dendritic bias observed in the experimental cells, as well as the protracted sensitive period for the production of biased dendrites (Perry and Maffei, 1988), it is likely that remodeling of the distal dendritic tree may play an ancillary role in shaping the population obtained in that study. In addition, the sensitive period for the production of dendritic bias in similar experiments made in the retina of the cat extends beyond the period of naturally occurring neuron death (Eysel et al., 1985). Nonetheless, the evidence that relatively shallow gradients of cell density are sufficient to create dendritic bias as long as they occur within, but not later than, the period of natural cell death (see above) strengthens the hypothesis of cell selection as an early determinant factor in the production of cell density–related biased dendrites.

A necessary requirement for the hypothesis of selection is the availability of ganglion cells with disoriented axons in early stages of development. We (Nakatani and Linden, 1991) therefore studied the distributions of axon orientations and axodendritic polarities of ganglion cells as projected to the retinal plane, in whole-mounted retinas of developing rats (see Figure 6.4).

It was found that newborn and adult rats have approximately the same proportions of ganglion cells with either disoriented axons or anomalous axodendritic polarity or both. Since newborn rats have approximately twice as many ganglion cells as adults, there is indeed a stock of cells with anomalous neurite geometry to account for the experimental results reviewed above and interpreted on the basis of cell selection. Further, during the first 4–5 days after birth, the

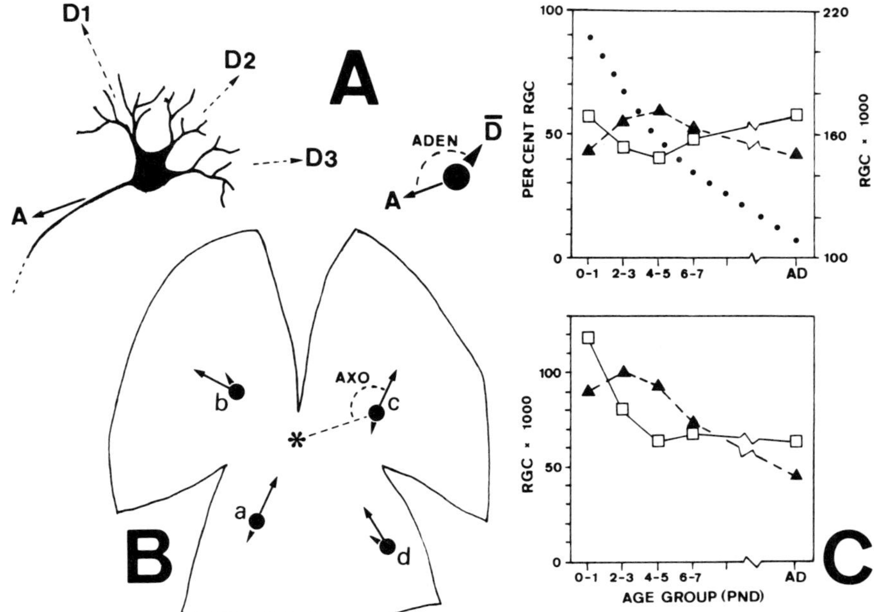

FIGURE 6.4. Geometry of the primary neurites of ganglion cells as projected to the plane of the retina. A. The direction of the mean vector ($\overline{\mathrm{D}}$) of the primary dendrites (D1–D3) was calculated and the angle formed with the axon hillock (A) was measured (ADEN). B. The angle between the axon direction and a line connecting the cell body with the optic disk was measured on drawings of whole-mounted retinas (AXO). Cells were scored as *regular* (a) or as *irregular* (b–d) according to the two parameters. C. The numbers of *regular* (open squares) and *irregular* (filled triangles) cells (lower graph) were estimated from the percentages of each type (upper graph) and the total number of ganglion cells (dotted line) at each age. Notice that newborn rats have excess *irregular* cells up to postnatal days 4–5. Mod. from Nakatani and Linden (1991).

number of anomalous ganglion cells remains stable, and decreases thereafter. Thus, mechanisms operating at distinct epochs during normal postnatal development appear to determine the selective loss of ganglion cells based on the geometry of both their primary dendrites and their axon.

As shown above (see Figure 6.2) following optic tract lesions made at birth the dendrites of ipsilaterally projecting ganglion cells located at the borders of the temporal crescent become clearly biased toward the depleted nasal retina (Lau et al., 1991; Linden and Perry, 1982). This bias was found to reflect a selective increase in the number of cells with both the initial portion of the axon directed toward the temporal periphery and the primary dendrites pointing to the more central, nasal retina, at the expense of the regular cells with centrally directed axon and peripherally directed primary dendrites (M. Nakatani and R. Linden, unpublished data). These data are also most easily explained by the hypothesis of cell selection. It should be noted that the optic tract lesions do not by themselves create an artificial bias as do the retinal lesions, and therefore the appearance of neurons with altered neuritic geometry must be secondary to the modified cell interactions within the retina.

The preceding data therefore suggest that the dendritic architecture of the mature population of retinal ganglion cells depends considerably upon the selection of neurons on the basis of the geometry of the dendritic arbors. This selection can be partly attributed to the operation of dendritic competition during the period of natural neuronal death. Additional remodeling of the distal dendrites at later stages (Ramoa et al., 1987), and highly specialized events like the late movement of cells away from the fovea that has been shown in the monkey (Kirby, 1991), would thus exert their shaping effect upon the dendrites of a preselected cell population.

MECHANISMS OF DENDRITIC COMPETITION

Little progress has been made in the search for the mechanisms of dendritic competition. It has been shown, however, that the survival of retinal ganglion cells *in vitro* depends on electrical activity (Lipton, 1986). Blockade of the voltage-dependent sodium channel with tetrodotoxin (TTX) produced the degeneration of about half of the ganglion cells located in cell clusters dissociated from retinas of postnatal rats. About half of the clustered cells had developed electrical activity after a few hours *in vitro*, in contrast with isolated ganglion cells which do not show spontaneous electrical activity nor respond to TTX with increased degeneration. It was then proposed that the cells that develop spontaneous activity as a result of contacts with other neurons within the clusters become dependent on this activity for survival. This could be related to the intraretinal control of neuron death, if the acquisition of electrical activity is determined by the growth of dendrites toward the afferents in the inner nuclear layer.

Culture medium conditioned for 1 day by monolayers of retinal cells prevented the degeneration in the presence of TTX (Lipton, 1986). Thus soluble molecules appear to be needed to protect the ganglion cells from degeneration produced by blockade of the electrical activity. Since conditioned medium obtained from TTX-treated cultures reportedly had no effect, it was proposed that activity was necessary for the production or release of soluble trophic molecules (Lipton, 1986).

Further work from the same laboratory showed that the vasoactive intestinal peptide (VIP) mimicks the effect of electrical activity. VIP is found in some amacrine cells in the normal retina of the rat. Addition of VIP to TTX-treated cultures blocked the induced degeneration of retinal ganglion cells in the clusters, and the effect was shown to be mediated through an increase in cAMP (Kaiser and Lipton, 1990). The overall data are consistent with the interpretation that the release of VIP by active amacrine cells helps sustain ganglion cells that become dependent on activity for survival. On the other hand, silent ganglion cells, either isolated or within clusters, did not seem to respond to VIP. The fact that VIP is effective in the presence of TTX shows that the electrical activity of the ganglion cells proper is not necessary for their survival, although it may still be necessary to trigger trophic interactions in the absence of TTX.

The data from Lipton's lab led to the suggestion that VIP may be an endogenous factor modulating normal cell death in the retina. It is not clear, however, why VIP seems to act exclusively upon the degeneration experimentally induced by TTX (see also Brenneman et al., 1990), and not on the spontaneous degeneration that occurs among ganglion cells maintained in control conditions *in vitro* (Kaiser and Lipton, 1990).

Recently we started to investigate the degeneration of ganglion cells in dissociated cell cultures of retinas from neonatal rats (Araujo, 1991; Araujo and Linden, 1990; Linden et al., 1991). The survival of ganglion cells identified by retrograde labeling with HRP was evaluated in the monolayers under various experimental conditions (see Figure 6.5). We found that ganglion cell survival was enhanced by culture medium previously conditioned by aggregates or explants from either rat (Araujo and Linden, 1990) or chick (Linden et al., 1991) retina. The trophic activity is nondyalizable and is sensitive to temperature.

Explants of rat retinas kept *in vitro* for different periods of time secreted the trophic activity as long as the ganglion cell and inner plexiform layers were reasonably well preserved. Explants or aggregates kept for up to 7 days *in vitro* secreted the trophic activity, although axotomized ganglion cells die within 3–4 days in the explants. Explants kept for 15 days in culture still had large numbers of photoreceptors in an outer nuclear layer, but the inner nuclear and ganglion cell layers were replaced by debris. Medium conditioned by the latter explants had no trophic activity, similar to medium conditioned by explants of cerebral hemispheres (Araujo, 1991).

The data indicate that retinal cells other than ganglion cells or photoreceptors, located in either the ganglion cell layer or the inner nuclear layer, or both, release a proteinaceous trophic factor capable of preventing ganglion cell degeneration *in vitro*. Further characterization and ultimately the study of the localization and expression of the trophic molecules will be needed to evaluate their possible role in the control of developmental neuronal death *in vivo,* as well as their relationship with VIP and the induced degeneration reported by Lipton and Kaiser. Nonetheless, the data suggest that ganglion cells in developing rats may compete within the retina for a neurotrophic factor of a similar nature as those implied in the target control of naturally occurring neuron death.

The experiments *in vitro* have not yet allowed the identification of the types of cells that secrete the trophic activity. In other systems, however, there is increasing evidence that the survival of developing neurons depends on the afferent supply. The evidence for afferent-dependency is based on experiments analogous to those that demonstrated the target-dependency of developing neurons (see above). Thus, early deafferentation produced increased cell death during the period of naturally occurring neuronal death (Clarke, 1985; Linden and Piñón, 1987; Okado and Oppenheim, 1984). Hyperinnervation reduced naturally occur-

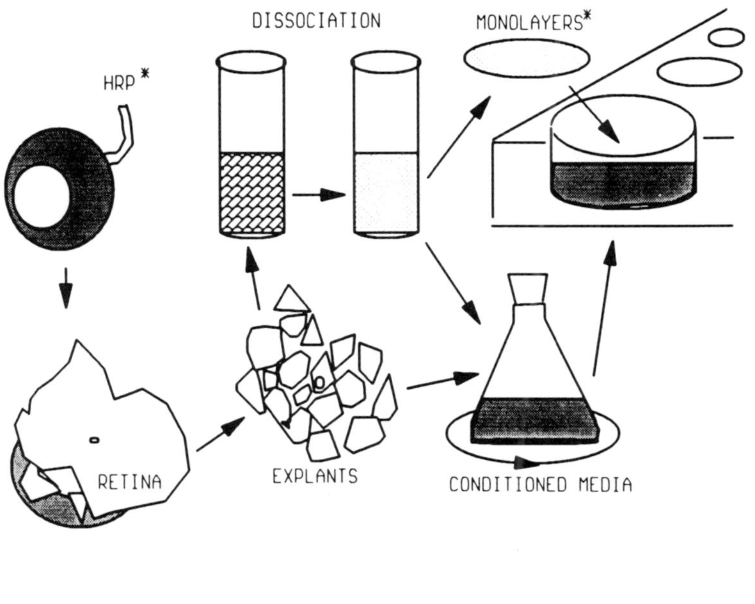

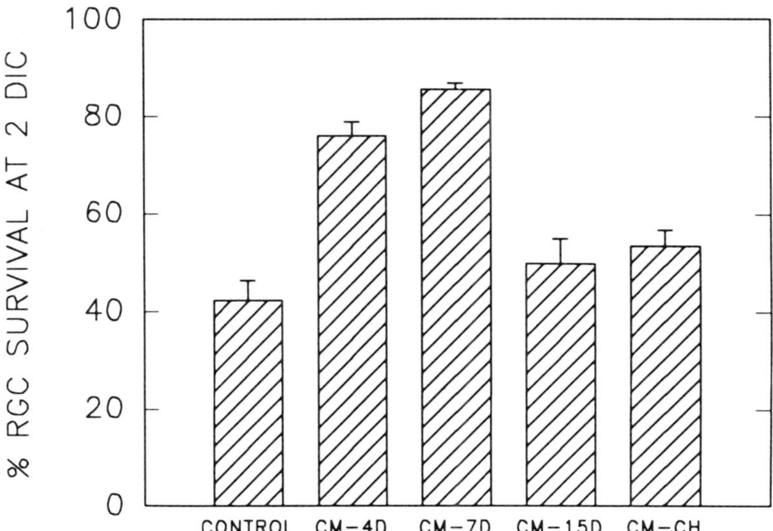

FIGURE 6.5. Survival of retinal ganglion cells *in vitro*. The cartoon depicts the methods used for testing the neurotrophic effects of culture media conditioned by cells of the retina. Asterisk indicates HRP-labeling of retinal ganglion cells. The graph shows the survival of HRP-labeled ganglion cells at 2 days in culture, in monolayers dissociated from the retina of neonatal rats and kept in the presence of various conditioned media. Notice that only the conditioned media from retinal explants maintained for 4–7 days *in vitro* had a neurotrophic effect, in contrast with either media from older explants or media conditioned by explants of cerebral hemispheres (CM-CH). Data from Araujo (1991).

ring neuron death (Cunningham et al., 1979). Removal of part of a neuron population reduced cell death among neurons that share the same afferents (Linden and Perry, 1982; Linden and Serfaty, 1985; see above). Finally, the number of neurons that survived naturally occurring neuronal death following partial deafferentation was linearly related to the number of afferent neurons (Linden and Renteria, 1988). Therefore, the intraretinal control of ganglion cell death may be dependent on the ganglion cell afferents, that is, on amacrine and/or bipolar cells. Further studies will be needed to test for the contribution of glial cells to the trophic interactions described above.

CONCLUSION

The available data are consistent with the currently proposed version of the principle of dendritic competition. Notwithstanding that much of the evidence is indirect, this principle offers a plausible explanation both for the intraretinal control of developmental cell death and for the observed reactions of dendritic trees to cell density gradients, either normal or experimentally induced by early lesions.

In the current review, emphasis has been placed on the idea that the dendrites are a vehicle for the trophic support of the retinal ganglion cells. Thus a favorable dendritic geometry would afford an advantage to a given neuron in the competition for survival. An alternative interpretation for the data on dendritic bias, namely, that the shape of the dendritic tree is itself modeled by competitive interactions not related to cell death, is also likely and the two possibilities are not mutually exclusive. Unfortunately, early markers of neurons destined to degenerate are still unavailable. Such markers will be needed to critically test these two alternative views and evaluate their relative contributions to the generation of the pattern of ganglion cell populations and their dendritic plexus in the mature retina.

Acknowledgments. The author's research has been supported by grants from the Conselho Nacional de Desenvolvimento Cientifico e Tecnológico, Financiàdora de Estudos e Projetos, and Fundação de Amparo à Pesquisa do Estado do Rio de Janeiro. The critical comments of Ben E. Reese, Elizabeth G. de Araujo, and Margarete Nakatani are gratefully acknowledged.

REFERENCES

Araujo EG (1991): *Fatores neurotróficos intraretinianos aumentam a sobrevivência de células ganglionares da retina in vitro*. D. Sc. thesis, Instituto de Biofisica da Universidade Federal do Rio de Janeiro, Rio de Janeiro

Araujo EG, Linden R (1990): Survival of retinal ganglion cells in vitro: The effect of conditioned medium from retinal cell aggregates. *Braz J Med Biol Res* 23:743–746

Barde YA (1988): What, if anything, is a neurotrophic factor? *Trends Neurosci* 11:343–346

Blaser PF, Catsicas S, Clarke PGH (1991): Limits to the dependence of developing neurons on protein synthesis in their axonal target territory. *Anat Embryol (Berl)* 184:15–24

Brenneman DE, Yu C, Nelson PG (1990): Multi-determinate regulation of neuronal survival—Neuropeptides, excitatory amino acids and bioelectric activity. *Intl J Dev Neurosci* 8:371–378

Clarke PGH (1985): Neuronal death during development in the isthmo-optic nucleus of the chick: Sustaining role of afferents from the tectum. *J Comp Neurol* 234:365–379

Cowan WM (1979): Selection and control in neurogenesis. In: *The Neurosciences: Fourth Study Program*, Schmitt FO, Worden FG, eds. Cambridge: MIT Press

Cunningham TJ, Huddleston C, Murray M (1979): Modification of neuron numbers in the visual system of the rat. *J Comp Neurol* 184:423–433

Dann JF, Buhl EH, Peichl L (1987): Dendritic maturation in cat retinal ganglion cells: A lucifer yellow study. *Neurosci Lett* 80:21–26

Dreher B, Robinson SR (1988): Development of the retinofugal pathway in birds and mammals: Evidence for a common "timetable." *Brain Behav Evol* 31:369–390

Dunlop S (1990): Early development of retinal ganglion cell dendrites in the marsupial *Setonix brachiurus,* Quokka. *J Comp Neurol* 293:425–447

Easter SS, Barald KF, Carlson BM (1988): *From Message to Mind.* Sunderland, USA Sinauer

Eysel UT, Peichl L, Wassle H (1985): Dendritic plasticity in the early postnatal feline retina: Quantitative characteristics and sensitive period. *J Comp Neurol* 242:134–145

Guillery RW (1988): Competition in the development of visual pathways. In: *The Making of the Nervous System*, Parnavelas JG, Stern CD, Stirling RV, eds. Oxford: Oxford University Press

Hughes A (1977): The topography of vision in mammals of contrasting life style: Comparative optics and retinal organization. In: *Handbook of Sensory Physiology*: Vol. 7, Part 5. *The Visual System in Vertebrates*, Crescitelli F, ed. Berlin: Springer-Verlag

Jacobson M (1991): *Developmental Neurobiology*. 3rd ed. New York: Plenum Press

Kaiser PK, Lipton SA (1990): VIP-mediated increase in cAMP prevents tetrodotoxin-induced retinal ganglion cell death in vitro. *Neuron* 5:373–381

Kirby MA (1991): Evidence for the radial migration of retinal ganglion cells (RGCs) during formation of the foveal depression in the primate. In: *Abstracts of the III World Congress of Neurosciences, IBRO*, Montreal, Canada

Lau KC, So KF, Tay D, Jen LS (1991): Elimination of transient dendritic spines in ipsilaterally projecting retinal ganglion cells in rats with unilateral thalamectomy. *Neurosci Lett* 121:255–258

Leventhal PG, Schall JD, Ault SJ (1988a): Extrinsic determinants of retinal ganglion cell structure in the cat. *J Neurosci* 8:2028–2038

Leventhal AG, Schall JD, Ault SJ, Provis JM, Vitek DJ (1988b): Class-specific cell death shapes the distribution and pattern of central projection of cat retinal ganglion cells. *J Neurosci* 8:2011–2027

Levi-Montalcini R (1987): The nerve growth factor: Thirty-five years later. *Science* 237:1154–1162

Lia B, Williams RW, Chalupa LM (1987): Formation of retinal ganglion cell topography during prenatal development. *Science* 236:848–851

Linden R (1987): Competitive interactions and regulation of developmental neuronal death in the retina. In: *Developmental Neurobiology of Mammals*, Chagas C, Linden R, eds, Vatican City: Pontifical Academy of Sciences

Linden R (1988): *Dinâmica de populações celulares no sistema nervoso em desenvolvimento*. Rio de Janeiro: Editora da Universidade Federal do Rio de Janeiro

Linden R (1992): *Dendritic competition in the developing retina: Ganglion cell density gradients and laterally displaced dendrites*. Submitted for publication

Linden R, Araujo EG, Pires RA (1991): Intraretinal neurotrophic factors prevent ganglion cell death *in vitro*. *Soc Neurosci Abstr*, 17:228

Linden R, Perry VH (1982): Ganglion cell death within the developing retina: A regulatory role for retinal dendrites? *Neuroscience* 7:2813–2827

Linden R, Piñón LGP (1987): Dual control by targets and afferents of developmental neuronal death in the mammalian central nervous system: A study in the parabigeminal nucleus of the rat. *J Comp Neurol* 266:141–150

Linden R, Renteria AS (1988): Afferent control of neuron numbers in the developing brain. *Dev Brain Res* 44:291–295

Linden R, Serfaty CA (1985): Evidence for differential effects of terminal and dendritic competition upon developmental neuronal death in the retina. *Neuroscience* 15:853–868

Lipton SA (1986): Blockade of electrical activity promotes the death of mammalian retinal ganglion cells in culture. *Proc Natl Acad Sci U S A* 83:9774–9778

Maffei L, Perry VH (1988): The axon initial segment as a possible determinant of retinal ganglion cell dendritic geometry. *Dev Brain Res* 41:185–194

McCall MJ, Robinson SR, Dreher B (1987): Differential retinal growth appears to be the primary factor producing the ganglion cell density gradient in the rat. *Neurosci Lett* 79:78–84

Nakatani M, Linden R (1991): Axon orientation and axo-dendritic polarity of retinal ganglion cells during postnatal development in the rat. *Braz J Med Biol Res*, 24:937–941

Okado N, Oppenheim RW (1984): Cell death of motoneurons in the chick embryo spinal cord: 9. The loss of motoneurons following removal of afferent inputs. *J Neurosci* 4:1639–1652

Oppenheim RW (1981): Neuronal cell death and some related regressive phenomena during neurogenesis: A selective historical review and progress report. In: *Studies in Developmental Neurobiology. Essarys in Honor of Viktor Hamburger*, Cowan WM, ed. New York: Oxford University Press

Oppenhiem RW (1991): Cell death during development of the nervous system. *Ann Rev Neurosci* 14:453–501

Perry VH (1979): The ganglion cell layer of the rat retina: A Golgi study. *Proc R Sco Lond (Biol)* 204:363–375

Perry VH, Henderson Z, Linden R (1983): Postnatal changes in retinal ganglion cell and optic axon populations in the pigmented rat. *J Comp Neurol* 219:356–368

Perry VH, Linden R (1982): Evidence for dendritic competition in the developing retina. *Nature* 297:683–685

Perry VH, Maffei L (1988): Dendritic competition: Competition for what? *Dev Brain Res* 41:195–200

Perry VH, Walker M (1980): Morphology of cells in the ganglion cell layer during development of the rat retina. *Proc R Soc Lond (Biol)* 208:433–445

Potts RA, Dreher B, Bennett MR (1982): The loss of ganglion cells in the developing retina of the rat. *Dev Brain Res* 3:481–486

Purves D, Lichtman JW (1985): *Principles of Neural Development*. Sunderland, USA Sinauer

Ramoa AS, Campbell G, Shatz CJ (1987): Transient morphological features of identified ganglion cells in living fetal and neonatal retina. *Science* 237:522–525

Ramoa AS, Campbell G, Shatz CJ (1988): Dendritic growth and remodeling of cat retinal ganglion cells during fetal and postnatal development. *J Neurosci* 8:4239–4261

Schall JD, Leventhal AG (1987): Relationships between ganglion cell dendritic structure and retinal topography in the cat. *J Comp Neurol* 257:149–159

Schall JD, Perry VH, Leventhal AG (1987): Ganglion cell dendritic structure and retinal topography in the rat. *J Comp Neurol* 257:160–165

Smith DE (1977): The effect of deafferentation on the development of brain and spinal nuclei. *Prog Neurobiol* 8:349–367

Sohal GS (1976): Effects of deafferentation on the development of the isthmo-optic nucleus in the duck *(Anas platyrhyncos). Exp Neurol* 50:161–173

Thoenen H (1991): The changing scene of neurotrophic factors. *Trends Neurosci* 14:165–170

Wassle H, Boycott BB (1991): Functional architecture of the mammalian retina. *Physiol Rev* 71:447–480

Williams RW, Bastiani MJ, Lia B, Chalupa LM (1986): Growth cones, dying axons, and developmental fluctuations in the fiber population of the cat's optic nerve. *J Comp Neurol* 246:32–69

Yamasaki EN (1990): *Morfologia de células ganglionares da retina do rato em desenvolvimento*. M.Sc. thesis, Instituto de Biofisica da Universidade Federal do Rio de Janeiro, Rio de Janeiro

7

Role of Postsynaptic Activity in Retinogeniculate Pattern Formation

MRIGANKA SUR, JONG-ON HAHM, AND MANUEL ESGUERRA

THE ISSUE

There is considerable evidence now that activity in retinal afferents plays an important role in the formation of connections between retinal ganglion cell axons and their target cells in the lateral geniculate nucleus (LGN). Thus, postnatal eyelid suture in kittens prevents retinogeniculate X and Y axon arbors from developing normally (Sur et al., 1982), and causes postsynaptic X and Y LGN cells to have abnormal structural and functional features (Friedlander et al., 1982; Sherman and Spear, 1982). Blocking action potentials in retinal afferents by infusing tetrodotoxin (TTX) intraocularly in postnatal kittens also alters the morphology of retinogeniculate axon arbors (Sur et al., 1985), and causes LGN cells to have abnormal physiological properties (Dubin et al., 1986). Infusion of TTX into the vicinity of the optic chiasm in prenatal kittens prevents the formation of eye-specific laminae (Shatz and Stryker, 1988), and causes retinogeniculate axon arbors to be abnormally large (Sretavan et al., 1988).

While the theme behind the experiments mentioned above has been to examine the effect of afferent activity on the development of afferent-target connections, in practice, in the retinogeniculate pathway, it is not a straightforward matter to separate the effect of afferent activity from that of target activity. This is because retinal afferents provide by far the major excitatory drive to LGN cells, and blocking or reducing the activity of retinal axons inevitably and severely reduces the activity of LGN cells themselves.

In the experiments described below, we have asked whether and how the activity of postsynaptic cells influences the patterns formed by retinal afferents in the developing ferret LGN, and hence the connections made by presynaptic afferents on LGN target cells. The ferret LGN contains, in addition to eye-specific laminae, a further segregation of retinal axons into "on" and "off" sublaminae. Our approach has been to examine how retinogeniculate transmission occurs in adult and developing ferrets, use specific blockers to reduce retinogeniculate transmission during development, and subsequently examine the effect on retinogeniculate axon arbors. We find that retinogeniculate transmission employs excitatory amino acids and their receptors, including N-methyl-D-aspartate (NMDA) and non-

NMDA receptors. Blocking NMDA receptors on LGN cells during the formation of "on" and "off" sublaminae disrupts the formation of the sublaminae. Blockade of NMDA receptors reduces postsynaptic activity as well as disrupts a mechanism for detecting temporal correlations between pre- and postsynaptic activity (Constantine-Paton et al., 1990); these results indicate, at the least, that the activity of postsynaptic cells plays a significant role in the development of retinogeniculate termination patterns.

THE NORMAL DEVELOPMENT OF RETINOGENICULATE PATTERNS

In ferrets, the segregation of retinogeniculate afferents into eye-specific laminae and "on" and "off" sublaminae occurs progressively in the first 3 weeks after birth. Prior to eye-specific segregation, axons from each eye terminate within the entire LGN (Linden et al., 1981), and individual optic tract axons have short fibrils that extend throughout the mediolateral extent of the nucleus (Hahm, 1991). During the first postnatal week, axons from the two eyes give rise to arbors within discrete, nonoverlapping laminae within the LGN, and eye-specific laminar segregation is essentially complete by the second postnatal week. The two major laminae are lamina A, which received input from the contralateral eye, and lamina A1, which receives input from the ipsilateral eye. Beginning the third postnatal week, arbors within laminae A and A1 confine themselves further into sublaminae (Hahm et al., 1991): the inner sublamina in each of the two eye-specific laminae retains input from "on" center afferents while the outer sublamina retains input from "off" center afferents. Thus, individual retinogeniculate axons form arbors within lamina A or A1 that are restricted to one inner or one outer sublamina (see Figure 7.1A and 7.1B; also Roe et al., 1989).

The third to fourth postnatal week in ferrets is the critical period in which to examine factors that influence the formation of sublaminae. Not only do individual retinogeniculate axons come to form focal arbors within them, sublaminae are also distinguished by clear, termination-free, zones in between. Assays for sublamina formation include (a) defining the projections from an entire eye to each LGN, by injecting an anterograde tracer such as WGA-HRP into the vitreous of one eye, and (b) reconstructing the terminations of single retinogeniculate axons, by bulkfilling individual axons with HRP in chunks of thalamus maintained *in vitro* (Hahm et al., 1991).

THE ROLE OF NMDA AND NON-NMDA RECEPTORS IN RETINOGENICULATE TRANSMISSION

Two kinds of experiment from our laboratory and others demonstrate that retinal ganglion cell axons employ an excitatory amino acid (such as glutamate) for transmission, and that retinogeniculate transmission is mediated by both NMDA and non-NMDA receptors on LGN cells. First, iontophoresis of specific NMDA

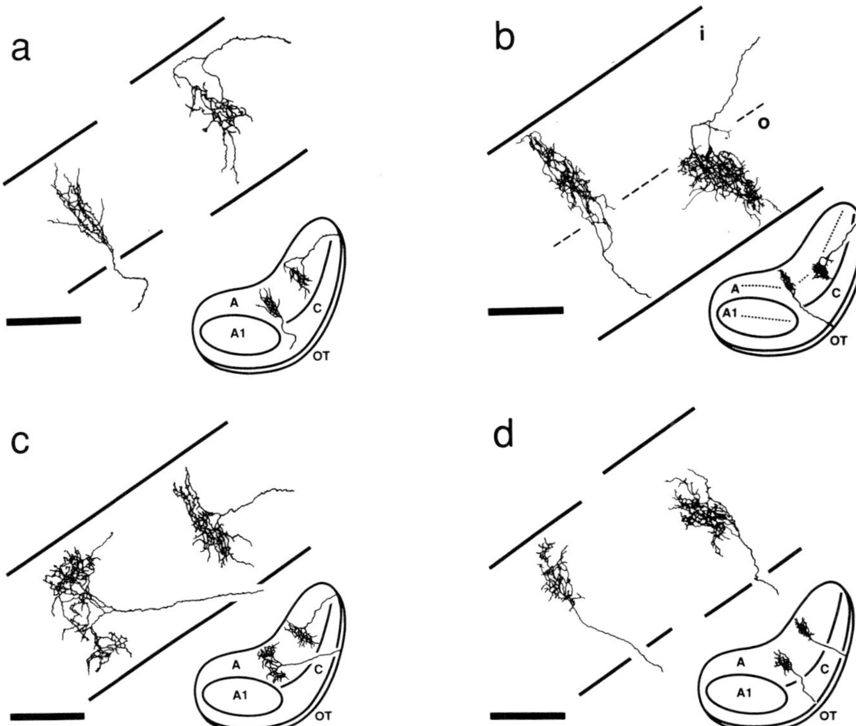

FIGURE 7.1. Retinogeniculate axon arbors in normal ferrets and in ferrets reared with infusion of d-APV into the LGN during the third postnatal week. Axons were labeled by removing the thalamus from the brain, maintaining it in artificial cerebrospinal fluid, and depositing HRP in the optic tract (OT) ventral to the LGN. Arbors were reconstructed from serial sections using camera lucida. A. Arbors in a normal 2-week-old animal. These arbors span nearly the entire height of lamina A (or A1). The inset shows, in a horizontal section of the LGN, the relative locations of laminae A and A1 and the position of each arbor within lamina A. Scale bar: 100 μm for all large axon drawings. B. Arbors in a normal 3-week-old animal. These arbors are now restricted to the height of an inner (i) or outer (o) sublamina within lamina A (or A1). C,D. Arbors in animals reared with infusion of d-APV from age 2–3 wk. Some arbors, shown in C, span the entire height of lamina A (or A1) and do not restrict in height in all. Other arbors, shown in D, are restricted in height and are either positioned appropriately, at the inner or outer half of lamina A, or inappropriately, in the middle of lamina A. Reprinted by permission from *Nature* 351:568–570. Copyright 1991 Macmillan Journals Ltd.

and non-NMDA receptor antagonists on LGN cells in adult cats reduces the visual responses of these cells (Heggelund and Hartveit, 1990; Kwon et al., 1991; Sillito et al., 1990a, 1990b). All LGN cells appear to have NMDA and non-NMDA receptors, though there is some variation in the proportion of these receptors employed in the visual responses of different cells (Kwon et al., 1991). Second, EPSPs recorded in LGN cells of adult ferrets *in vitro* following optic tract stimulation are reduced in both amplitude and duration by NMDA and non-

NMDA antagonists (Esguerra et al., 1992). Even the postsynaptic response to a single optic tract stimulus is mediated significantly by NMDA receptors. Moreover, in the LGN of developing ferrets, particularly at ages at which eye-specific laminae and "on" and "off" sublaminae are forming, retinogeniculate EPSPs can be recorded (White and Sur, 1991); both NMDA and non-NMDA receptors contribute significantly to these EPSPs (C. A. White, M. Esguerra, and M. Sur, unpublished observations).

Thus, retinal ganglion cell axons communicate synaptically with LGN cells throughout the time that retinal axon arbors are being shaped into their adult termination patterns. Retinogeniculate synaptic transmission is a critical means by which LGN cells come to be active during development. It is reasonable to postulate that the activity of these target cells plays a role in the formation of retinogeniculate termination patterns. We have started to examine such a hypothesis by using specific antagonists to block or reduce retinogeniculate transmission during development.

EFFECT OF NMDA ANTAGONISTS ON RETINOGENICULATE ARBORS

We have examined the effect on sublaminar development of blocking NMDA receptors in the LGN of ferrets between 2 and 3 weeks of age (Hahm et al., 1991). The NMDA antagonists d-APV and MK-801, delivered via osmotic minipumps to achieve concentrations that block NMDA receptors at retinogeniculate synapses *in vitro*, appears to prevent the formation of sublaminae. Labeling an eye's projection zone with WGA-HRP in 3-week-old ferrets reared with blockade of NMDA receptors shows that while eye-specific laminae are clearly visible, "on" and "off" sublaminae, that are normally present by 3 weeks, now can fail to form. We have examined single retinal axon arbors in animals reared with d-APV; these differ systematically from normal 3-week arbors. Some arbors span the entire height of lamina A or A1 and fail to restrict at all into sublaminae (see Figure 7.1C). Other arbors develop to normal size, and some are positioned appropriately at the inner or outer half of each lamina. Still other normal-sized arbors, however, are inappropriately located within the LGN, being positioned in the middle of lamina A or A1 (see Figure 7.1D). Quantitative measures of arbors indicate that arbors in animals reared with NMDA receptor blockade are similar in overall size to normal arbors, but differ significantly in sublaminar restriction. Such arbors form the cellular basis for the lack of sublamination observed in the entire eye's projection to the LGN.

NMDA receptors have been postulated to form the substrate for a Hebbian mechanism for development (Constantine-Paton et al., 1990), by providing a means for detecting temporal correlations between pre- and postsynaptic activity and allowing the entry of calcium into the postsynaptic cell. Intracellular calcium can then act via second messenger systems to modulate the efficacy of synapses. It is possible that NMDA receptors on LGN cells play such a specific developmental role. However, the retinogeniculate pathway also employs NMDA receptors in synaptic transmission, and hence in regulating the overall activity of postsynaptic cells. There are several different ways by which a cells's activity level (e.g., its membrane potential) can regulate intracellular signals, independent of NMDA receptors. It therefore becomes complicated to separate a specifically developmental role for NMDA receptors from a more general role in setting activity levels per se of LGN cells (cf. Shatz, 1990).

Regardless of whether NMDA receptors play a specific role in development or a general role by mediating synaptic transmission, our results suggest that the activity of postsynaptic cells is crucial for the formation of retinogeniculate afferent patterns. It is possible that postsynaptic cells influence presynaptic afferents via a retrograde signal (Dumuis et al., 1988; Gally et al., 1990; Garthwaite et al., 1988, 1989; Schuman and Madison, 1991), and it is of considerable interest at this time to identify such signals and their specific effect on retinogeniculate axon arbors.

POSTSYNAPTIC ACTIVITY AND RETINOGENICULATE DEVELOPMENT

One way to begin addressing whether NMDA receptors play a specific developmental role is to examine the effect of blocking non-NMDA receptors on retinogeniculate pattern formation. Blocking non-NMDA receptors reduces synaptic transmission to most LGN cells, thereby reducing their activity levels; such reduction in activity might also affect NMDA receptors secondarily due to the voltage-dependence of these receptors. Thus, blocking non-NMDA receptors does not provide an unequivocal answer to whether NMDA receptors are critical. At the same time, non-NMDA receptors may not be that distinct from NMDA receptors, at least in their developmental role. For example, recent evidence indicates that non-NMDA receptors can also allow entry of calcium into neurons (Gilbertson et al., 1991; Iino et al., 1990), and hence can provide specific developmental signals as well. Furthermore, the voltage dependence of NMDA receptors may play much less of a role *in vivo*, at least in conditions where there is ambient glutamate (Sah et al., 1989) or where the background activity of cells is high, as in the retinogeniculate pathway (Kwon et al., 1992).

A second way to examine the issue is to block NMDA receptors but maintain postsynaptic activity at a high level: if NMDA receptors play a critical role, sublaminae should not form; if postsynaptic activity per se is the critical element, sublaminae should form. Such experiments would need to involve a combination of NMDA receptor block and either direct excitation of LGN cells or a reduction of inhibition on them, and are a major focus of our present work.

Acknowledgments. We thank Cheryl White for her comments on the manuscript. Supported by grants from the National Institutes of Health.

REFERENCES

Constantine-Paton M, Cline H, Debski E (1990): Patterned activity, synaptic convergence, and the NMDA receptor in developing visual pathways. *Ann Rev Neurosci* 13:129–154

Dubin M, Stark L, Archer S (1986): A role for action-potential activity in the development of neuronal connections in the kitten retinogeniculate pathway. *J Neurosci* 6:1021–1036

Dumuis A, Sebben M, Haynes L, Pin J-P, Bockaert J (1988): NMDA receptors activate the arachidonic cascade system in striatal neurons. *Nature* 336:68–70

Esguerra M, Kwon YH, Sur M (1992): Retinogeniculate EPSPs recorded intracellularly in the ferret lateral geniculate nucleus *in vitro*: Role of NMDA receptors. *Visual Neurosci* 8:545–555

Friedlander MJ, Stanford LR, Sherman SM (1982): Effects of monocular deprivation on the structure-function relationship of individual neurons in the cat's lateral geniculate nucleus. *J Neurosci* 2:321–330

Gally J, Montague PR, Reeke G Jr, Edelman G (1990): The NO hypothesis: Possible effects of a short-lived, rapidly diffusible signal in the development and function of the nervous system. *Proc Natl Acad Sci U S A* 87:3547–3551

Garthwaite J, Charles S, Chess-Williams R (1988): Endothelium-derived relaxing factor release on activation of NMDA receptors suggests role as intercellular messenger in the brain. *Nature* 336:385–388

Garthwaite J, Garthwaite G, Palmer RMJ, Moncada S (1989): NMDA receptor activation induces nitric-oxide synthesis from arginine in rat-brain slices. *Eur J Pharmacol* 172:413–416

Gilbertson TA, Scobey R, Wilson M (1991): Premeation of calcium ions through non-NMDA glutamate channels in rod bipolar cells. *Science* 251:1613–1615

Hahm JO (1991): *Influence of the target on development of the ferret retinogeniculate projection*. Unpublished Ph.D. dissertation, Massachusetts Institute of Technology

Hahm JO, Langdon RB, Sur M (1991): Disruption of retinogeniculate afferent segregation by antagonists to NMDA receptors. *Nature* 351:568–570

Heggelund P, Hartveit E (1990): Neurotransmitter receptors mediating excitatory input to cells in the cat lateral geniculate nucleus: 1. Lagged cells. *J Neurophysiol* 63:1347–1360

Kwon YH, Esguerra M, Sur M (1991): NMDA and non-NMDA receptors mediate visual responses of neurons in the cat's lateral geniculate nucleus. *J Neurophysiol* 66:414–428

Kwon YH, Nelson SB, Toth LJ, Sur M (1992): Effect of stimulus contrast and size on NMDA receptor activity in the cat lateral geniculate nucleus. *J Neurophysiol* (in press)

Linden D, Guillery R, Cucchiaro J (1981): The dorsal lateral geniculate nucleus of the normal ferret and its postnatal development. *J Comp Neurol* 203:189–211

Lino M, Ozawa S, Tsuzuki K (1990): Permeation of calcium through excitatory amino acid receptor channels in cultured rat hippocampal neurones. *J Physiol* 424:151–165

Roe AW, Garraghty PE, Sur M (1989): Terminal arbors of single On-center and Off-center X and Y retinal ganglion cell axons within the ferret's lateral geniculate nucleus. *J Comp Neurol* 288:208–242

Sah P, Hestrin S, Nicoll RA (1989): Tonic activation of NMDA receptors by ambient glutamate enhances excitability of neurons. *Science* 246:815–818

Schuman EM, Madison DV (1991): A requirement for the intercellular messenger nitric oxide in long-term potentiation. *Science* 254:1503–1506

Shatz CJ (1990): Impulse activity and the patterning of connections during CNS development. *Neuron* 5:745–756

Shatz CJ, Stryker M (1988): Prenatal tetrodotoxin infusion blocks segregation of retinogeniculate afferents. *Science* 242:87–89

Sherman SM, Spear P (1982): Organization of visual pathways in normal and visually deprived cats. *Physiol Rev* 62:738–855

Sillito AM, Murphy PC, Salt TE (1990a): The contribution of the non-N-methyl-D-aspartate group of excitatory amino acid receptors to retinogeniculate transmission in the cat. *Neuroscience* 34:273–280

Sillito AM, Murphy PC, Salt TE, Moody CI (1990b): Dependence of retinogeniculate transmission in cat on NMDA receptors. *J Neurophysiol* 63:347–355

Sretavan D, Shatz C, Stryker M (1988): Modification of retinal ganglion cell axon morphology by prenatal infusion of tetrodotoxin. *Nature* 336:468–471

Sur M, Garraghty P, Stryker M (1985): Morphology of physiologically identified retinogeniculate axons in cats following blockade of retinal impulse activity. *Soc Neurosci Abstr* 11:805

Sur M, Humphrey AL, Sherman SM (1982): Monocular deprivation affects X- and Y-cell retinogeniculate terminations in cats. *Nature* 300:183–185

White CA, Sur M (1991): Development of membrane and synaptic properties in neurons of the ferret LGN. *Neurosci Abstr* 17:1136

8

Generation of Cell Diversity in the Mammalian Visual Cortex

MARLA B. LUSKIN

COMPOSITION, PROLIFERATION AND DETERMINATION OF CELLS IN THE CEREBRAL CORTEX

One of the hallmarks of the central nervous system (CNS) of all mammalian species is that it contains a diverse population of specialized cell types and specific sets of connections. The cerebral cortex is the ultimate example of such specificity where a unique pattern of afferent and efferent connections distinguish a large number of individual cortical areas, each containing a heterogeneous array of neuronal and glial subtypes. The visual cortex, the topic of this chapter, is one such area. To ensure the emergence of such complexity, the developing nervous system has somehow devised a way to regulate the number of cells generated, their distribution, their course of differentiation, and the connections they form. This chapter is concerned with a consideration of how the immense variety of cells belonging to the cerebral cortex, and visual cortex in particular, arises during the course of development. Any acceptable explanation of how cells acquire their identity must account for the observation that all of the cells of the cerebral cortex—both neurons and glia—are derived from a seemingly homogeneous layer of neuroepithelial cells surrounding the immature cerebral ventricles, known as the ventricular zone. The question this raises is whether the cells of the ventricular zone form a homogeneous population, or whether it might contain distinct sets of progenitor cells that give rise exclusively to one type of cell: either all neurons, all astrocytes, or all oligodendrocytes. This also raises the related question of if and when lineage restrictions occur among the population of progenitor cells producing the cells of the mature cerebral cortex. An analysis of the phenotype of the progeny of individual progenitor cells of the ventricular zone would shed light on these questions. As will be discussed below, the adaptation of recombinant retroviruses as tools to study cell lineage has placed such an analysis of the genealogical relationships of cells of the cerebral cortex within the realm of possibility.

This chapter will review the cellular composition of the cerebral cortex and aspects of its development that have become understood utilizing a variety of experimental approaches, including ^{3}H-thymidine birthdating methods and electron microscopy. To a large extent, the visual cortex has served as a model for these studies because more is known about it than any other cortical area. However, it is worth emphasizing that the early events in telencephalic development are likely to be similar for all neocortical areas. Therefore, the present discussion will not be restricted to the acquisition of cell identity in the visual cortex, but rather, where applicable, it will draw upon studies dealing with other cortical areas in an attempt to arrive at some general principles of cortical development. I will mainly discuss studies from my own laboratory that in recent years have concentrated on an analysis of the lineage relationships of cells in the cerebral cortex involving the use of retroviral lineage tracers. These studies have revealed that by the onset of cortical neurogenesis the lineages for neurons, astrocytes, and oligodendrocytes have diverged, suggesting a role for lineage in the determination of cell fate.

Determination of Cell Identity in the Nervous System

Lineage relationships among the cells of the mammalian CNS have been of interest to neurobiologists ever since His (1889) first claimed to distinguish different populations of cells in the ventricular zone. As a means to determine what governs the pathway of differentiation a particular cell takes, in recent years much attention has focused on gaining a better understanding of the relative roles of lineage-derived and environmentally-derived information (for review, see McConnell, 1988). In invertebrates, such as *C. elegans*, in which the lineage of the entire nervous system has been elucidated, lineage is believed to play an important role in defining the differentiated phenotype of neurons (Kenyon, 1985; Sulston et al., 1983). In contrast, the differentiation of cells in the *Drosophila* eye ommatidium follows a stereotypical pattern of development that is now known to be regulated by cell-cell interactions (Banerjee and Zipursky, 1990). Moreover, a number of studies on the development of the vertebrate peripheral nervous system have underscored the significance of environmental factors in dictating the neurotransmitter phenotype of cells (Anderson, 1989; Doupe et al., 1985; Patterson, 1978; Potter et al., 1986). Despite these advances, much less is known about the mechanisms that regulate cell identity in the mammalian CNS, largely because progress in this area has been hampered by the lack of a suitable way to trace the geneology of *individual* cells. Nonetheless, Raff and co-workers, in an elegant series of experiments, have determined some of the rules by which glial cells in the rat optic nerve seem to acquire their phenotype (for review, see Miller et al., 1989; Raff, 1989). But it is not known whether these same mechanisms can explain glial diversification in the cortex, where analogous populations of glia have not been documented, and where one must also explain how the multitude of neuronal cell types arise. In the last few years, studies aimed at unraveling lineage relationships in the cerebral cortex have been undertaken by taking advantage of a recently introduced method (Price et al., 1987; Sanes et al., 1986) that enlists replication-defective recombinant retroviruses as heritable markers (Luskin et al., 1988; Price and Thurlow, 1988; Walsh and Cepko, 1988).

The advent of recombinant retroviruses several years ago has made lineage analysis in the cerebral cortex feasible principally because the engineered retroviruses have been endowed with desirable properties. Replication-defective recombinant retroviruses can be utilized as heritable lineage markers because, like wild type retroviruses, they are able to infect a dividing cell, integrate into its DNA, and be passed on to all its progeny. Additionally, due to the excision of essential parts of the wild type viral genome (see Sanes, 1989, for more details), they are unable to replicate on their own and infect unrelated cells. Moreover, the recombinant retroviruses, which are most useful as lineage tracers, have had added to their genome an exogenous gene, such as the *E. coli.* β-galactosidase gene, whose expression can be detected histochemically and immunohistochemically in all the descendants of an infected progenitor cell. Hence, the application of recombinant retroviruses as a tool for analyzing cell lineage overcomes many of the problems encountered with using other lineage tracers in the mammalian CNS. Foremost, the retroviral lineage tracer (1) does not dilute out as a function of increasing numbers of cell divisions; (2) is restricted to all the progeny of an infected cell; (3) can be introduced into small cells, such as ventricular zone cells of the telencephalon that are extremely difficult to inject intracellularly with a marker *in vivo*; and (4) can be introduced into precursor cells at virtually any stage of development.

Histogenesis of the Cerebral Cortex

Thymidine birthdating studies and ultrastructural observations have demonstrated that during the development of the mammalian cerebral cortex, neurons generated in a proliferative zone lining the ventricles subsequently migrate radially through an intermediate zone to form the cortical plate (Angevine and Sidman, 1961; Berry and Rogers, 1965; Sauer, 1935; Shimada and Langman, 1970; Sidman and Rakic, 1973). There they undergo differentiation and form layers, each of which has a distinct subset of neuronal phenotypes. Furthermore, birthdating studies have established that the neurogenesis of cells of the deep layers precedes that of the more superficial layers in a systematic fashion referred to as the inside-first, outside-last pattern of neurogenesis (e.g., Caviness and Sidman, 1973; Luskin and Shatz, 1985; Rakic, 1974). Glial cells are also derived from precursors in the ventricular zone, but unlike neurons, after leaving the ventricular zone glial cells retain the capacity to divide within the subependymal zone, and throughout the white and grey matter of the cerebral cortex (Ichikawa and Hirata, 1982; Mares and Bruckner, 1978; Privat, 1975; Skoff, 1980). Despite the wealth of information derived from these studies, fundamental questions about the spatial and temporal properties of cell proliferation and migration, and the mechanisms that regulate the number of cells generated, have gone unanswered. For instance, it is not clear when and where most glial cells originate, or what proportion arise prenatally and what proportion arise postnatally. Nor is it known how much glial cell division occurs in the cerebral cortex proper in the form of *in situ* division. The introduction of recombinant retroviruses combined with recent technological advances in the development of low-light-level confocal video microscopy now make it possible to directly study dynamic aspects of neurogenesis, gliogenesis, and migration in the developing telencephalon *in situ*, and headway in these areas can be anticipated in the coming years.

The question of whether all the neurons derived from a single precursor cell in the ventricular zone follow the same route to their destination, and whether they do so by migrating along radial glial fibers, as suggested by ultrastructural observations (Meller and Tetzloff, 1975; Peters and Feldman, 1973; Rakic, 1972, 1988), has received renewed attention. An inference from this idea is that clonally related neurons remain in close proximity and radially arrayed in the mature cortex (Rakic, 1988). We have recently obtained results that lend support to this model (Luskin et al., 1988). Our studies demonstrated that in the developing telencephalon, clonally related cells often form radial arrays that span the distance from ventricular zone to cortical plate, and that neuronal clones generally remained radially oriented in the mature cerebral cortex. On the other hand, Walsh and Cepko (1988) reported that neuronal clones are not radially oriented, but that clonally related neurons become dispersed in the tangential dimension of the cortex, casting some doubt on the model proposed by Rakic (1988). However, neither study compared the distribution of radial glial fibers to the disposition of clonally related neurons. One explanation, then, for the seemingly contradictory results is that if clonally related neurons follow radial glial fibers, they could become distributed tangentially to the same extent that radial glial fibers spread out as they defasciculate and become deflected upon leaving the intermediate zone and enter the cortical plate (Misson et al., 1988, 1991). This issue has recently been addressed by Austin and Cepko (1990). They reported that the pattern of migration of clonally related cells from the ventricular zone to the cortical plate in mice corresponded to the distribution of radial glial fibers, and that in the medial and dorsal aspect of the cortex the migration was usually in the radial dimension, agreeing with our earlier report (Luskin et al., 1988). An even more decisive way to determine whether the descendants of a progenitor cell migrate along the same fascicle or nearby fascicles of radial glial fibers is to chart the changing positions of the members of a clone in relation to radial glial fibers in living slices of telencephalon over an extended period of time.

Considerations of the Emergence of Separate Neuronal and Glial Lineages

For over 100 years cell lineage in the cerebral cortex has been investigated, but until recently, as discussed above, there was no way to directly trace cell lineage in mammals (Price et al., 1987; Sanes et al., 1986). Nonetheless, several different models of cell lineage have emerged and are now being tested. His, among the first investigators to propose a scheme for neuronal-glial lineage, suggested, based on the histological appearance of cells, that there was a population of germinal cells in the ventricular zone that divide and produce only neurons, whereas glial cells were believed to be derived from a distinct class of spongioblast cells (His, 1889). Subsequent analysis underscored the problem with relying on morphological criteria to distinguish between different types of precursor cells; Sauer (1935) demonstrated that the observed differences between these two presumed types of cells in the ventricular zone simply reflect the changing morphologies of cells at different stages in the mitotic cycle. On the other hand, the apparent homogeneity of cells in the ventricular zone led some investigators to hypothesize that a single uncommitted class of precursor cells gives rise to both neurons and glia (Schaper, 1897; reviewed in Levitt et al., 1981).

This latter model of a multipotential cell in the ventricular zone has been challenged by several more recent studies of the embryonic telencephalon that have used antibodies to cell-type specific markers to distinguish between different populations of precursor cells. These studies concluded that the precursor cells of the ventricular and subventricular zone, another layer of proliferating cells circumscribing the ventricular zone, are not a homogeneous population. For example, Levitt and colleagues (1981, 1983) found separate populations of GFAP-positive and GFAP-negative cells in the ventricular zone of the primate telencephalon, and argued that neuronal and glial precursors coexist at early stages of cortical development. However, their results did not enable them to distinguish which cell types GFAP-negative cells might give rise to besides neurons, or if GFAP-positive cells could give rise to cells other than astroyctes. In another study, LeVine and Goldman (1988b) observed that, in tissue sections of the neonatal rat forebrain, some cells of the subventricular zone stained with both antibodies to GD3, a marker of immature neuroectoderm, and galactocerebroside or carbonic anhydrase, markers for oligodendrocytes (LeVine and Goldman 1988a, 1988b). Other cells were GD3-negative and stained with antibodies to vimentin; they reasoned these were giving rise to astrocytes. From these observations, they suggested that the lineages of astrocytes and oligodendrocytes diverge by early postnatal life. Because LeVine and Goldman's analysis began after neurogenesis had ceased, there was no way they could investigate the lineage of neurons. Furthermore, there is no way to determine which of the immunostained cells are derived from the same precursor cell. However, in another series of studies, Vaysse and Goldman (1990) demonstrated directly using retroviral lineage tracing combined with cell-type specific markers *in vitro*, that the lineages for astrocytes and oligodendrocytes have diverged by the time of birth (after the period of neurogenesis) in the developing rat striatum. This finding is in agreement with our *in vitro* lineage analysis of dissociated cells from the prenatal telencephalon of the mouse (Luskin et al., 1988), in which we also used retroviral-mediated gene transfer. Our *in vivo* studies, discussed below, have corroborated this result of separate lineages for neurons and glia (Barfield et al., 1990; Luskin, 1990). Moreover, this work has provided the first direct evidence that the lineage of neurons, astrocytes, and oligodendrocytes diverge early in development (Barfield, 1990; Luskin, 1990; Luskin et al., 1988). However, other attempts to uncover how cell diversity arises in the mammalian CNS using cultured cells have provided evidence for a common precursor for neurons and glia in the septal region (Temple, 1989), albeit a small proportion of the dividing cells were bipotential. Presumably, there must be multipotential precursor cells in the developing telecephalon as well, but at a developmental stage preceding that at which we have begun tracking cell lineage to date.

Aspects of Neuronal Diversity in the Mammalian Cerebral Cortex

As a prelude to determining the lineage relationships between the diverse array of neurons in the cerebral cortex, the cellular components must be distinguished from one another in a meaningful way. To a large extent this has been accomplished. The multitude of neurons in the cerebral cortex have been characterized by their laminar position, morphology, physiology, and pattern of projections (for review, see Gilbert, 1983). Furthermore, correlations have been reported between these features used to classify cells and their neurochemical identity. Broadly speaking, though, on the basis of morphological and functional criteria, the immense population of neurons in the visual cortex can be subdivided into two types: pyramidal neurons or projection neurons, and nonpyramidal neurons or interneurons (Lin et al., 1986; Wise, 1985). It has been suggested that the pyramidal neurons primarily utilize the excitatory amino acids L-glutamate and L-aspartate as neurotransmitters (Baughman and Gilbert, 1981; Conti et al., 1987; Parnavelas et al., 1989; Streit, 1984), whereas virtually all of the nonpyramidal neurons contain the inhibitory amino acid, gamma-amino-butyric acid (GABA) (Houser et al., 1983). However, a variety of peptides, mosts significantly somatostatin, vasoactive intestinal polypeptide, cholecystokinin, and neuropeptide Y, have been colocalized with GABA, defining subpopulations of nonpyramidal neurons (Jones and Hendry, 1986). Furthermore, each of these subpopulations has been distinguished by its ultrastructural and synaptic features, as well as by its peptide content (Parnavelas et al., 1989; Peters, 1985). In addition, different subtypes of GABAergic cells have been distinguished on the grounds that they have different cell surface molecules (Naegele et al., 1988). Collectively, these studies suggest that there are at least two types of pyramidal neurons, and several types of nonpyramidal neurons for which the lineage must be accounted.

Despite our lack of understanding about what controls the identity of cortical neurons, and in particular, their transmitter phenotype, many other features of their development have been elucidated using immunocytochemistry alone or in combination with ^{3}H-thymidine (Fairen, 1986; Meyer and Wahle, 1988; Miller, 1985, 1986; Parnavelas et al., 1988; Wolff et al., 1984). Together these studies have defined the patterns of neurogenesis and differentiation for various transmitter-containing cell types, and have demonstrated that within a layer, the pyramidal and nonpyramidal cells are born simultaneously (Luskin and Shatz, 1985; Miller, 1985). The question that naturally arises is whether the transmitter phenotype of a neuron is dictated by lineage-derived factors or by factors in its environment. At present there is little direct support for either alternative in the visual cortex. A first step in evaluating the role that lineage-derived factors play in the determination of phenotype is to ask whether the same progenitor cell gives rise to pyramidal and nonpyramidal neurons, and even whether the different subtypes of nonpyramidal cells have a separate lineage. Findings from our previous study did not exclude either possibility (Luskin et al., 1988). This study showed that a single progenitor cell can contribute neurons to several laminae, consistent with both the existence of a bipotential progenitor cell for pyramidal and nonpyramidal neurons and with the scenario that the progeny of some progenitor cells are all pyramidal cells, whereas other progenitor cells bear all nonpyramidal cells. Our more recent studies summarized below support the latter possibility (Luskin, 1990; Parnavelas et al., 1990, 1991).

LINEAGE RELATIONSHIPS IN THE CEREBRAL CORTEX ANALYZED WITH A RECOMBINANT RETROVIRUS ENCODING THE REPORTER GENE *E. COLI* β-GALACTOSIDASE

Appearance of lacZ-Positive Cells at the Light and Electron Microscopic Level

In order to analyze the phenotypic composition of clonally related cells originating from progenitor cells of the telencephalon we developed a technique to deliver retroviral lineage tracers to dividing cells of the ventricular zone, as shown in Figure 8.1. For these investigations we have used the rat as our experimental animal for several reasons. First, the cell types in the rat cerebral cortex, and visual cortex in particular, are well characterized. Second, recombinant retroviruses

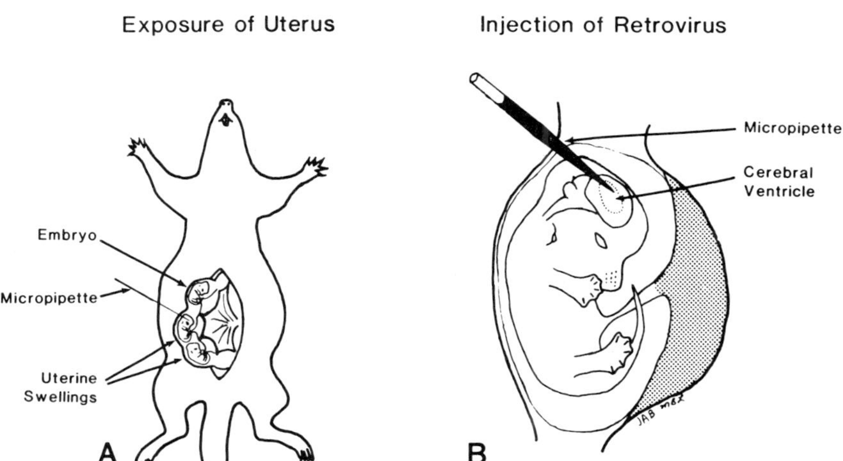

FIGURE 8.1. Method for introducing retrovirus to dividing cells of the telencephalic ventricular zone of an E15 or E16 rat embryo. In order to make injections of recombinant retrovirus into the cerebral ventricles of rat embryos, a laparotomy is performed and the uterus exposed as shown in **A**. The uterus contains several embryos arranged bilaterally in 2 rows (right half exposed), each supported by a separate placenta. **B.** An enlargement of one uterine swelling. It shows the placement of a micropipette in the cerebral ventricle of one hemisphere once it has penetrated the uterine membrane, amniotic cavity, and skull. Recombinant retrovirus (0.5–1.0 μl) is injected directly into the cerebral ventricles, providing it with access to proliferating cells of the ventricular zone. Shaded area represents placenta.

have been made that are specific to rodents—none exist for cat or monkey. Third, given the similarity in early cortical development between primate and nonprimate animals, most likely the findings in rodents can be applied to cortical development in higher mammals, including human development.

Initially, we were interested in examining whether the same progenitor cell gives rise to neurons and glia, or if there are separate progenitor cells giving rise to neurons or glia, but not both, at a time when or before postmitotic neurons destined for the cerebral cortex are being generated. We reasoned that the onset of neurogenesis might be a likely time for lineage restrictions to be manifested, because not only is proliferation occurring then, but more importantly, some neurons are becoming postmitotic. The onset of neurogenesis in the rat visual cortex is embryonic day 15 (E15) or 16 (Raedler and Raedler, 1978). Thus, to give the retrovirus access to dividing cells of the ventricular zone intraventricular injections of BAG retrovirus were made. The BAG retrovirus carries the reporter gene *E. coli* β-galactosidase (see Price et al., 1987, for a description of its construction), abbreviated *lacZ*, and by convention its gene product is referred to as lacZ.

The brains of injected embryos were subsequently perfused postnatally, sectioned on a Vibratome, and incubated in X-Gal, which turns the cells expressing β-galactosidase blue. To permit subsequent ultrastructural analysis, the sections were embedded in plastic before the distribution of lacZ-positive cells was charted in order to identify discrete clusters of closely spaced lacZ-positive cells, considered to be clones for the purpose of our analysis. Operationally, we restricted our analysis to clusters of lacZ-positive cells that were separated from any other stained cell by at least 500 μm in a plane parallel to the pial surface, although usually clusters were more widely distributed in the cortex. The presumption that clusters of lacZ-positive cells represent clones is bolstered by the findings of Austin and Cepko (1990) in which the number of clusters of lacZ-positive cells in the mouse telencephalon depended on the amount of virus injected into the cerebral ventricles, but the number of cells per cluster remained comparable. Galileo et al., (1990) have also provided evidence that discrete groups of lacZ-positive cells represent clones, although their studies were in the optic tectum of the chick. However, there is still some uncertainty about whether a cluster of cells constitutes a clone. Other types of experiments must resolve this issue.

At the light microscopic level labeled cells in the adult rat show considerable variation in the staining of their processes. A relatively small proportion of cells show quite extensive process staining sufficient to classify them as neuronal or nonneuronal (astroglial or oligodendroglial). However, the vast majority of lacZ-positive cells show only incomplete or no process staining. While glia could frequently be distinguished from neurons, there was no satisfactory way in the majority of cases to distinguish between pyramidal and nonpyramidal cells, and between astrocytes and oligodendrocytes. The inability to determine the identity of cells at the light microscopic level prompted us to undertake an ultrastructural analysis of the phenotype of lacZ-positive cells. Ultrastructure can be used to unambiguously ascertain phenotype for the major cell types of the cortex. Fortunately, this was a manageable, though laborious endeavor because the β-galactosidase histochemical reaction product is electron dense (Bunge et al., 1989). In our experiments the reaction product is predominantly associated with the nuclear membrane and endoplasmic reticulum, and sometimes with the Golgi apparatus and mitochondria. Furthermore, the staining does not obscure our ability to visualize axon terminals synapsing upon labeled cells. This turned out to be quite fortunate, because the most meaningful way to distinguish between pyramidal and nonpyramidal cells is by the type of synapses they receive on their cell soma. Pyramidal neurons receive exclusively symmetrical axosomatic synapses (Colonnier, 1981; Peters, 1985), whereas both symmetrical and asymmetrical synapses occur on the soma of nonpyramidal neurons.

The findings discussed below are based on 28 clones in which every cell (192) contributing to a clone was serially sectioned for ultrastructural analysis. Figure 8.2 shows a representative example of a cell from each type of clone encountered. Briefly, neurons could be identified by their axosomatic synapses, extension of dendrites from their cell soma, and abundant endoplasmic reticulum (see Figure 8.2A and 8.2B) (Parnavelas et al., 1989; Peters, 1985; Peters et al., 1976). The most characteristic feature of astrocytes at the ultrastructural level is the presence of glial filaments (see Figure 8.2D), although not detectable in every cell (Peters et al., 1976). In addition, astrocytes were frequently located adjacent to blood vessels, had an electron lucent cytoplasm, and exhibit an irregular cytoplasmic outline weaving in and around other profiles in the tissue. Finally, the most revealing features of oligodendrocytes are their dense clumps of nuclear heterochromatin (see Figure 8.2C) and cytoplasmic granularity (Peters et al., 1976; Vaughan, 1984). Collectively, these well-established features allowed us to distinguish the phenotype of lacZ-positive cells. Thus, when the morphological phenotype of individual cells within a clone was compared following injections of retrovirus at the onset of neurogenesis, we observed a striking similarity among the phenotype of cells within a clone. With few exceptions, when cells were classified as either astrocytes, oligodendrocytes, pyramidal neurons, or nonpyramidal neurons, each clone contained a homogeneous population of cells.

Lineage Relationships between Pyramidal and Nonpyramidal Neurons

In the 15 clusters of lacZ-positive cells classified as neuronal clones, every cell had the fine structural features characteristic of neurons (see above). These clones typically contained between 2 and 9 cells. In order to investigate the lineage relationships between neuronal subtypes, the cells within neuronal clones were further classified as either pyramidal or nonpyramidal neurons according to widely acknowledged ultrastructural features (Colonnier, 1981; Jones and Powell, 1970; Parnavelas et al., 1977; Peters, 1985). Therefore, we took advantage of the characteristic differences between the morphology and ultrastructure of pyramidal and nonpyramidal cells to conclusively distinguish between them (Parnavelas et al., 1991). First, in the mammalian visual cortex pyramidal neurons vary widely in

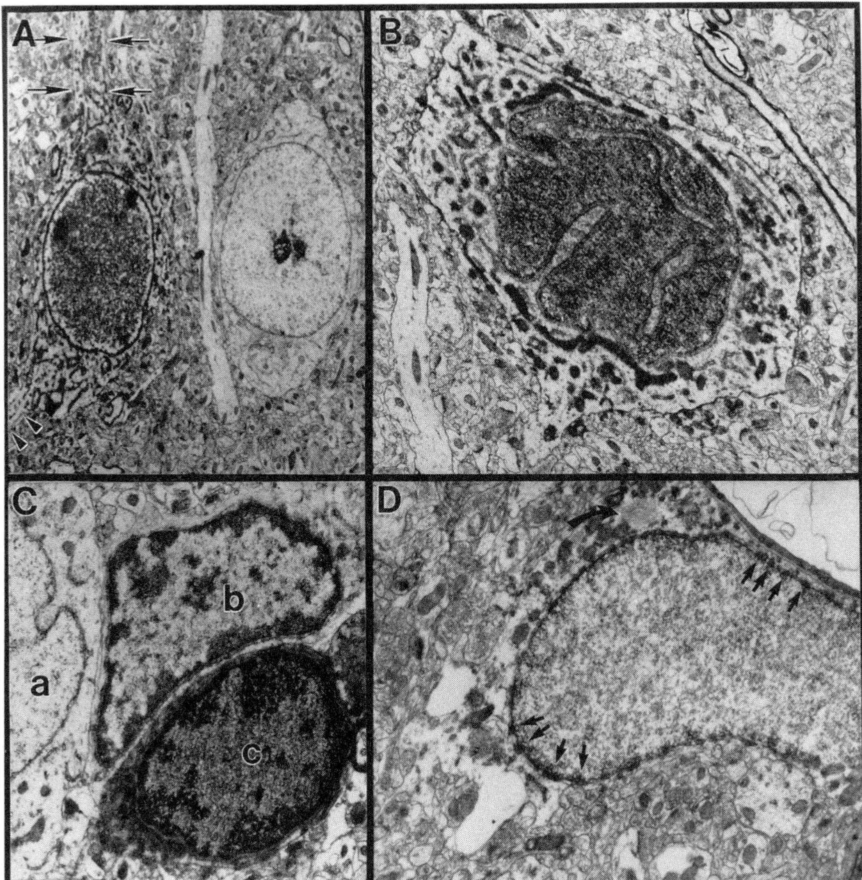

FIGURE 8.2. Representative examples of the ultrastructural appearance of each of the major cell types in the visual cortex expressing the inherited lineage marker *E. coli* β-galactosidase. Virtually all clones examined in the adult cerebral cortex resulting from an injection of recombinant retrovirus at the onset of neurogenesis (E15 or E16) were composed of a single cell type; they contained either **(A)** all pyramidal neurons, **(B)** all nonpyramidal neurons, **(C)** all oligodendrocytes, or **(D)** all astrocytes. Each cell type has a distinctive appearance when histochemically stained with the substrate X-Gal to reveal *E.coli* β-galactosidase, and can be easily distinguished from unlabeled cells. The reaction product is predominantly associated with the nuclear membrane, and endoplasmic reticulum when present. For example, the cell on the right in **A** is an unlabeled pyramidal neuron, whereas the cell on the left is a lacZ-positive pyramidal neuron. Notice that the reaction product extends into the proximal processes of neurons, indicated by arrows (apical dendrite) and arrowheads (basal dendrite). Also compare in **C**, the appearance of an unlabeled neuron (*a*) and unlabeled oligodendrocyte (*b*) to a lacZ-positive oligodendrocyte (*c*). The small arrows in **D** designate deposits of reaction product around the nuclear membrane, and the curved large arrow points to a collection of intermediate glial filaments indicative of astrocytes. See text for a more complete description of the ultrastructural features characteristic of each cell type and the distribution of β-galactosidase reaction product unique to each cell type. **A**, x5,500; **B**, x9,200; **C**, x12,000; **D**, x13,500.

size, but on average have a larger soma diameter than nonpyramidal neurons (Peters, 1985). Nonetheless, we would argue that it would not be prudent to use soma size as a decisive criteria to judge phenotype because the two populations are not mutually exclusive with respect to the size of the soma. Furthermore, pyramidal and nonpyramidal neurons generally differ in regards to their ultrastructure. For example, pyramidal neurons usually have dispersed heterochromatin in an otherwise pale, round nucleus (see Figure 8.2A, 8.3C, and 8.3D), whereas the nucleus of nonpyramidal cells is more electron-dense and irregular in outline (see Figure 8.2B and 8.4C). Similarly, there are subtle, but significant, differences in the fine structure appearance of their cytoplasm, illustrated in Figures 8.2, 8.3, and 8.4; nonpyramidal neurons often exhibit parallel arrays of granular endoplasmic reticulum. Despite the noticeable differences between pyramidal and nonpyramidal cells, as mentioned above, we relied on the types of synapses impinging upon the soma of lacZ-positive cells to unequivocally distinguish between pyramidal and nonpyramidal neurons. The soma of pyramidal neurons receive exclusively symmetrical synapses (Colonnier, 1981; Peters, 1985), whereas both symmetrical and asymmetrical synapses contact the soma of nonpyramidal neurons.

Nearly all of the neuronal clones examined at the ultrastructural level were homogeneous with respect to cell type; Figures 8.3 and 8.4 exemplify the features of pyramidal and nonpyramidal cell clones, respectively. Figure 8.3 is an illustration of a clone containing exclusively pyramidal neurons, as well as exclusively symmetrical axosomatic synapses. Conversely, Figure 8.4 offers an example of a clone containing exclusively nonpyramidal cells. In this, as in all nonpyramidal cell clones, all the neurons received a mixture of symmetrical and asymmetrical axosomatic synapses. As evident, there can be a great deal of variability in the process staining between cells in a clone (compare cells *a* and *b* in Figure 8.4). Despite this variability, the ultrastructure and distribution of reaction product was remarkably similar for the two cells. In fact, members of a clone usually displayed a similar disposition and density of reaction product. This consistency in the appearance of the labeling serves an ancillary evidence that a cluster of cells is indeed clonally related. In addition, a striking difference between lacZ-positive pyramidal and nonpyramidal cells is that the nucleoplasm of nonpyramidal cells becomes especially electron-dense, allowing these cells to be more readily classified. It is not clear what accounts for the intense staining in nonpyramidal cell nuclei. The dense staining of the pyramidal cell nucleus in Figure 8.2A is unusual. We judged it to be a pyramidal cell because of its apical and basal dendrites and, more importantly, absence of asymmetrical axosomatic synapses.

Only one clone containing exclusively neurons was composed of a mixture of neuronal cell types: one pyramidal neuron and one nonpyramidal neuron. Although the definition of a clone dictated that these two cells be considered part of the same clone (i.e., within 500 μm of each other), alternative possibilities should not be overlooked. For instance, they may have arisen from two distinct progenitor cells that happened to be close to one another in the ventricular zone. On the other hand, they may be the product of a bipotential progenitor cell. If the site of integration of the *lacZ* gene could be shown to be the same in both cells, one would feel more convinced that they are clonally related.

While pyramidal neurons and nonpyramidal neurons occur in every cellular layer of the rat visual cortex, the pyramidal cells outnumber the nonpyramidal cells by approximately 2:1. This is intriguing in light of the finding that, of the limited number of clones analyzed so far, clones composed of pyramidal neurons contain more cells on average than clones composed of nonpyramidal neurons (mean

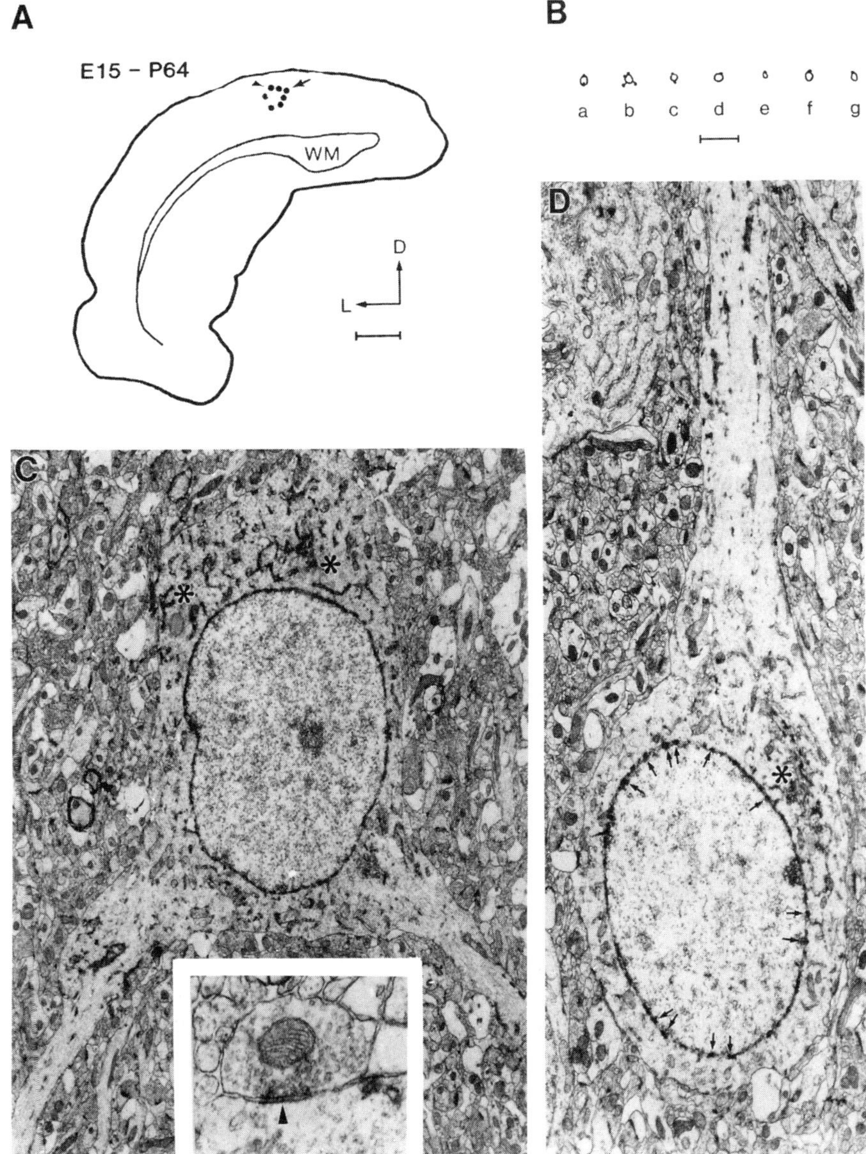

number: 5.6 cells vs 3.0 cells). One possibility to account for this difference is that there may be twice as many pyramidal cell progenitor cells than nonpyramidal cell progenitor cells. However, other mechanisms could also sculpt the final number of each cell type, such as cell death or differences in the timing of the mitotic cycle for each progenitor cell type.

Lineage Relationships between Neurons and Glia

Of the nearly 30 clones we have analyzed fully at the ultrastructural level, resulting from the injection of retrovirus at the onset of neurogenesis (ED15 or ED16), all but two contained only one type of cell—either they were all neurons, all astrocytes, or all oligodendrocytes (see Figure 8.2 for appearance of lacZ-positive astrocytes and oligodendrocytes). No clones were identified that contained both astrocytes and oligodendrocytes, although homogeneous clones were found in the grey and white matter of the cerebral cortex (Barfield et al., 1990; Luskin, 1990). Furthermore, even in the two mixed neuronal/glial clones encountered, the different cell types were not intermingled with one another. Rather, they were spatially separated from each other, though they fell within the spatial criteria that grouped them together as one clone. As in the case of the mixed neuronal clone, we have no absolute way to verify whether the neurons and glia are from the same progenitor cell or from two different ones whose offspring migrated to the same location in the cortex. Overall, we conclude from these findings that from the onset of neurogenesis there are separate progenitor cells for neurons, astrocytes, and oligodendrocytes.

FIGURE 8.3. Illustration of a clone composed exclusively of 7 pyramidal neurons. This group of neurons was situated in the visual cortex (area 17) of a rat brain injected with retrovirus at E15 and fixed at P64. **A**. The clone was reconstructed by superimposing the positions of the lacZ-positive neurons (dots) in two consecutive camera lucida drawings of 100 μm thick sections. D, dorsal; L, lateral; WM, white matter. Scale bar: 1 mm. **B**. Camera lucida drawings of the 7 neurons that composed this clone. At the light microscopic level, all cells exhibited a relatively small amount of blue reaction product surrounding a pale nucleus. A few (e.g., cells *b, c*) showed rudimentary processes extending from their somata, although none of the cells displayed sufficient morphological features to allow classification into one of the major types of cortical neurons. Scale bar: 50 μm. **C, D**. However, when the neurons were examined by electron microscopy, they all showed ultrastructural features typical of pyramidal cells, in particular only symmetrical axosomatic synapses. Electron micrographs of two of the neurons are shown in **C** and **D; C** shows cell *b* (arrow in **A**) and **D** shows cell *c* (arrowhead in **A**). In both, the β-galactosidase histochemical reaction product is associated with the nuclear membrane, appearing as discrete "puffs" (arrows in **D**), and with a number of organelles in the cytoplasm (large asterisks). In **C** and **D** pia is toward the top of the page. Inset illustrates, at higher magnification, one of the symmetrical synapses (arrowhead) that was identified in another section through the same cell. **C**, x7400; **D**, x8600; inset x42000. (Modified from Parnavelas et al., 1991).

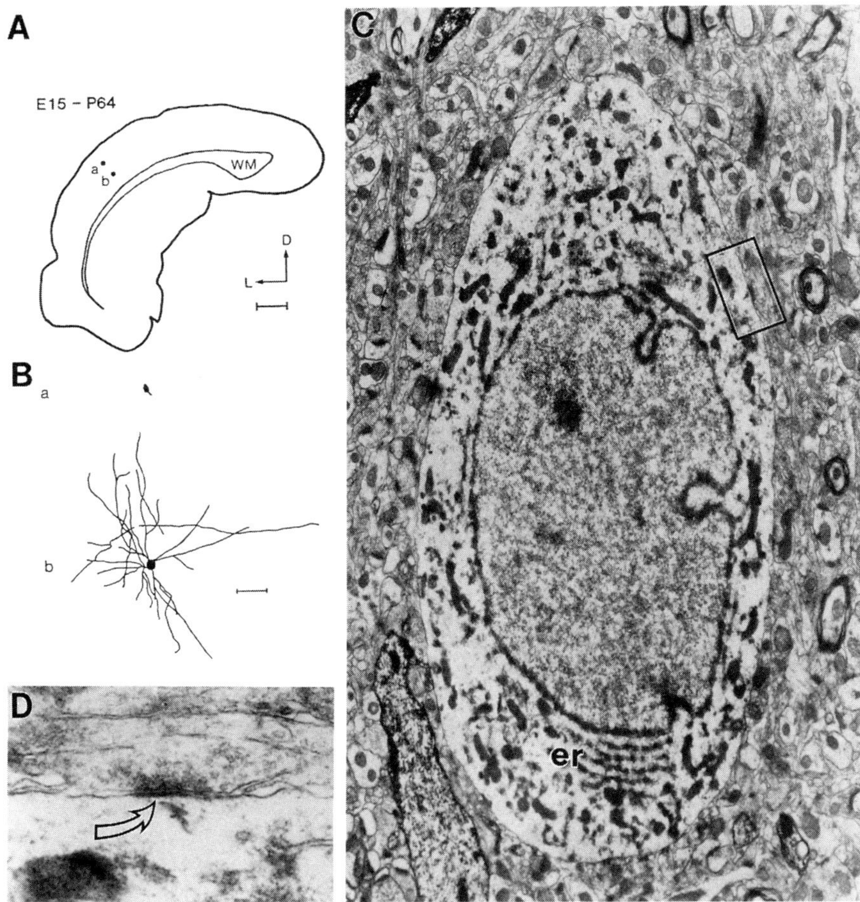

FIGURE 8.4. Example of a two-cell nonpyramidal clone examined from the visual cortex (area 18a) of a rat brain injected with retrovirus at E15 and fixed at P64. **A**. This clone was reconstructed by superimposing the positions of the 2 lacZ-positive cells (*a* and *b*) in three consecutive 100 μm thick Vibratome sections. D, dorsal; L, lateral; WM, white matter. Scale bar: 1 mm. **B**. Camera lucida drawings of the two clonally related neurons. The phenotype of cell *a* could not be ascertained by its light microscopic appearance. It showed a densely stained soma giving rise to one short process, while cell *b* revealed a very extensive dendritic field characteristic of multipolar nonpyramidal neurons in the rat visual cortex. Scale bar: 50 μm. **C,D**. However, upon examination at the ultrastructural level, it was revealed that cell *a*, like cell *b*, was indeed a nonpyramidal neuron. Cell *a* showed an indented nucleus containing clumps of heterochromatin, a rich complement of organelles, including stacked endoplasmic reticulum (**er**) laden with reaction product, and a number of both symmetrical and asymmetrical axosomatic synapses. The presence of an asymmetrical axosomatic synapse, an exclusive feature of nonpyramidal neurons in the cortex, is indicated by a box in **C**, and illustrated at higher magnification (arrow) in **D. C**, x10000; **D**, x42000. (Modified from Parnavelas et al., 1991).

Neuronal and glial clones were different from each other in a number of ways. Most significantly, on average, glial clones tended to be larger than neuronal clones, and frequently the cells within glial clones occurred close together, forming small balls of cells within the cortex. As mentioned above, all neuronal clones contained less than 10 cells, whereas glial clones commonly contained more than 10 cells, including one clone of astrocytes containing 32 cells. The large number of cells per clone and the tight packing of cells within a glial clone suggests that some of the lacZ-positive astrocytes and oligodendrocytes are the product of cells that divided *in situ* after migrating away from the ventricular zone.

CONCLUSIONS

Two major conclusions follow from these studies showing that when the members of a clone are classified at the ultrastructural level, they constitute a homogeneous population of cells. First, by the onset of neurogenesis for the cerebral cortex, the lineages for neurons and glia have diverged, and moreover, there are separate progenitor cells for astrocytes, oligodendrocytes, and neurons (see Figure 8.5). Second, by the onset of neurogenesis, the lineages for pyramidal and nonpyramidal neurons have diverged (see Figure 8.5). Future experiments should determine if there are separate progenitor cells for the different subtypes of nonpyramidal cells, which can be distinguished according to their neurochemical identity. Furthermore, experiments in which retroviral lineage tracers are administered at younger developmental timepoints, either *in vivo* or *in vitro*, should reveal multipotential progenitor cells and when lineage restrictions between neurons and glia are first manifested.

The findings given would tend to suggest that not every part of the mammalian CNS utilizes the same strategy for the determination of cell fate. In some parts, lineage seems not to be critically involved, and instead, cell-cell interactions or position-dependent cues have been shown to play an important role in governing the fate of cells. For instance, in the rat retina, progenitor cells remain multipotential and give rise to more than one cell type up to their last division (Turner and Cepko, 1987). On the other hand, evidence from transplantation studies have shown that the laminar fate of a cortical neuron is determined during or just following its final mitosis in the ventricular zone (McConnell and Kaznowski, 1991). Thus, it appears that the developmental potential of a progenitor cell or its progeny becomes progressively restricted as development proceeds. Decisions about cell phenotype appear to be made before a neuron's laminar fate (position) is slated. The determination of when and where various restrictions occur, possibly mediated by cell-cell interactions, will most likely, in time, lead to the time, lead to the identification of genes that are differentially regulated and expressed as cells become committed to a specific pathway of differentiation.

Acknowledgments. I wish to thank Alice Schmid for her helpful comments on the manuscript, and Ms. Eileen Breding for her much-appreciated photographic assistance.

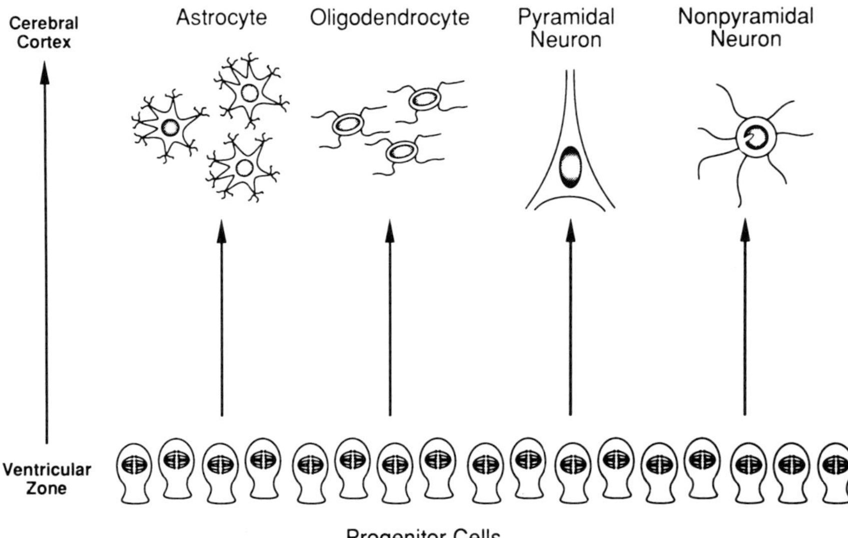

FIGURE 8.5 Separate progenitor cells in the ventricular zone give rise to sets of clonally related astrocytes, oligodendrocytes, and neurons in the rat telencephalon. All of the cells of the cerebral cortex are derived from a layer of progenitor cells, the ventricular zone, surrounding the cerebral ventricles. When a retroviral lineage tracer is used to infect progenitor cells of the telencephalon at the onset of neurogenesis for the cerebral cortex, most of the resulting clusters of clonally related cells in the adult cortex are composed of a homogeneous population of cells. They contain either all astrocytes, all oligodendrocytes, all pyramidal neurons, or all nonpyramidal neurons. This suggests that by the onset of neurogenesis for the cerebral cortex, the lineages for neurons, astrocytes, and oligodendrocytes have diverged, and moreover that the lineages for pyramidal and nonpyramidal neurons have diverged.

This work was partially supported by grants from the Alfred P. Sloan Foundation, Pew Memorial Trust, National Institutes of Health, the March of Dimes Birth Defects Foundation, Research Committee of Emory University.

REFERENCES

Anderson DJ (1989): The neural crest cell lineage problem: Neuropoiesis? *Neuron* 3:1–12

Angevine JB, Sidman RL (1961): Autoradiographic study of cell migration during histogenesis of cerebral cortex in the mouse. *Nature* 192:766–768

Austin CP, Cepko CL (1990): Cellular migration patterns in the developing mouse cerebral cortex. *Development* 110:713–732

Banerjee U, Zipursky SL (1990): The role of cell-cell interaction in the development of the *Drosophila* visual system. *Neuron* 4:177–187

Barfield JA, Parnavelas JG, Luskin MB (1990): Separate progenitor cells give rise to neurons, astrocytes and oligodendrocytes in the rat cerebral cortex. *Neurosci Abstr* 16:1272

Baughman RW, Gilbert CD (1981): Aspartate and glutamate as possible neurotransmitters in the visual cortex. *J Neurosci* 1:427–439

Berry M, Rogers AW (1965): The migration of neuroblasts in the developing cerebral cortex. *J Anat* 99:691–709

Bunge MB, Wood PM, Tynar LB, Bates ML, Sanes JR (1989): Perineurium originates from fibroblasts: Demonstration *in vitro* with a retroviral marker. *Science* 243: 229–231

Caviness VS, Jr (1982): Neocortical histogenesis in normal and Reeler mice: A development study based upon ^{3}H-thymidine autoradiography. *Dev Brain Res* 4:293–302

Caviness VS. Jr, Sidman RL (1973): Time of origin of corresponding cell classes in the cerebral cortex of normal and Reeler mutant mice: An autoradiographic analysis. *J Comp Neurol* 148:141–151

Colonnier M (1981): The electron-microscopic analysis of the neuronal organization of the cerebral cortex. In: *The Organization of the Cerebral Cortex*, Schmitt FO, Worden FG, Adelman G, Dennis SG, eds. Cambridge: MIT Press

Conti F, Rustioni A, Petrusz P, Towle AC (1987): Glutamate-positive neurons in the somatic sensory cortex of rats and monkeys. *J Neurosci* 7:1887–1901

Doupe AJ, Landis SC, Patterson PH (1985): Environmental influences in the development of neural crest derivatives: Glucocorticoids, growth factors, and chromaffin cell plasticity. *J Neurosci* 5:2119–2142

Fairen A, Cobas A, Fonseca M (1986): Times of generation of glutamic acid decarboxylase immunoreactive neurons in mouse somatosensory cortex. *J Comp Neurol* 251:67–83

Galileo DS, Gray GE, Owens GC, Majors J, Sanes JR (1990): Neurons and glia arise from a common progenitor in chick optic tectum: demonstration with two retroviruses and cell-type specific antibodies. PNAS 87:458–462

Gilbert CP (1983): Neurocircuitry of the visual cortex. *Ann Rev Neurosci* 6:217–247

Hendry SHC, Jones EG (1986): Reduction in number of immunostained GABAergic neurons in deprived-eye dominance columns of monkey area 17. *Nature* 320:750–753

His W (1889): Die neuroblasten und deren Entstehung im embryonalen. *Marke Abn Phys Cl Kgl Sach ges Wiss* 15:313–372

Houser CR, Hendry SHC, Jones EG, Vaughn JE (1983): Morphological diversity of immunocytochemically identified GABA neurons in the monkey sensory-motor cortex. *J Neurocytol* 12:617–638

Ichikawa M, Hirata Y (1982): Morphology and distribution of postnatally generated glial cells in the somatosensory cortex of the rat: An autoradiographic and electron microscopic study. *Dev Brain Res* 4:369–377

Jones EG, Hendry SHC (1986): Peptide containing neurons in primate cerebral cortex. In: *Neuropeptides in Neurologic and Psychiatric Diseases*, Martin JB, Barchas JD, eds. New York: Raven Press

Jones EG, Powell TPS (1970): Electron microscopy of the somatic sensory cortex of the cat: 1. Cell types and synaptic organization. *Philos Trans R Soc Lond (Biol)* 257:1–11

Kenyon C (1985): Cell lineage and the control of *Caenorhabditis elegans* development. *Philos Trans R Soc Lond (Biol)* 312:21–38

LeVine SM, Goldman JE (1988a): Embryonic divergence of oligodendrocyte and astrocyte lineages in developing rat cerebrum. *J Neurosci* 8:3992–4006

LeVine SM, Goldman JE (1988b): Spatial and temporal patterns of oligodendrocyte differentiation in rat cerebrum and cerebellum. *J Comp Neurol* 277:441–455

Levitt P, Cooper ML, Rakic P (1981): Coexistence of neuronal and glial precursor cells in the cerebral ventricular zone of the fetal monkey: An ultrastructural immunoperoxidase analysis. *J Neurosci* 1:27–39

Levitt P, Cooper ML, Rakic P (1983): Early divergence and changing proportions of neuronal and glial precursor cells in the primate cerebral ventricular zone. *Dev Biol* 96:472–484

Lin C-S, Lu SM, Schmechel DE (1986): Glutamic acid decarboxylase and somatostatin immunoreactivities in rat cerebral cortex. *J Comp Neurol* 244:369–383

Luskin MB (1990): Generation of cell diversity in the cerebral cortex: Lineage relationships. *Eur Neurosci Assoc Abstr* 22:4

Luskin MB, Parnavelas JG, Barfield TA: Neurons, astrocytes and oligodendrocytes of the rat cerebral cortex originate from separate progenitor cells: on ultrastructural analysis of clonally related cells. (submitted)

Luskin MB, Pearlman A, Sanes JR (1988): Cell lineage in the cerebral cortex of the mouse studied *in vivo* and *in vitro* with a recombinant retrovirus. *Neuron* 1:635–647

Luskin MB, Shatz CJ (1985): Pattern of neurogenesis of the cat's primary visual cortex. *J Comp Neurol* 242:611–631

Mareš V, Brückner G (1978): Postnatal formation of non-neuronal cells in the rat occipital cerebrum: An autoradiographic study of the time and space pattern of cell division. *J Comp Neurol* 177:519–528

McConnell SK (1988): Development and decision-making in the mammalian cerebral cortex. *Brain Res Rev* 13:1–23

McConnell SK, Kaznowski CE (1991): Cell cycle dependence of laminar determination in developing neocortex. *Science* 254:282–285

Meller K, Tetzlaff W (1975): Neuronal migration during the early development of the cerebral cortex: A scanning electron microscope study. *Cell Tissue Res* 163:313–325

Meyer G, Wahle P (1988): Early postnatal development of cholecystokinin-immunoreactive structures in the visual cortex of the cat. *J Comp Neurol* 276:360–386

Miller MW (1985): Cogeneration of retrogradely labeled corticocortical projection and GABA-immunoreactive local circuit neurons in cerebral cortex. *Dev Brain Res* 23:187–192

Miller MW (1986): The migration and neurochemical differentiation of gamma-aminobutyric acid (GABA)-immunoreactive neurons in rat visual cortex as demonstrated by a combined immunocytochemical-autoradiographic technique. *Dev Brain Res* 28: 41–46

Miller RH, ffrench-Constant C, Raff MC (1989): The macroglial cells of the rat optic nerve. *Ann Rev Neurosci* 12:517–34

Misson J-P, Austin CP, Takahashi T, Cepko C, Caviness VS Jr (1991): The alignment of migrating neural cells in relation to the murine neopallial radial glial fiber system. *Cerebral Cortex* 1:221–229

Misson J-P, Edwards MA, Yamamoto M, Caviness VS Jr (1988): Identification of radial glial cells within the developing murine central nervous system: Studies based upon a new immunohistochemical marker. *Dev Brain Res* 44:95–108

Naegele JR, Arimatsu Y, Schwartz P, Barnstable CJ (1988): Selective staining of a subset of GABAergic neurons in cat visual cortex by monoclonal antibody VC1.1. *J Neurosci* 8:79–89

Parnavelas JG, Barfield JA, Frank E, Luskin MB (1991): Separate progenitor cells give rise to pyramidal and nonpyramidal neurons in the rat telencephalon. *Cerebral Cortex* 1:463–468

Parnavelas JG, Barfield JA, Luskin MB (1990): Lineage relationships of pyramidal and nonpyramidal neurons in the rat cerebral cortex. *Neurosci Abstr* 16:1272

Parnavelas JG, Dinopoulos A, Davies SW (1989): The central visual pathways. In:

Handbook of Chemical Neuroanatomy, vol. 7, Integrated systems of the CWS, Pt II (Bjorklund A, Hokfelt T, Swanson SW, eds), pp 1–164. Amsterdam: Elsevier

Parnavelas JG, Papadopoulos GC, Cavanagh ME (1988): Changes in neurotransmitters during development. In: *Cerebral Cortex*, vol. 7, Jones EG, Peters A, eds. New York: Plenum Press

Parnavelas JG, Sullivan K, Lieberman AR, Webster KE (1977): Neurons and their synaptic organization in the visual cortex of the rat: Electron microscopy of Golgi preparations. *Cell Tissue Res* 183:499–517

Patterson PH (1978): Environmental determination of autonomic neurotransmitter functions. *Ann Rev Neurosci* 1:1–17

Peters A (1985): The visual cortex of the rat. In: *Cerebral Cortex*, vol. 3, Jones EG, Peters A, eds. New York: Plenum Press

Peters A, Feldman M (1973): The cortical plate and molecular layer of the late rat fetus. *Z Anat Entwickl-Gesch* 141:3–37

Peters A, Palay SL, Webster de F (1976): *The Fine Structure of the Nervous System.* Philadelphia: Saunders

Potter DD, Landis SC, Matsumoto SG, Furshpan EJ (1986): Synaptic functions in rat sympathetic neurons in microcultures: 2. Adrenergic/cholinergic dual status and plasticity. *J Neurosci* 6:1080–1098

Price J, Thurlow L (1988): Cell lineage in the rat cerebral cortex: A study using retroviral-mediated gene transfer. *Development* 104:473–482

Price JL, Turner D, Cepko C (1987): Lineage analysis in the vertebrate nervous system by retrovirus mediated gene transfer. *PNAS U S A* 84:156–160

Privat A (1975): Postnatal gliogenesis in the mammalian brain. *Int Rev Cytol* 40:281–323

Raedler E, Raedler A (1978): Autoradiographic study of early neurogenesis in rat neocortex. *Anat Embryol (Berl)* 154:267–284

Raff MC (1989): Glial cell diversification in the rat optic nerve. *Science* 243:1450–1455

Rakic P (1972): Mode of cell migration to the superficial layers of monkey neocortex. *J Comp Neurol* 145:61–84

Rakic P (1974): Neurons in the rhesus monkey visual cortex: Systematic relation between time of origin and eventual disposition. *Science* 183:425–427

Rakic P (1988): Specification of cerebral cortical areas. *Science* 241:170–176

Sanes JR (1989): Analyzing cell lineage with a recombinant retrovirus. *Trends Neurosci* 12:21–28

Sanes JR, Rubenstein JLR, Nicolas J-F (1986): Use of a recombinant retrovirus to study post-implantation cell lineage in mouse embryos. *EMBO J* 5:3133–3142

Sauer FC (1935): Mitosis in the neural tube. *J Comp Neurol* 62:377–405

Schaper A (1897): The earliest differentiation in the central nervous system of vertebrates. *Science* 5:430–431

Shimada M, Langman J (1970): Cell proliferation, migration and differentiation in the cerebral cortex of the golden hamster. *J Comp Neurol* 139:227–244

Sidman RL, Rakic P (1973): Neuronal migration, with special reference to the developing human brain: A review. *Brain Res* 62:1–35

Skoff RB (1980): Neuroglia: A re-evaluation of their origin and development. *Pathol Res Pract* 168:279–300

Streit P (1984): Glutamate and aspartate as transmitter candidates for systems of the cerebral cortex. In: *Cerebral Cortex*, vol. 2, Jones EG, Peters A, eds. New York: Plenum Press

Sulston J, Schierenberg E, White JG, Thomson JN (1983): The embryonic cell lineage of the nematode *Caenorhabditis elegans*. *Dev Biol* 100:64–119

Temple S (1989): Division and differentiation of isolated CNS blast cells in microculture. *Nature* 340:471–473

Turner D, Cepko C (1987): A common progenitor for neurons and glia persists in rat retina late in development. *Nature* 328:131–136

Vaughan DW (1984): The structure of neuroglial cells. In: *Cerebral Cortex*, vol. 2, Jones EG, Peters A, eds. New York: Plenum Press

Vaysse PJ-J, Goldman JE (1990): A clonal analysis of glial lineages in neonatal forebrain development *in vitro*. *Neuron* 5:227–235

Walsh C, Cepko CL (1988): Clonally related cortical cells show several migration patterns. *Science* 241:1342–1345

Wise SP (1985): Co-localization of GABA and neuropeptides in the cerebral cortex. *Trends Neurosci* 8:92–95

Wolff JR, Bottacher H, Zetzshe T, Oertel WH, Chronwall BM (1984): Development of GABAergic neurons in rat visual cortex as identified by glutamate decarboxylase-like immunoreactivity. *Neurosci Lett* 47:207–212

9

Different Developmental Strategies of the Telencephalic Commissures: A Comparison between the Ontogeneses of Visual Callosal Connections and of Olfactory Commissural Connections in Rodents

ROBERTO LENT

INTRODUCTION

In many areas of experimental biology new concepts receive so much attention from researchers in the field that they soon begin to be considered undoubted truths of universal value. In neuroscience, one such truism is the concept that the development of cortical connectivity *necessarily* employs a dual strategy in which a sequence of progressive events is followed by a sequence of regressive mechanisms that sculpt the final cortical hardware.

Progressive events are understood as those that involve the numerical increase of elements such as cells, fibers, synapses, and so forth. Regressive mechanisms are conceived as those involving a decrease—usually selective—of these same elements. The progressive events are assumed to provide the brain with an oversized (or "exuberant") set of circuits, far more numerous than necessary for adult life. This exuberant hardware would then be gradually trimmed by regressive mechanisms—cell ceath, axonal elimination, and others—resulting in the selective survival of those circuits that fulfill more appropriately the needs of the individual and of the species.

The occurrence of this dual strategy is supported by experimental evidence collected from studies of neocortical development, particularly—but not exclusively—of the visual cortex and its interhemispheric connections. However, recent studies on the development of interhemispheric connections of paleocortical regions have uncovered the existence of another, much simpler, ontogenetic strategy, composed predominantly, if not entirely, of progressive phenomena.

In this review, it is proposed that while the dual strategy is characteristic of neocortical projections, the paleocortex employs the singular ontogenetic strategy. These different developmental strategies of the cerebral cortex are illustrated using as typical examples the interhemispheric projections of the primary visual area and of the olfactory cortex of laboratory rodents.

THE DEVELOPMENT OF INTERHEMISPHERIC CONNECTIONS OF THE VISUAL CORTEX THROUGH THE CORPUS CALLOSUM

Visual Callosal Projections in Adults

The visual cortex of mammals is composed of many different cytoarchitectonic areas, each of them containing at least one complete representation of the visual field (Drager, 1975; Montero, 1981; Tiao and Blakemore, 1976; Tusa et al., 1981; Van Essen, 1985; Wagor et al., 1980). The primary visual cortex (V1) receives visual input from the retina mostly through the dorsal lateral geniculate nucleus of the thalamus (Dean, 1990; Dursteler et al., 1979; Peters and Feldman, 1976; Sefton and Dreher, 1985) and initiates at least two divergent streams of visual processing involving the other visual areas of the ipsilateral cortex (Gattass et al., 1990; Ungerleider and Mishkin, 1982). In rodents, V1 comprises a cytoarchitectonic field numbered 17 by Krieg (1946), and named Oc1 by Zilles (1985).

The interhemispheric projections arising in V1 course through the corpus callosum in all the Eutherian mammals. In Metatherians, such as the opossum, the callosum is absent and neocortical interhemispheric communication is conveyed by an enlarged anterior commissure (Ebner and Myers, 1967; Granger et al., 1985; Martin, 1967). The visual callosal projections of adult mammals are organized according to certain structural principles (Innocenti, 1986; Lent and Schmidt, 1992), among which one in particular has been relevant for revealing the strategy employed by this projection system during development. That is the incomplete, albeit corticotopic, representation of V1 in the corpus callosum. Only the border of V1 with V2 (in rodents, this latter area has been numbered 18a by Krieg, and named Oc2L by Zilles) receives afferents and projects fibers through the corpus callosum (Cipolloni and Peters, 1979; Cusick and Lund, 1981; Lund and Lund, 1970; Olavarria and Van Sluyters, 1985; Rhoades and Dellacroce, 1980; Yorke and Caviness, 1975; Záborsky and Wolff, 1982; but also see Dursteler et al., 1979). This border region corresponds to a partial representation of the visual field that comprises about 30° off the vertical meridian in rodents (Rhoades and Dellacroce, 1980; Thomas and Espinoza, 1987), and less in other species (Innocenti, 1980; McCourt et al., 1990; Payne, 1990; Swadlow, 1977).

Callosal neurons are mostly pyramidal (Dursteler et al., 1979; Yorke and Caviness, 1975), although other cell types also contribute fibers to the corpus callosum (Hughes and Peters, 1990). In rodents, most visual callosal cells are located in equivalent numbers in layers II–III and V (Dursteler et al., 1979; Jacobson and Trojanowski, 1974; Yorke and Caviness, 1975), but in higher mammals there is a tendency for them to predominate in the supragranular layers (Segraves and Rosenquist, 1982; Shatz, 1977; Van Essen et al., 1982; Winfield et al., 1975). There are also cells with callosal axons in layer IV (Diao and So, 1991; Hughes and Peters, 1990).

The Development of Visual Callosal Projections: A Dual Strategy of Progressive and Regressive Events

The crucial questions for the developmental neurobiology of visual callosal projections are: How is this mature pattern of connections achieved? Which strategies are employed for the morphogenesis of the system?

The first, distinguishable stage in the development of visual callosal neurons is possibly the interruption in the proliferation of their precursors during embryogenesis. In rodents, this occurs during the second week of gestation (Lent et al., 1990), together with the birth of neurons that will subsequently migrate to layers V, IV, and III of the occipital cortex (Angevine and Sidman, 1961; Berry and Rogers, 1965; Caviness, 1982; Crossland and Uchwat, 1982; Raedler and Raedler, 1978).

It is possible that callosal axogenesis in the visual cortex may start precociously during cell migration, since this has been shown to occur in the primate prefrontal cortex (Schwartz et al., 1991; but see Hernit et al., 1990). It is also possible that the transitory neurons of the subplate (Uylings et al., 1990) do elongate callosal fibers before those of layer V, as demonstrated in cat embryos (Chun et al., 1987). These phenomena, however, still need direct documentation in rodents.

The first callosal axons can be seen approaching the midsagittal plane in rodent embryos a few days before birth (Floeter and Jones, 1985; Silver et al., 1982). In hamsters, midplane crossing occurs during the last few hours of gestation (Lent et al., 1990; Norris and Kalil, 1990). These first callosal axons arise from frontal and parietal cortices. Those from the occipital cortex reach the midline a few days later (Fish et al., 1990; Lent et al., 1990; Mooney et al., 1984). In the hamster, most callosal development takes place postnatally. Pioneer fibers have been detected leading the growth front of callosal axons during midplane crossing (Hogan and Berman, 1990; Lent et al., 1990), but their significance remains to be established. After commissuration, callosal fibers elongate rapidly and soon occupy the prospective white matter of the opposite hemisphere down to the rhinal fissure. Anterograde labeling with horseradish peroxidase and wheat germ agglutinin has shown that the occipital callosal axons do not enter the cortical gray matter as soon as they arrive at its deeper border. Instead, they remain within the white matter for some days before innervating the appropriate target regions (Lent et al., 1990; Olavarria and Van Sluyters, 1985). This delay before innervation had previously been shown for other cortical sectors, and became known as the callosal "waiting period" (Wise and Jones, 1976), an inappropriate expression implying a lack of functional activity that is probably not true. The actual existence of a waiting period for all callosal fibers has been recently challenged by the use of more sensitive techniques of anterograde axonal tracing (Fish et al., 1990; but see Hedin-Pereira et al., 1991).

Immediately after birth, when the first set of callosal fibers have crossed the midplane, visual cortex neurons with callosal axons are already numerous and, most importantly, they distribute homogeneously throughout the whole extent of the occipital cortex (Lent et al., 1990; Mooney et al., 1984; Olavarria and Van Sluyters, 1985). During the first two postnatal weeks, the number of callosal neurons in the visual cortex increases noticeably, suggesting a gradual recruitment of neurons whose fibers course through the corpus callosum. Indeed, during this period, axon counts in the callosal tract of postnatal rats indicate a steep increase in the total number of fibers (Gravel et al., 1990).

Until this point of postnatal development, all the events have been progressive: proliferation and neurogenesis followed by migration, and elongation of callosal axons arising from an increasing number of cells. A regressive phenomenon first occurs when some callosal fibers start arborizing within the cortex opposite the V1/V2 border, while the others situated at the core of V1 are selectively eliminated. This phenomenon was first inferred in kittens from the observation that the initially exuberant callosal zone of origin in the neonate is replaced by that typical of the adult, that is, restricted to the border between V1 and V2 (Innocenti et al., 1977). This observation has since been confirmed in other species including

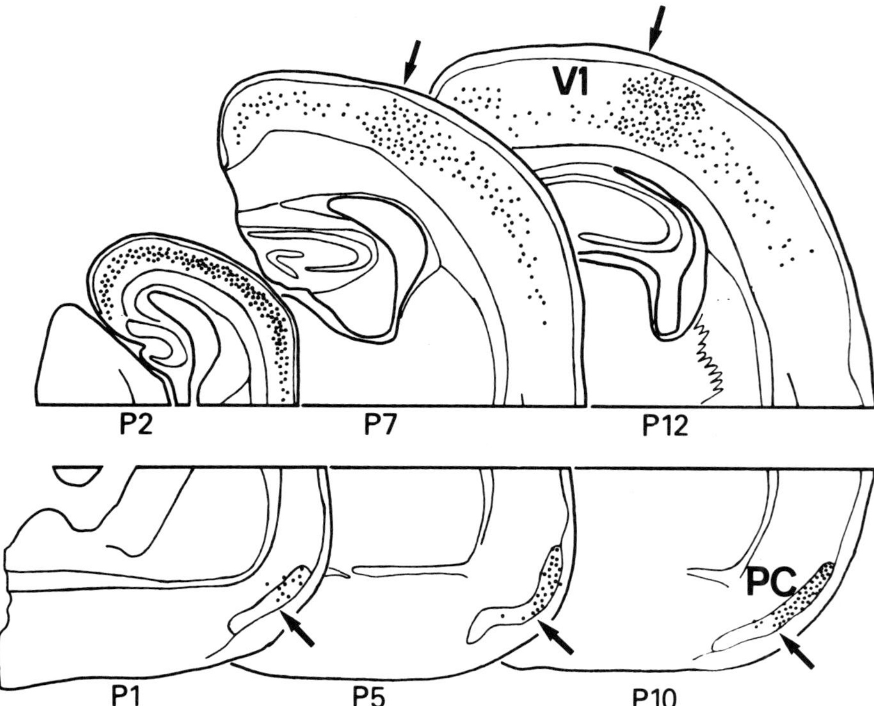

FIGURE 9.1. Schematic representation of the development of visual callosal connections in the hamster (top), as compared with that of olfactory connections through the anterior commissure (bottom). The arrows point to the borders between callosal/commissural and acallosal/acommissural cortical regions.

rodents (see Figure 9.1) (Chow et al., 1981; Lent et al., 1990; Olavarria and Van Sluyters, 1985), and in other areas of the neocortex (Cabana and Martin, 1985; Feng and Brugge, 1983; Ivy and Killackey, 1981). That selective elimination of fibers, rather than cell death, is the mechanism underlying this regressive phenomenon, was shown by Innocenti (1981) and by O'Leary et al. (1981) after elegant experiments involving long-lasting fluorescent tracers. More recently, elimination of callosal axons was directly demonstrated by quantitative ultrastructural studies of the total number of fibers in the callosal tract of cats and primates (Berbel and Innocenti, 1988; Koppel and Innocenti, 1983; LaMantia and Rakic, 1990). A similar study, however, failed to confirm this phenomenon in rats (Gravel et al., 1990). But it is possible that the period of callosal fiber elimination in rats overlaps that of fiber growth. The net result of axon counts, in this case, would conceal the former phenomenon.

Other progressive events such as synaptogenesis (Blue and Parnavelas, 1983) and myelination (Lent and Jhaveri, 1992; Sturrock, 1980) occur when the mature pattern of the visual callosal projections is stabilized. Also, it cannot be excluded—although not directly demonstrated for callosal neurons—that other regressive phenomena shown to occur in the neocortex, such as cell death (Ferrer et al., 1990; Finlay and Slattery, 1983; Heuman and Leuba, 1983) and synaptic elimination (Rakic et al., 1986) do exist for callosal neurons and afferents as well.

THE DEVELOPMENT OF INTERHEMISPHERIC CONNECTIONS OF THE OLFACTORY CORTEX THROUGH THE ANTERIOR COMMISSURE

Commissural Connections of the Piriform Cortex in Adults

The olfactory cortex of mammals is comprised of different cytoarchitectonic areas that contain a representation of the nasal epithelium. Among these areas, the piriform cortex (PC) is the largest (Switzer et al., 1985). In rodents, it is subdivided, on cytoarchitectonic and hodologic grounds (Haberly and Price, 1978a), into two main sectors: one anterior (PCa), the other posterior (PCp).

Many differences distinguish the PC from V1 (see Table 9.1). One difference is that while V1 is a neocortical area with the six layers characteristic of this phylogenetically recent cortex, the PC is part of the paleocortex, and as such has a simpler, three-layered structure. The second difference is that while V1 receives sensory input from the retina through a thalamic relay, the PC receives olfactory input from the nasal epithelium through a relay at the olfactory bulb. Finally, the third important difference is the fact that the interhemispheric connections of the PC of Eutherian mammals are contained in the anterior commissure, while those of V1 course through the corpus callosum.

Despite these differences, there is one structural similarity between rodent V1 and PC that is essential for the purpose of comparing their ontogenetic strategies. While rostral PCa contains commissural neurons covering its entire mediolateral extent, only the dorsolateral half of caudal PCa and PCp contains commissural

TABLE 9.1 A comparison between the visual cortex (V1) and the piriform cortex (PC) in rodents.

V1	PC
Structurally isocortical (6-layered)	Structurally allocortical (3-layered)
Receives sensory input from retina through thalamus	Receives sensory input from nasum through olfactory bulb
Highly topographical organization of afferents	Highly divergent organization of afferents
Interhemispheric connections through corpus callosum	Interhemispheric connections through anterior commissure
Incomplete representation of callosal connections (restricted to V1–V2 border)	Incomplete representation of commissural connections (restricted to dorsolateral half)

cells and receives commissural afferents (Haberly and Price, 1978a; Jouandet and Hartenstein, 1983; Lent and Guimarães, 1991). Concerning the distribution of commissural cells and afferents, this region is analogous to the border between V1 and V2 (see figure 9.1). In contrast, the ventromedial sector is acommissural, resembling the acallosal core of V1. Due to the high degree of divergence characteristic of the olfactory system (Davis et al., 1978; Haberly and Price, 1978a, 1978b), it is unlikely that this organization involves a partial representation of the sensory surface, as is the case for the visual callosal projections.

Similar to other association neurons in the piriform cortex, those projecting through the anterior commissure are mostly pyramidal (Haberly and Bower, 1984; Haberly and Price, 1978a), but there is some evidence that other cell types may be involved in interhemispheric communication (Jouandet and Hartenstein, 1983). Most are located in layer II of the PC, but a few can be found in layer III (Haberly and Price, 1978a).

The Development of Piriform Commissural Projections: A Singular Strategy of Progressive Events

The neurogenesis of anterior commissure cells has not been directly studied until this moment. That of the rat piriform cortex as a whole, however, has been described by Bayer (1986), and was shown to occur earlier than the genesis of occipital neocortex (also see Derer et al., 1977).

The first commissural fibers in hamster embryos were seen to approach the midsagittal plane about 2 days before callosal axons (Pires-Neto and Lent, 1991). They have a tortuous trajectory in the ipsilateral hemisphere, until they reach the midplane on the following day, embryonic day 14 (ED14). At this moment, fibers are more numerous within the prospective anterior commissure, forming a compact bundle at the midline. From the growing front at the midplane, a few pioneer axons are discerned in the opposite hemisphere, about 500 μm or more ahead of the follower fibers (Pires-Neto and Lent, 1991). These fibers are tipped by structures resembling growth cones. Pioneers and followers are seen to elongate rapidly within the contralateral hemisphere. In 24 hr the three limbs of the anterior commissure are filled with fibers that occupy the whole rostrocaudal extent of the external capsule and lateral white matter. Immediately after arriving at the deeper border of layer III of the PC, fibers initiate a profuse arborization within cortical layers III and Ib. These observations indicate an important difference between the growth of anterior commissure fibers from the PC, and that of the visual callosal axons. For the bulk of commissural fibers, no waiting period (longer than 12 hr) can be noticed before they innervate the target piriform layers in the opposite hemisphere. For the callosal projections, however, it appears that most fibers wait several days within the white matter before innervating the cortex, although some do elongate sprouts into the cortical layers (Fish et al., 1990; Hedin-Pereira et al., 1991). These observations also indicate that in both cases pioneer fibers lead the way toward the target regions into the opposite hemisphere. The development of AC projections in rats has been studied with radiolabeled aminoacids (Schwob and Price, 1984), but perhaps due to lower precision of this tracing technique, arborization of commissural axons in layer Ib was seen to start only on PD9.

At birth, when a great number of commissural fibers have already arborized within the PC layers, the corresponding commissural neurons are still few, mostly located in layer II (Lent and Guimarães, 1991). However, differently from the visual callosal neurons, in neonates the tangential pattern of distribution of PC commissural neurons is equivalent to the adult pattern (see Figure 9.1). AC neurons are found covering the mediolateral extent of rostral PCa, but more caudally they occupy only the dorsolateral half of the PC, sparing the prospective acommissural sector at the medioventral half of the cortex. Thus, the qualitative evidence indicates that there is no exuberant distribution of anterior commissure neurons within the PC at birth.

Observation of retrogradely labeled commissural neurons in the piriform cortex at different postnatal ages indicates that while the number of AC neurons increases with age, their tangential and radial distribution does not change significantly (see Figure 9.1). This was confirmed by counting commissural neurons (Lent and Guimarães, 1991).

Taken together, both qualitative and quantitative evidence suggest that the development of the interhemispheric projections of the PC through the anterior commissure involves only a progressive sequence of events, lacking the regressive mechanisms that characterize the ontogenesis of the visual cortex.

So far only preliminary data have been produced by use of the electron microscope, in agreement with this conclusion. Guadaño-Ferraz and colleagues (1991) have counted fibers from electron micrographs in the anterior commissure of developing rats, and Sturrock (personal communication) has done the same in mice. Both failed to detect any decline in the number of axons during postnatal development. In contrast, LaMantia and Rakic (1984) have reported an overproduction followed by elimination of fibers in the anterior commissure of monkeys. In primates, however, the anterior commissure accommodates a large complement of neocortical fibers, a fact that hinders comparison with the development of the AC in rodents (see below). The same rationale has to be used to qualify the observation by Cabana and Martin (1985) that anterior commissure projections in the opossum undergo the dual strategy typical of the callosal projections of Eutherian mammals.

Other progressive mechanisms occur when the numerical stabilization of the commissural projections is advanced. This is the case of myelination (Lent and Jhaveri, 1992; Sturrock, 1976). As for regressive mechanisms different from the selective elimination of fibers, cell death and synaptic elimination cannot be excluded, but so far they have not been demonstrated in the piriform cortex.

In light of the available results, it can be concluded that the interhemispheric projections of the PC undergo a singular, entirely progressive, developmental strategy, that contrasts with the dual, progressive-regressive strategy adopted by V1 interhemispheric projections through the corpus callosum.

DIFFERENT DEVELOPMENTAL STRATEGIES OF INTERHEMISPHERIC CONNECTIONS: THE PHYLOGENESIS OF CORTICAL ONTOGENESIS

The relationship between ontogeny and phylogeny is an issue that has attracted the interest of biologists for more than a century (for a recent review, see Northcutt, 1990), but the interpretations for it remains highly controversial. For all investigators in the field, however, the comparative study of different ontogenies is the most fruitful approach to shed light on this problem. It is of interest, then, to analyze the developmental strategies employed by the cerebral cortex under an evolutionary optics. In order to be more specific, this can be done for one particular set of cortical circuits, for example, the interhemispheric projections.

Two different approaches can be used. The first is to compare the development of interhemispheric connections of extant animals having different phylogenetic ancestry, and the second is to select one particular taxon, and compare the ontogenesis of the interhemispheric connections of cortical regions having different evolutionary origins.

Both approaches must take into account an important fact and its corollary. The fact is that the interhemispheric connections of homologous cortical regions may take different pathways in different animals. This is the case of the interhemispheric connections of the neocortex. In Metatherians, they course through the anterior commissure together with those originating from the paleocortex (Ebner and Myers, 1967; Granger et al., 1985; Martin, 1987). In lower Eutherians, such as rodents, these neocortical projections course mostly through the corpus callosum (Jacobson and Trojanowski, 1974; Yorke and Caviness, 1975), a very small part taking the anterior commissure (Horel and Stelzner, 1981; Jouandet and Hartenstein, 1983). In higher Eutherians, such as cats and monkeys, on the other hand, large sectors of temporal neocortex project through the anterior commissure, together with paleocortical regions (Jouandet, 1982; Jouandet et al., 1984; Jouandet and Gazzaniga, 1979; Pandya et al., 1973). The corollary of the above fact is that each telencephalic commissure accommodates interhemispheric projections of different evolutionary history. The corpus callosum, for instance, contains not only neocortical projections, but also fibers from medial and lateral sectors of the transitional periallocortex (Reep and Winans, 1982; Vogt et al., 1981), as well as claustral projections (Squatrito et al., 1980). The anterior commissure, on the other hand, as mentioned above, contains paleocortical connections together with a complement of fibers from the neocortex that may vary pronouncedly from one taxon to another.

In this context, one may compare the developmental strategy of rodent callosal connections of V1, with that of the anterior commissure connections of PC (see Figure 9.2 and 9.3), as described in the preceding sections. The main conclusion of this comparison is that the development of the interhemispheric connections of the paleocortex adopts a singular, progressive strategy, while that of the neocortex employs a dual, progressive-regressive strategy. It appears, then, that the regressive mechanisms of sculpting the final pattern of interhemispheric connections have evolved together with the neocortex itself, from a previous, less flexible, strategy characteristic of older cortex. It is not obligatory, however, for

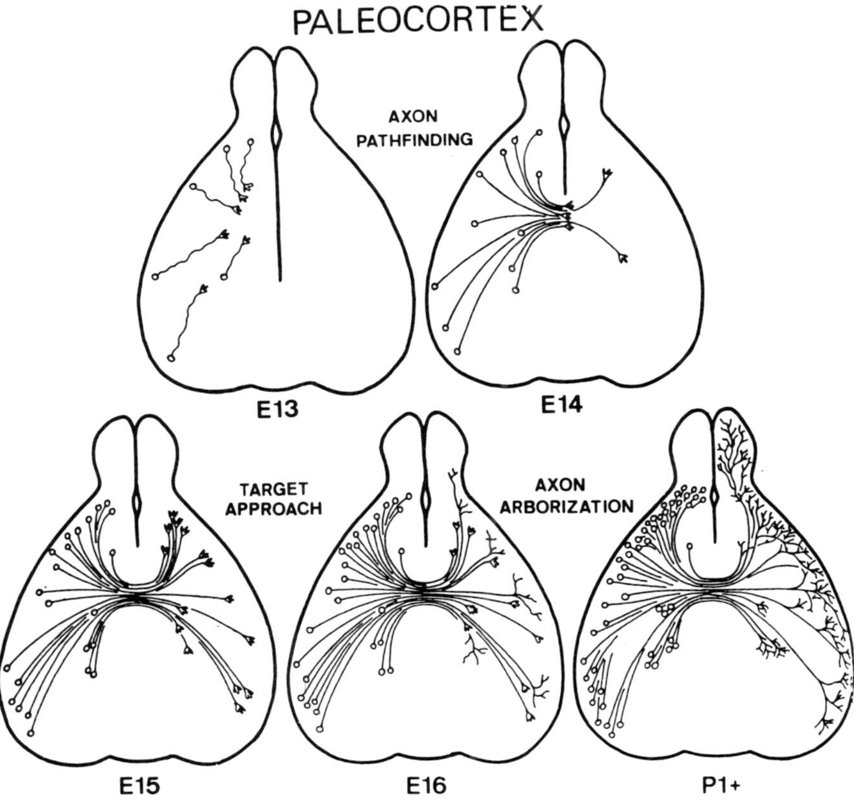

FIGURE 9.2. Schematic representation of the essential stages that characterize the singular developmental strategy of the interhemispheric connections of the paleocortex. Ages below each drawing refer to hamsters. Compare with Figure 9.3.

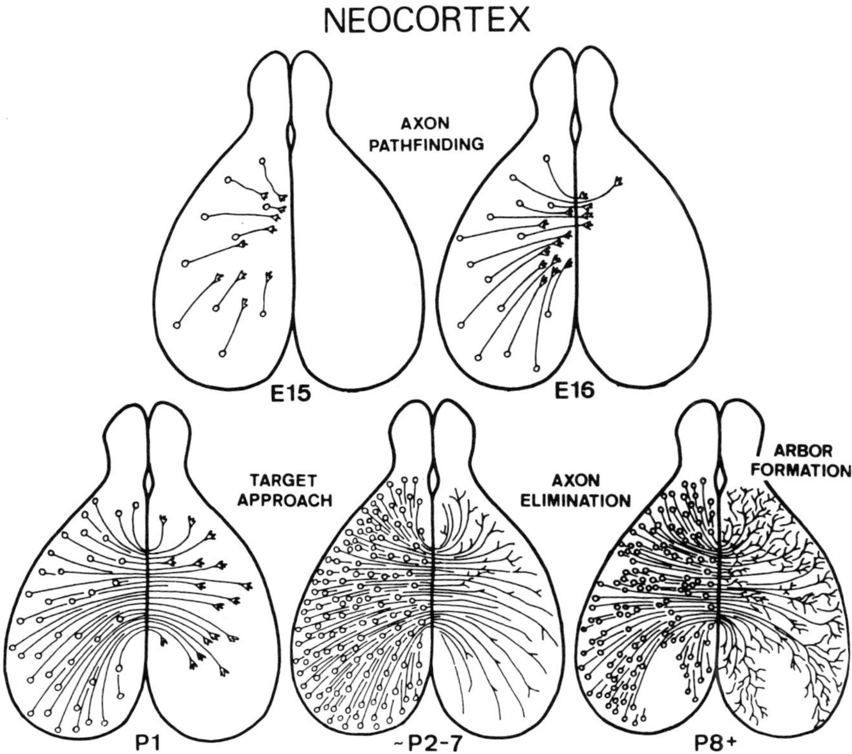

FIGURE 9.3. Schematic representation of the essential stages that characterize the dual ontogenetic strategy of the interhemispheric connections of the neocortex. Ages below each drawing refer to hamsters. Compare with Figure 9.2.

all neocortical regions to rely on this strategy, as shown by recent work in visual callosal projections of developing primates (Chalupa et al., 1989; Dehay et al., 1988).

Progressive events are subject to modulation by environmental factors during development. This is a potential source of individual variation. Growing callosal fibers, for instance, can be diverted ipsilaterally to form an aberrant bundle when the midline is split surgically in newborn hamsters (Lent, 1984), or when mouse fetuses are gamma-irradiated (Lent and Schmidt, 1986; Schmidt and Lent, 1987). Anterior commissure axons, in addition, can be made to invade the terminal territory of the olfactory bulb in layer I of PC, when the bulb is ablated in newborn rats (R. Lent and E.C.P. Moraes, unpublished data). Progressive mechanisms are also subject to genetic change. In this case, this may be considered as a source of variation among taxa. Genetic mutations capable of influencing these mechanisms, for example, may have been at the base of evolutionary change in interhemispheric connections (Katz et al., 1983). When a novel region of interhemispheric fusion appeared rostrally to the lamina terminalis in Eutherians, genetic change must probably have induced growing neocortical axons to deflect dorsally and medially toward the corpus callosum, instead of ventrally and laterally toward the anterior commissure.

Regressive events, on the other hand, have been proposed as additional generators of variance among individuals in somatic time (Edelman, 1987), and among taxa in the evolutionary time scale (Ebbesson, 1980, 1984; Finlay et al., 1987). This hypothesis is in accordance with the fact that the elimination of exuberant callosal connections can be modulated by manipulation of environmental factors. Visual callosal connections, for instance, can be modified when the eyes are disaligned during development (Elberger et al., 1983; Innocenti and Frost, 1979; Lund et al., 1978; Lund and Mitchell, 1979) and when the cortex is precociously deprived of retinal input (Cusick and Lund, 1981; Innocenti and Frost, 1980) or even of form vision (Innocenti and Frost, 1979). Similar anomalies in the callosal connections have been observed in naturally strabismic animals, such as Siamese cats (Shatz, 1977). Recently, it has been shown also that thyroid hormone serum levels influence the ontogenetic elimination of visual callosal connections (Gravel et al., 1990). It is conceivable, therefore, that regressive mechanisms have been a favourable evolutionary acquisition of neocortical systems, capable of enriching the plastic repertoire of cortical connections and thus increasing individual variation.

SUMMARY

A comparison between the development of the interhemispheric projections of the visual cortex, and those of the olfactory cortex, leads to the conclusion that while the former adopts a dual ontogenetic strategy, the latter employs a much simpler, singular strategy. The singular strategy consists of a sequence of progressive events, namely neuron proliferation, axon growth and arborization, synaptogenesis, and eventually the stabilization of the projection pattern. Through the dual strategy, on the other hand, these same progressive events result in an oversized projection that is trimmed subsequently before stabilization of the mature pattern. Since the olfactory cortex (paleocortex) precedes the visual cortex (neocortex) in evolution, it is conceivable to conclude that regressive mechanisms may have appeared as an additional source of individual variance through epigenetic influence during the development of cortical circuits.

Acknowledgments. The author is indebted to Beth C. P. Moraes and Antonia L. Carvalho for excellent technical support. This work was funded by various grants from Financiadora de Estudos e Projetos, Conselho Nacional de Desenvolvimento Cientĭfico e Tecnológico, and Fundacão de Amparo `a Pesquisa do Estado do Rio de Janeiro.

REFERENCES

Angevine JB Jr, Sidman RL (1961): Autoradiographic study of cell migration during histogenesis of cerebral cortex in the mouse. *Nature* 192:766–768

Bayer SA (1986): Neurogenesis in the rat primary olfactory cortex. *Int J Dev Neurosci* 4:251–271

Berbel P, Innocenti GM (1988): The development of the corpus callosum in cats: A light- and electron-microscopic study. *J Comp Neurol* 276:132–156

Berry M, Rogers AW (1965): The migration of neuroblasts in the developing cerebral cortex. *J Anat* 99:691–709

Blue ME, Parnavelas JG (1983): The formation and maturation of synapses in the visual cortex of the rat: 1. Quantitative analysis. *J Neurocytol* 12:697–712

Cabana T, Martin GF (1985): The development of commissural connections of somatic motor-sensory areas of neocortex in the North American opossum. *Anat Embryol* 171:121–128

Caviness VS Jr (1982): Neocortical histogenesis in normal and Reeler mice: Developmental study based on [^{3}H]-thymidine autoradiography. *Dev Brain Res* 4:293–302

Chalupa LM, Killackey HP, Snider CJ, Lia B (1989): Callosal projection neurons in area 17 of the fetal rhesus monkey. *Dev Brain Res* 46:303–308

Chow KL, Baumbach HD, Lawson R (1981): Callosal projections of the striate cortex in the neonatal rabbit. *Exp Brain Res* 42:122–126

Chun JJ, Nakamura MJ, Shatz CJ (1987): Transient cells of the developing mammalian telencephalon are peptide immunoreactive neurons. *Nature* 325:617–620

Cipolloni PB, Peters A (1979): The bilaminar and banded distribution of callosal terminals in the posterior neocortex of the rat. *Brain Res* 176:33–47

Crossland WJ, Uchwat CJ (1982): Neurogenesis of the central visual pathways in the golden hamster. *Dev Brain Res* 5:99–103

Cusick CG, Lund RD (1981): The distribution of the callosal projection to the occipital visual cortex in rats and mice. *Brain Res* 214:239–259

Cusick CG, Lund RD (1982): Modification of visual callosal projections in rats. *J Comp Neurol* 212:385–398

Davis BJ, Macrides F, Youngs WM, Schneider SP, Rosene DL (1978): Efferents and centrifugal afferents of the main and accessory olfactory bulbs in the hamster. *Brain Res Bull* 3:59–72

Dean P (1990): Sensory cortex: Visual perceptual functions. In: *The Cerebral Cortex of the Rat*, Kolb B, Tees RC, eds. Cambridge: MIT Press

Derer P, Caviness VS Jr, Sidman RL (1977): Early cortical histogenesis in the primary olfactory cortex of the mouse. *Brain Res* 123:27–40

Diao Y-C, So K-F (1991): Dendritic morphology of visual callosal neurons in the golden hamster. *Brain Behav Evol* 37:1–9

Drager U (1975): Receptive fields of single cells and topography in mouse visual cortex. *J Comp Neurol* 160:269–290

Dursteler MR, Blakemore C, Garey LJ (1979): Projections to the visual cortex in the golden hamster. *J Comp Neurol* 183:185–204

Ebbesson SOE (1980): The parcellation theory and its relation to interspecific variability in brain organization, evolutionary and ontogenetic development, and neuronal plasticity. *Cell Tissue Res* 213:179–212

Ebbesson SOE (1984): Evolution and ontogeny of neural circuits. *Behav Brain Sci* 7:321–331

Ebner FF, Myers RE (1967): Afferent connections to neocortex in the opossum (*Didelphis virginiana*). *J Comp Neurol* 129:241–248

Edelman GM (1987): *Neural Darwinism. The Theory of Neuronal Group Selection*. New York: Basic Books

Elberger AJ, Smith EL III, White JM (1983): Spatial dissociation of visual inputs alters the origin of the corpus callosum. *Neurosci Lett* 35:19–24

Feng JZ, Brugge JF (1983): Postnatal development of auditory callosal connections in the kitten. *J Comp Neurol* 214:416–426

Ferrer I, Bernet E, Soriano E, Del Rio T, Fonseca M (1990): Naturally occurring cell death in the cerebral cortex of the rat and removal of dead cells by transitory macrophages. *Neuroscience* 39:451–458

Finlay BL, Slattery M (1983): Local differences in the amount of early cell death in neocortex predict adult local specializations. *Science* 219:1349–1351

Finlay BL, Wikler KC, Sengelaub DR (1987): Regressive events in brain development and scenarios for vertebrate brain evolution. *Brain Behav Evol* 30:102–117

Fish SE, Mooney RD, Rhoades RW (1990): Development of occipital callosal axons in hamster: Anterograde tracing with Di-I. *Soc Neurosci Abstr* 16:495

Floeter MK, Jones EG (1985): The morphology and phased outgrowth of callosal axons in the fetal rat. *Dev Brain Res* 22:7–18

Gattass R, Rosa MGP, Sousa APB, Piñón MCG, Fiorani M Jr, Neuenschwander S (1990): Cortical streams of visual information processing in primates. *Braz J Med Biol Res* 23:375–394

Granger EM, Masterton RB, Glendenning KK (1985): Origin of interhemispheric fibers in acallosal opossum (with a comparison to callosal origins in rat). *J Comp Neurol* 241:82–98

Gravel C, Sasseville R, Hawkes R (1990): Maturation of the corpus callosum of the rat: 2. Influence of thyroid hormones on the number and maturation of axons. *J Comp Neurol* 291:147–161

Guadaño-Ferraz A, Berbel P, Balboa R, Innocenti GM (1991): The development of the anterior commissure in normal and hypothyroid rats. *Eur J Neurosci (Suppl)*4:225

Haberly LB, Bower JM (1984): Analysis of association fiber system in piriform cortex with intracellular recording and staining techniques. *J Neurophysiol* 51:90–112

Haberly LB, Price JL (1978a): Association and commissural fiber systems of the olfactory cortex of the rat: 1. Systems originating in the piriform cortex and adjacent areas. *J Comp Neurol* 178:711–740

Haberly LB, Price JL (1978b): Association and commissural fiber systems of the olfactory cortex of the rat: 2. Systems originating in the olfactory peduncle. *J Comp Neurol* 181:781–808

Hedin-Pereira C, Jhaveri S, Lent R (1991): Development of callosal axon arbors in hamsters. *Soc Neurosci Abstr* 17:740

Hernit CS, Van Sluyters RC, Murphy KM (1990): Visual callosal development in neonatal rats: Do migrating or undifferentiated cells have an interhemispheric axon? *Soc Neurosci Abstr* 16:803

Heuman D, Leuba G (1983): Neuronal death in the development and aging of the cerebral cortex of the mouse. *Neuropathol Appl Neurobiol* 9:297–311

Hogan D, Berman NEJ (1990): Growth cone morphology, axon trajectory and branching patterns in the neonatal rat corpus callosum. *Dev Brain Res* 53:283–287

Horel JA, Stelzner DJ (1981): Neocortical projections of the rat anterior commissure. *Brain Res* 220:1–12

Hughes CM, Peters A (1990): Morphological evidence for callosally projecting nonpyramidal neurons in rat visual cortex. *Anat Embryol* 182:591–604

Innocenti GM (1980): The primary visual pathway through the corpus callosum: Morphological and functional aspects in the cat. *Arch Ital Biol* 118:124–188

Innocenti GM (1981): Growth and reshaping of axons in the establishment of visual callosal connections. *Science* 212:824–827

Innocenti GM (1986): General organization of callosal connections in the cerebral cortex. In: *Cerebral Cortex*, Jones EG, Peters A. eds. New York: Plenum Press

Innocenti GM, Fiore L, Caminiti R (1977): Exuberant projections into the corpus callosum from the visual cortex of newborn cats. *Neurosci Lett* 4:237–242

Innocenti GM, Frost DO (1979): Effects of visual experience on the maturation of the efferent system to the corpus callosum. *Nature* 280:231–234

Innocenti GM, Frost DO (1980): The postnatal development of visual callosal connections in the absence of visual experience or of the eyes. *Exp Brain Res* 39:365–375

Ivy GO, Killackey HP (1981): The ontogeny of distribution of the callosal projection neurons in the rat parietal cortex. *J Comp Neurol* 195:367–389

Jacobson S, Trojanowski JQ (1974): The cells of origin of the corpus callosum in rat, cat, and rhesus monkey. *Brain Res* 74:149–155

Jouandet ML (1982): Neocortical and basal telencephalic origins of the anterior commissure of the cat. *Neuroscience* 7: 1731–1752

Jouandet ML, Garey LJ, Lipp HP (1984): Distribution of the cells of origin of the corpus callosum and the anterior commissure in the marmoset monkey. *Anat Embryol* 169:45–59

Jouandet ML, Gazzaniga MS (1979): Cortical field of origin of the anterior commissure of the rhesus monkey. *Exp Neurol* 66:381–397

Jouandet ML, Hartenstein V (1983): Basal telencephalic origins of the anterior commissure of the rat. *Exp Brain Res* 50:183–192

Katz MJ, Lasek RJ, Silver J (1983): Ontophyletics of the nervous system: Development of the corpus callosum and evolution of axon tracts. *Proc Natl Acad Sci U S A* 80:5936–5940

Koppel H, Innocenti GM (1983): Is there a genuine exuberancy of callosal projections in development? A quantitative electron microscopic study in the cat. *Neurosci Lett* 41:33–40

Krieg WJS (1946): Connections of the cerebral cortex: 1. The albino rat. A topography of the cortical areas. *J Comp Neurol* 84:221–275

LaMantia AS, Rakic P (1984): The number, size, myelination, and regional variation of axons in the corpus callosum and anterior commissure of the developing rhesus monkey. *Soc Neurosci Abstr* 10:1081

LaMantia AS, Rakic P (1990): Axon overproduction and elimination in the corpus callosum of the developing rhesus monkey. *J Neurosci* 10:2156–2175

Lent R (1984): Neuroanatomical effects of neonatal transection of the corpus callosum in hamsters. *J Comp Neurol* 223:548–555

Lent R, Guimarães RZP (1991): Development of paleocortical projections through the anterior commissure of hamsters adopts progressive, not regressive, strategies. *J Neurobiol* 22:475–498

Lent R, Hedin-Pereira C, Menezes JRL, Jhaveri S (1990): Neurogenesis and development of callosal and intracortical connections in the hamster. *Neuroscience* 38:21–37

Lent R, Jhaveri S (1992): Myelination of the cerebral commissures in the hamster, as revealed by a monoclonal antibody specific for oligodendrocytes. *Dev Brain Res* 66:193–201

Lent R, Schmidt SL (1986): Dose-dependent occurrence of the aberrant longitudinal bundle in the brains of mice born acallosal after prenatal gamma-irradiation. *Dev Brain Res* 25:127–132

Lent R, Schmidt SL (1992): The ontogenesis of the cortical commissures and the determination of brain asymmetries. *Progr Neurobiol*, in press

Lund JS, Lund RD (1970): The termination of callosal fibers in the paravisual cortex of the rat. *Brain Res* 17:25–45

Lund RD, Mitchell DE (1979): Asymmetry in the visual callosal connections of strabismic cats. *Brain Res* 167:176–179

Lund RD, Mitchell DE, Henry GH (1978): Squint-induced modification of callosal connections in cats. *Brain Res* 144:169–172

Martin GF (1967): Interneocortical connections in the opossum, *Didelphis virginiana*. *Anat Rec* 157:607–616.

McCourt ME, Thalluri J, Henry GH (1990): Properties of area 17/18 border neurons contributing to the visual transcallosal pathway in the cat. *Visual Neurosci* 5:83–98

Montero VM (1981): Comparative studies on the visual cortex. In: *Cortical Sensory Organisation: Vol. 2 Multiple Visual Areas*, Woolsey CN, ed. Clifton, NJ: Humana Press.

Mooney RD, Rhoades RW, Fish SE (1984): Neonatal superior colliculus lesions alter visual callosal development in hamster. *Exp Brain Res* 55:9–25

Norris C, Kalil K (1990): Morphology and cellular interactions of growth cones in the developing corpus callosum. *J Comp Neurol* 293:268–291

Northcutt RG (1990): Ontogeny and phylogeny: A re-evaluation of conceptual relationships and some applications. *Brain Behav Evol* 36:116–140

Olavarria J, Van Sluyters RC (1985): Organization and postnatal development of callosal connections in the visual cortex of the rat. *J Comp Neurol* 239:1–26

O'Leary DDM, Stanfield BB, Cowan WM (1981): Evidence that the early postnatal restriction of cells of origin of the callosal projection is due to the elimination of axonal collaterals rather than to the death of neurons. *Dev Brain Res* 1:607–617

Pandya DN, Karol EA, Lele PP (1973): The distribution of the anterior commissure in the squirrel monkey. *Brain Res* 49:177–180

Payne BR (1990): Function of the corpus callosum in the representation of the visual field in cat visual cortex. *Visual Neurosci* 5:205–211

Peters A, Feldman ML (1976): The projection of the lateral geniculate nucleus to area 17 of the rat cerebral cortex: 1. General description. *J Neurocytol* 5:63–84

Pires-Neto MA, Lent R (1991): Pioneer fibers in the anterior commissure of hamster embryos. *Braz J Med Biol Res*, 24:1067–1070

Raedler E, Raedler A (1978): Autoradiographic study of early neurogenesis in rat neocortex. *Anat Embryol* 154:267–284

Rakic P, Bourgeois JP, Eckenhoff MF, Zecevic N, Goldman-Rakic P (1986): Concurrent overproduction of synapses in diverse regions of the primate cerebral cortex. *Science* 232:232–235

Reep RL, Winans SS (1982): Efferent connections of dorsal and ventral agranular insular cortex in the hamster, *Mesocricetus auratus*. *Neuroscience* 7:2609–2635

Rhoades RW, Dellacroce DD (1980): Visual callosal connections in the golden hamster. *Brain Res* 190:248–254

Schmidt SL, Lent R (1987): The effects of prenatal irradiation on the development of cerebral cortex and corpus callosum of the mouse. *J Comp Neurol* 264:193–204

Schwartz ML, Rakic P, Goldman-Rakic PS (1991): Early phenotype expression of cortical neurons: Evidence that a subclass of migrating neurons have callosal axons. *Proc Natl Acad Sci U S A* 88:1354–1358

Schwob JE, Price JL (1984): The development of axonal connections in the central olfactory system of rats. *J Comp Neurol* 223:177–202

Sefton AJ, Dreher B (1985): Visual system. In: *The Rat Nervous System: Vol. 1. Forebrain and Midbrain*, Paxinos G, ed. Sydney, Australia: Academic Press

Segraves MA, Rosenquist AC (1982): The distribution of the cells of origin of callosal projections in cat visual cortex. *J Neurosci* 2:1079–1089
Shatz CJ (1977): Anatomy of interhemispheric connections in the visual system of Boston, Siamese, and ordinary cats. *J Comp Neurol* 173:497–518
Silver J, Lorenz SE, Wahlsten D, Coughlin J (1982): Axonal guidance during development of the great cerebral commissures: Descriptive and experimental studies, *in vivo*, on the role of preformed glial pathways. *J Comp Neurol* 210:10–29
Squatrito S, Battaglini PP, Galletti C, Sanseverino ER (1980): Projections from the visual cortex to the contralateral claustrum of the cat revealed by an anterograde axonal transport method. *Neurosci Lett* 19:271–275
Sturrock RR (1976): Development of the mouse anterior commissure: 1. A comparison of myelination in the anterior and posterior limbs of the anterior commissure of the mouse brain. *Zbl Vet Med C* 5:54–67
Sturrock RR (1980): Myelination of the mouse corpus callosum. *Neuropathol Appl Neurobiol* 6:415–420
Swadlow HA (1977): Relationship of the corpus callosum to visual areas I and II of the awake rabbit. *Exp Neurol* 57:516–531
Switzer RC, De Olmos J, Heimer L (1985): Olfactory system. In: *The Rat Nervous System: Vol 1. Forebrain and Midbrain*, Paxinos G, ed. Sydney, Australia: Academic Press
Thomas HC, Espinoza SG (1987): Relationships between interhemispheric cortical connections and visual areas in hooded rats. *Brain Res* 417:214–224
Tiao YC, Blakemore C (1976): Funtional organization in the visual cortex of the golden hamster. *J Comp Neurol* 168:459–482
Tusa RJ, Palmer LA, Rosenquist AC (1981): Multiple cortical visual areas in cat. In: *Cerebral Localization in Somatic, Visual and Auditory Systems*, Woolsey CN, ed. Clifton, NJ: Humana Press
Ungerleider LG, Mishkin M (1982): Two cortical visual systems. In: *Analysis of Behavior*, Ingle DJ, Goodale MA, eds. Cambridge: MIT Press
Uylings HBM, Van Eden CG, Parnavelas JG, Kalsbeek A (1990): The prenatal and postnatal development of rat cerebral cortex. In: *The Cerebral Cortex of the Rat*, Kolb B, Tees RC, eds. Cambridge: MIT Press
Van Essen DC (1985): Functional organization of primate visual cortex. In: *Cerebral Cortex, vol. 3*. Jones EG, Peters A, eds. New York: Plenum Press
Van Essen DC, Newsome WT, Bixby JL (1982): The pattern of interhemispheric connections and its relationship to extrastriate visual areas in the macaque monkey. *J Neurosci* 2:265–283
Vogt BA, Rosene DL, Peters A (1981): Synaptic termination of thalamic and callosal afferents in cingulate cortex of the rat. *J Comp Neurol* 201:265–283
Wagor E, Mangini NJ, Pearlman AL (1980): Retinotopic organization of striate and extrastriate visual cortex in the mouse. *J Comp Neurol* 193:187–202
Winfield DA, Gatter KC, Powell TPS (1975): Certain connections of the visual cortex of the monkey shown by the use of horseradish peroxidase. *Brain Res* 92:456–461
Wise SP, Jones EJ (1976): The organization and postnatal development of the commissural projections of the rat somatic sensory cortex. *J Comp Neurol* 168:313–344
Yorke CH, Caviness VS (1975): Interhemispheric neocortical connections of the corpus callosum in the normal mouse: A study based on anterograde and retrograde methods. *J Comp Neurol* 164:233–246
Záborsky L, Wolff JR (1982): Distribution patterns and individual variations of callosal connections in the albino rat. *Anat Embryol* 165:213–232
Zilles K (1985): *The Cortex of the Rat: A Stereotaxic Atlas*. Berlin: Springer-Verlag

Section 2

Maturity

10

Organization of Catecholaminergic Amacrine Cells in the Rhesus Monkey Retina

ANDREW P. MARIANI AND JAN NORA HOKOÇ

MORPHOLOGY OF CATECHOLAMINERGIC AMACRINE CELLS IN THE RHESUS RETINA

Dopamine-containing amacrine cells were among the first of many different neurotransmitter phenotypes identified in the macaque retina owing to their blue-green, formaldehyde-induced fluorescence (Ehinger, 1966; Laties and Jacobowitz, 1966). They are neurons with cell bodies in the innermost row of the inner nuclear layer, processes that arborize in the outermost stratum of the inner plexiform layer, and fine, radially oriented fibers that course in the inner nuclear layer (see Figure 10.1A), and have been well-characterized with regard to their morphology (Nguyen-Legros et al., 1984), distribution across the retina (Mariani et al., 1984), and synaptic organization (Hokoç and Mariani, 1987). Although the appearance of dopaminergic neurons varies somewhat between species (Ehinger, 1982), there was, with the exception of perhaps two or three vertebrate species (Hadjiconstantinou et al., 1984; Keyser et al., 1987), little evidence for morphological variability among this transmitter phenotype in a single retina, including the rhesus monkey's.

However, with immunohistochemistry and antibodies directed against tyrosine hydroxylase (TH), the rate-limiting enzyme in the catecholamine synthetic pathway, a second, morphologically distinct type of catecholamine (CA) neuron has been found in the rhesus monkey retina (Mariani and Hokoç, 1988). These newly identified cells were termed type 2 CA amacrine cells to distinguish them from the previously known dopaminergic cells, now termed type 1 CA cells (described briefly above). The type 2 CA amacrine cells are more numerous than type 1 cells. Their cell bodies, smaller in diameter than those of the type 1 cells, are located in the inner nuclear, inner plexiform, and ganglion cell layers. The processes of the type 2 CA cells always arborize in the center of the inner plexiform layer (see Figure 10.1B), and type 2 cells contain less than one-third the TH that type 1 CA cells contain.

The two types of TH-immunoreactive amacrine cell in the rhesus monkey retina were described from the indirect immunofluorescent method as follows. Type 1 CA amacrine cells possess relatively large cell bodies almost always in the

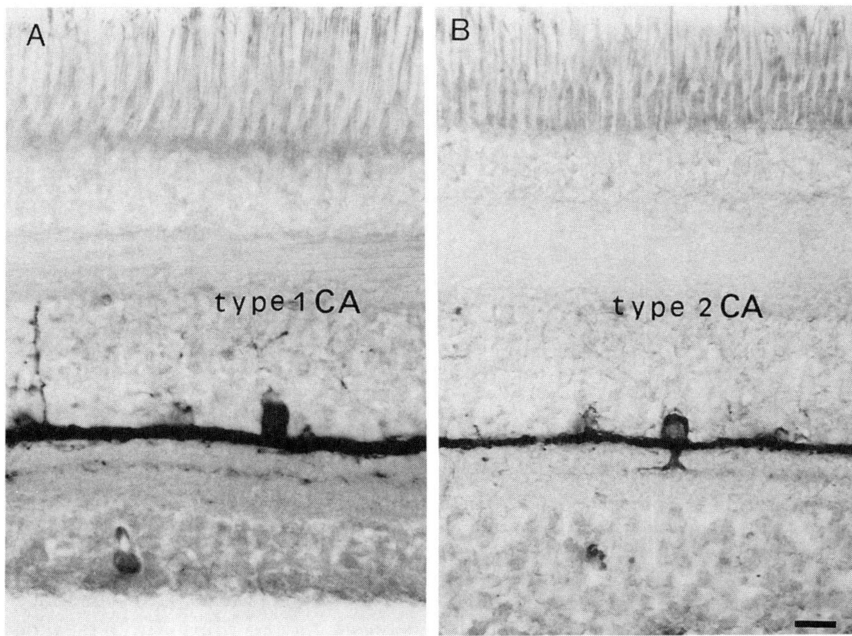

FIGURE 10.1. Light micrographs of TH-immunoreactive amacrine cells in the rhesus monkey retina visualized by the PAP method. A. A type 1 CA amacrine cell with a strongly immunoreactive cell body in the inner nuclear layer. Processes of the type 1 cells arborize in the outermost stratum of the inner plexiform layer and fine fibers that originate from these processes penetrate the inner nuclear layer. B. The cell body of a type 2 CA amacrine cell gives rise to a process that enters the inner plexiform layer where it branches and arborizes in the center of this synaptic region. Bar = 20 μm.

innermost row of the inner nuclear layer with processes that densely arborize in the outermost stratum of the inner plexiform layer, and fine, radially oriented fibers in the inner nuclear layer. Type 2 CA amacrine cells, on the other hand, have smaller cell bodies in the inner nuclear layer, the inner plexiform layer, and the ganglion cell layer. Their sparsely branching processes ramify in the center of the inner plexiform layer.

SYNAPTIC ORGANIZATION OF TYPE 1 CA AMACRINE CELLS

Electron microscopy showed synapses from other neuronal elements onto type 1 CA immunoreactive processes and these input synapses accounted for about one-fifth of all synapses involving this cell types. The most common synaptic input was from bipolar cell axon terminals at ribbon synapses (see Figure 10.2A). Other synapses to this cell type were from other nonimmunoreactive amacrine cells, and less frequently from other TH-immunoreactive profiles.

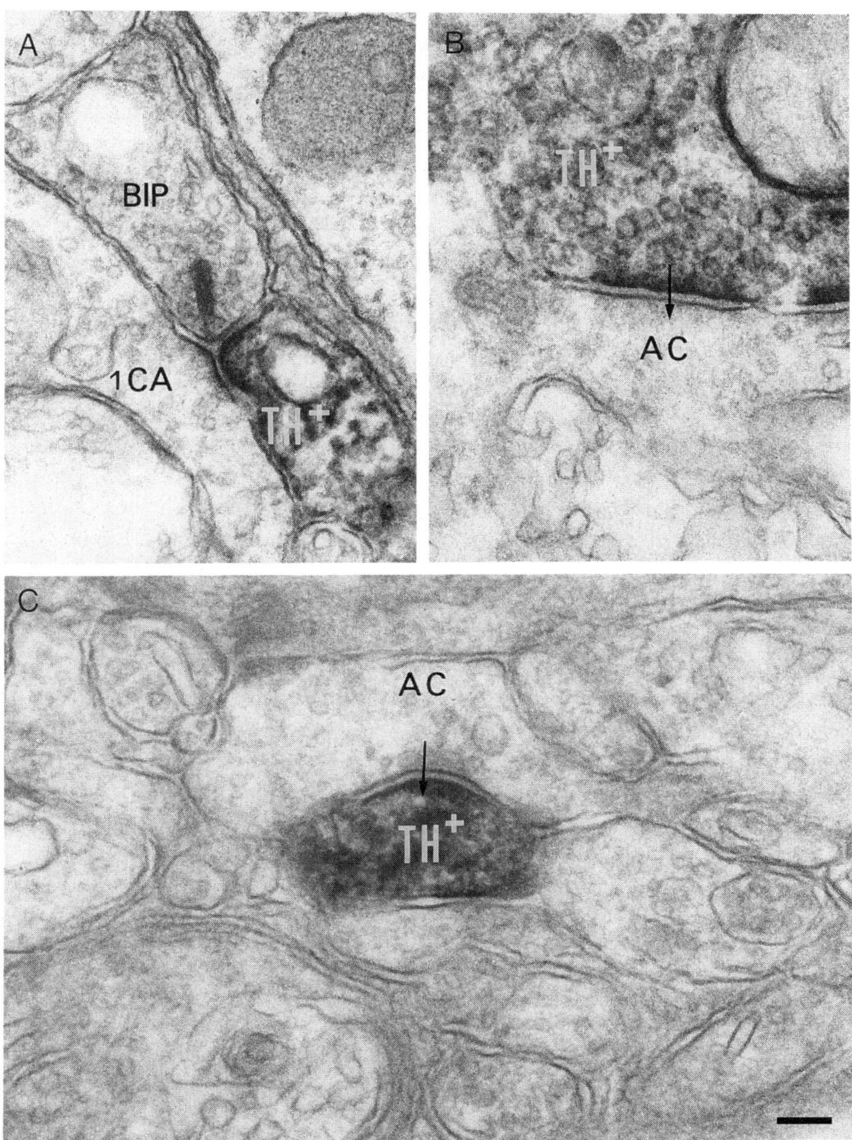

FIGURE 10.2. Electron micrographs of TH-immunoreactive (TH^+) synapses in rhesus retina. A. Input synapses to the type 1 CA cells are primarily from bipolar cell axon terminals (BIP) at ribbon synapses, and output synapses (B) are primarily onto other amacrine cells (AC). C. Type 2 CA amacrine cells have input synapses primarily from other amacrine (AC) cells, but very few identifiable output synapses. BAR = 0.1 μm.

Synaptic output of the type 1 CA amacrine cells accounted for about four-fifths of all synapses involving this cell type. These output synapses were primarily onto nonimmunoreactive amacrine cell processes (see Figure 10.2B). Much less frequently found were synapses onto other TH-immunoreactive profiles as well as reciprocal synapses back onto bipolar cell axon terminals. The quantitative results of the synaptic organization of the type 1 CA amacrine cells are summarized in Table 10.1.

SYNAPTIC ORGANIZATION OF TYPE 2 CA AMACRINE CELLS

Although type 2 CA cells contain less than one-third the TH as the type 1 cells (Mariani and Hokoç, 1988), the osmicated HRP reaction product in these cells was easily identifiable in the electron microscope.

Synapses onto type 2 CA amacrine cells accounted for 106/112 (95%) of all synapses involving TH immunoreactivity in the center of the inner plexiform layer. Amacrine cells provided the majority of the synaptic input (72/106 or 68%) to the TH-containing type 2 processes (see Figure 10.2C). Other synapses onto the type 2 CA amacrines, that is, the remainder of their synaptic input, were from bipolar cell axon terminals. These accounted for 32/106 (30%) of the synaptic input onto the TH-containing processes in the center of the inner plexiform layer. The synaptic endings of the bipolar cells were clearly identified by the presence of synaptic ribbons surrounded by synaptic vesicles. The ribbons approached a point bisecting the presynaptic ridge. The type 2 CA, TH-immunoreactive processes were most often one of the two postsynaptic processes that constitute the "dyad." In only a few instances (n = 2), both profiles of the dyad opposite the presynaptic bipolar cell axon terminal were TH-containing type 2 CA amacrine processes.

The synaptic output, that is, synapses of the type 2 CA amacrine cells onto other neuronal elements, amounted to only 6/112 (5%) of all synapses involving TH immunoreactivity in the center of the inner plexiform layer. All of the synaptic output of the type 2 CA amacrines appeared to be onto nonimmunoreactive amacrine cell processes. No synapses of the TH-containing processes onto bipolar cell axon terminals, other TH-immunoreactive profiles, or onto ganglion cell

TABLE 10.1 Synaptic "Input" and "Output" of Type 1 CA Amacrine Cells

	% of total (n = 148)
Synapses onto Type 1 CA Amacrines	"input" = 22% (n = 32)
Bipolar Cell Axon Terminals = 53% (n = 17)	
Nonimmunoreactive Amacrine Cells = 31% (n = 10)	
TH+ Amacrine Cells = 16% (n = 5)	
Synapses made by Type 1 CA Amacrines	"output" = 78% (n = 116)
Nonimmunoreactive Amacrine Cells = 92% (n = 107)	
TH+ Amacrines Cells = 4% (n = 5)	
Bipolar Cell Axon Terminals = 3% (n = 4)	

dendrites were observed. Data enumerating the synaptic input and output of the type 2 CA amacrine cells are listed in Table 10.2.

This pattern of synaptic organization contrasts sharply with that of the type 1 CA or dopaminergic amacrine cells with processes that arborize in the outermost stratum of the inner plexiform layer. Although both the type 1 and type 2 CA amacrine cells have qualitatively similar synaptic organizations in that they receive input synapses from bipolar cells and amacrines, and output onto amacrine cells, they are different, quantitatively. About one-fourth of the synapses involving the type 1 CA cells are input synapses from bipolar cells and amacrine cells (Hokoç and Mariani, 1987). Therefore, type 1 and 2 CA amacrine cells differ with regard to their synaptic organization (see Figure 10.2) as well as morphology and immunohistochemistry.

POSSIBLE FUNCTIONAL ROLES OF PRIMATE CATECHOLAMINERGIC AMACRINE CELLS

What are the roles of the catecholaminergic amacrine cells in the primate retina? In order to answer this question, it is useful to review what is already known about the role of dopamine in the retina in general. The most thorough studies on the function of dopamine in the vertebrate retina come from teleosts. In this class of vertebrate, dopamine is contained in neurons known as interplexiform cells. Interplexiform cells resemble conventional amacrine cells in their morphology and synaptic organization in the inner plexiform layer, but they have processes that ascend through the inner nuclear layer and reach the outer plexiform layer where they arborize and synapse on horizontal cells (Dowling and Ehinger, 1975). In the inner plexiform layer, interplexiform cells receive synaptic input from amacrine cells (Dowling and Ehinger, 1975), bipolar cells (Yazulla and Zucker, 1988) and centrifugal fibers (Zucker and Dowling, 1987) that contain LHRH and FRMF-amide (Stell et al., 1984). In the outer plexiform layer, dopamine has been demonstrated to uncouple the extensive gap junctions that electrically couple like types of horizontal cells through a cyclic AMP-mediated process (Teranishi et al., 1983) that reduces gap junction channel open probability (McMahon et al., 1989).

Similar effects of dopamine on the gap junctions of horizontal cells in turtle retina have been reported (Piccolino et al., 1984). In turtles, however, dopamine

TABLE 10.2 Synpatic "Input" and "Output" of Type 2 CA Amacrine Cells in Rhesus Monkey Retina

	% of total (n = 112)
Synapses onto Type 2 CA Cells	"input" = 95% (n = 106)
Nonimmunoreactive Amacrine Cells = 68% (n = 72)	
Bipolar Cell Axon Terminals = 32% (n = 34)	
Synapses Made by Type 2 CA Cells	"output" = 5% (n = 6)
Nonimmunoreactive Amacrine Cells = 100% (n = 6)	

is thought to be released only from amacrine cells with arborizations confined to the inner plexiform layer (Nguyen-Legros et al., 1985). If, therefore, dopamine is to regulate gap junctions of horizontal cells in the normal intact turtle retina, the effect must be paracrine, that is, due to the diffusion of dopamine some distance from its release site in the inner plexiform layer to the outer plexiform layer.

Other roles for dopamine in the fish retina also suggest a paracrine function for this transmitter/modulator. In the outermost retina of fish, dopamine appears to mimic the effect of light with regard to pigment dispersion in the retinal pigmented epithelium and induces the otherwise light adaptive movement of rods and cones (Dearry and Burnside, 1986). These effects upon the pigment epithelium and photoreceptors can only come from dopamine released by the plexus of processes in the outer plexiform layer as there are no direct synapses upon the photoreceptors or pigment epithelium cells. The shedding of photoreceptor outer segment discs as part of the daily photoreceptor renewal process may involve dopamine acting in concert with other transmitters/modulators (Iuvone, 1988).

The presence of two distinct dopaminergic amacrine cell types in the macaque retina suggests distinct functional roles for these two catecholaminergic pathways. The type 1 CA amacrine cells of the primate retina have been well studied with regard to their morphology (Mariani and Hokoç, 1988; Nguyen-LeGros et al., 1984), spatial distribution (Mariani et al., 1984), and synaptic organization (Hokoç and Mariani, 1987). The type 1 cells are the previously well-known dopaminergic amacrine cells. They receive the majority of their synaptic input from a type of cone bipolar proposed to be the giant bistratified bipolar cell (Mariani, 1983), and in turn are presynaptic primarily to other nonimmunoreactive amacrine cells (Hokoç and Mariani, 1987). The synaptic input accounts for about one-fifth of all the synapses involving the type 1 CA cells, while the synaptic output is four-fifths. It seems likely that the proposed giant bistratified bipolar cell input provides a convergent input signal of a photopic condition to the low density type 1 CA cells that in turn broadcast a divergent dignal to other amacrine cells. These other amacrine cells, as speculated earlier (Mariani et al., 1984), are likely to be the narrow field, bistratified rod amacrines that link the rod and cone pathways in the inner plexiform layer by gap junctions (Kolb and Famiglietti, 1983). Like fish retina, where dopamine uncouples gap junctions in the outer plexiform layer (Teranishi et al., 1983), it was suggested that dopamine in the primate retina uncoupled the rod and cone systems in the inner plexiform layer (Mariani et al., 1984). It is here suggested that this uncoupling occurs during photopic conditions in order to improve the resolution of the cone system by not allowing the cone-mediated signals to spill into the more sensitive but poorer resolving rod pathway (via the gap junctions). The function and role of the fine radially oriented fibers that arise from the type 1 CA amacrines and course in the inner nuclear layer are unknown. The synapses, if any, made by these dopaminergic processes have not been studied in the retinas of Old World monkeys, but in New World primates they are known to arborize in the outer plexiform layer and synapse on horizontal cells (Dowling and Ehinger, 1975). The present study has demonstrted by electron microscopy that the type 2 CA amacrine cells receive synaptic input from both amacrine cells and bipolar cells. Curiously, although the synaptic input is rather easily found, the output synapses, or synapses of the type 2 CA amacrine cells, onto other neuronal elements are rarely found. Discerning a role for the type 2 CA cells on the basis of a relative absence of output synapses is difficult, but there are a number of possibilities.

A direct and primarily synaptic function for this cell type suggests a high degree of convergence, summating a signal received from other amacrine and bipolar cells over a relatively wide area and providing a narrow, "pinpoint" output. Another example of this type, with any identifiable function, is not known for amacrine cells in the retina.

An alternate explanation for the presence of the type 2 CA amacrines in the rhesus retina is that they express transiently TH immunoreactivity. Such a transient expression of the catecholaminergic phenotype is well known from other parts of the nervous system (Teitleman et al., 1981), and has been more recently described for certain amacrine cells in the developing postnatal cat retina (Wang et al., 1990) as well, where the type 2 CA amacrines have also been identified, but found to stop expressing immunohistochemically detectable levels of TH by age 3 weeks. Such an explanation for the transient expression of TH by the type 2 CA cells in the rhesus monkey retina is unlikely since this and a previous study (Mariani and Hokoç, 1988) were performed on considerably more mature specimens. Moreover, assays of dopamine, one of its metabolites, and TH indicate that these compounds increase postnatally, up to about age 1 month, to reach levels that do not change significantly therafter for the entire life span of the animal (Iuvone et al., 1989). In addition, a second catecholaminergic phenotype has recently been described in adult rabbit retina (Tauchi et al., 1990).

The most plausible role for the type 2 cells that takes into account both their highly unusual pattern of synaptic organization and some of the proposed functions of dopamine in the vertebrate retina (see above), is to provide neuromodulatory control due to a paracrine action concomitant with the nonsynaptic release of dopamine. This can account for the relative absence of synaptic output for this cell types. The placement of the dendritic arbors of the type 2 CA amacrines suggests their effect may be centered in the inner plexiform layer. Alternatively, this positioning of the dendritic tree could be necessary for the cells to receive the "correct" synaptic input, and the effects of dopamine secreted by the type 2 CA amacrines more widespread throughout the retina.

Acknowledgments. The work reported here was performed while the authors were members of the Laboratory of Neurophysiology, National Institute of Neurological Disorders and Stroke, NIH, Bethesda, MD, originally reported in publications from that laboratory.

REFERENCES

Dearry A, Burnside B (1986): Dopaminergic regulation of cone retinomotor movement in isolated teleost retinas: 1. Induction of cone contraction is mediated by D2 receptors. *J Neurochem* 46:1006–1021

Dowling JE, Boycott BB (1966): Organization of the primate retina: Electron microscopy. *Proc R Soc Lond (Biol)* 166:80–111
Dowling JE, Ehinger B (1975): Synaptic organization of the amine-containing interplexiform cells of the goldfish and cebus monkey retinas. *Science* 188:270–273
Ehinger B (1966): Adrenergic nerves to the eye and to related structures in man and in the cynomolgus monkey (*Macaca irus*). *Invest Ophthalmol* 5:42–52
Ehinger B (1982): Neurotransmitter systems in the retina. *Retina* 2:305–321
Hadjiconstantinou M, Mariani AP, Panula P, Joh TH, Neff NH (1984): Immunohistochemical evidence for epinephrine-containing retinal amacrine cells. *Neuroscience* 13:547–551
Hokoç JN, Mariani AP (1987): Tyrosine hydroxylase immunoreactivity in the rhesus monkey retina reveals synapses from bipolar cells to dopaminergic amacrine cells. *J Neurosci* 7:2785–2793
Iuvone PM (1988): Dopamine: A light-adaptive modulator of melatonin synthesis in the frog retina. In: *Dopaminergic Mechanisms in Vision*, Bodis-Wollner I, Piccolino M, eds. New York: Liss
Iuvone PM, Tigges M, Fernandes A, Tigges J (1989): Dopamine synthesis and metabolism in rhesus monkey retina: Development, aging, and the effects of monocular deprivation. *Visual Neurosci* 2:465–471
Keyser KT, Karten HJ, Katz B, Bohn MC (1987): Catecholaminergic horizontal and amacrine cells in the ferret retina. *J Neurosci* 7:3996–4004
Kolb H, Famiglietti EV Jr (1983): Rod and cone pathways in the inner plexiform layer of cat retina. *Science* 186:47–49
Laties AM, Jacobowitz D (1966): A comparative study of the autonomic innervation of the eye in monkey, cat and rabbit. *Anat Rec* 156:383–396
Mariani AP (1983): Giant bistratified bipolar cells in monkey retina. *Anat Rec* 206:215–220
Mariani AP, Hokoç JN (1988): Two types of tyrosine hydroxylase immunoreactive amacrine cell in the rhesus monkey retina. *J Comp Neurol* 276:81–91
Mariani AP, Kolb H, Nelson R (1984): Dopamine-containing amacrine cells of rhesus monkey retina parallel rods in spatial distribution. *Brain Res* 322:1–7
McMahon DG, Knapp AG, Dowling JE (1989): Horizontal cell gap junctions: Single-channel conductance and modulation by dopamine. *Proc Nat Acad Sci, U S A* 86:7639–7643
Nguyen-Legros J, Botteri C, Phuc LH, Vigny A, Gay M (1984): Morphology of primate's dopaminergic amacrine cells as revealed by TH-like immunoreactivity on retinal flat-mounts. *Brain Res* 295:145–153
Nguyen-Legros J, Versaux-Botteri C, Vigny A, Raux N (1985): Tyrosine hydroxylase immunohistochemistry fails to demonstrate dopaminergic interplexiform cells in the turtle retina. *Brain Res* 339:323–328
Piccolino M, Neyton J, Gerschenfeld HM (1984): Decrease of gap junction permeability induced by cyclic adenosine 3′,5′-monophosphate in horizontal cells of the turtle retina. *Proc Nat Acad Sci U S A* 79:3671–3675
Stell WK, Walker SE, Chohan KS, Ball AK (1984): The goldfish nervus terminalis: A luteinizing hormone-releasing hormone and molluscan cardioexcitatory peptide immunoreactive olfactoretinal pathway. *Proc Nat Acad Sci U S A* 81:940–944
Tauchi M, Madigan NK, Masland RH (1990): Shapes and distributions of the catecholamine-accumulating neurons in the rabbit retina. *J Comp Neurol* 293:178–189
Teitleman G, Gershon MD, Rothman TP, Joh T-H, Reis DJ (1981): Proliferation and distribution of cells that transiently express a catecholaminergic phenotype during development in mice and rats. *Dev Biol* 86:348–355

Teranishi T, Negishi K, Kato S (1983): Dopamine modulates S-potential and dye-coupling between external horizontal cells in carp retina. *Nature* 301:243–246

Wang H-H, Cuenca N, Kolb H (1990): Development of morphological types and distribution patterns of amacrine cells immunoreactive to tyrosine hydroxylase in the cat retina. *Visual Neurosci* 4:159–175

Yazulla S, Zucker CL (1988): Synaptic organization of dopaminergic interplexiform cells in the goldfish retina. *Visual Neurosci* 1:13–29

Zucker CL, Dowling JE (1987): Centrifugal fibers synapse on dopaminergic interplexiform cells in teleost retina. *Nature* 330:166–168

11

Neurotransmitter Drugs that Affect Vertebrate Eye Movements

MICHAEL ARIEL

OVERVIEW

Application of pharmacological agents within the visual system has been used both to treat ocular disease and to investigate normal mechanisms of visual function and neural development. This chapter will describe experiments that use GABAergic and glutamatergic drug application to understand the underlying mechanisms through which the visual system influences oculomotor behaviors. This discussion will describe some results using intracranial drug injection but will mainly focus on drug effects following intravitreal injections.

Because there are several oculomotor behaviors and several parallel pathways by which visual information streams from the retina through the brain, let us generalize somewhat. The visual cortex is involved in perception of color and form, although the cortex also computes the position and direction of visual stimuli. Stimulus position is required by the superior colliculus for the control of saccades, and stimulus direction and speed are used by brainstem structures (the accessory optic system and pretectum) to stabilize retinal images (minimize retinal slip). A constant retinal slip results in ocular nystagmus, with the fast phase eye movements recentering the gaze to allow for more slow phase movement.

Thus, the eye movements of animals trained to detect a target may be modified by drugs that affect retino-geniculo-cortical visual pathways. Saccades to certain positions in visual space may be influenced by chemical agents that alter the output of the superior colliculus. Finally, nystagmus may result from drugs that upset the processing of the retinal slip velocity of the global visual image.

Irrespective of the oculomotor behavior, these experiments use a common rationale: mimic a visual neuron's neurotransmitter release or block its transmitter's effect on postsynaptic cells. The protocols for these experiments are also similar. First, a pharmacological agent is chosen that has a known effect on a specific cell surface synaptic receptor of a specific neuron type that is near the site of drug application. Second, a dose is selected such that the cell's response will be modulated without nonspecific effects on adjacent nerve cells or axons. Third, prior to drug application, normal oculomotor behavior is characterized by measuring eye movements that occur spontaneously, reflexively, or in response to training. Fourth, following these control measurements, the drug is administered and eye movements are again measured to detect drug-related changes. Fifth, oculomotor behavior is monitored to determine the time course of recovery from the drug's effect.

As with the analysis of any behavioral effect of a pharmacological manipulation, many difficulties arise in interpreting these results. First, the effects of the drug must be distinguished from any effects of drug administration (e.g., transient effects of the anesthesia, ocular irritation caused by puncture of the globe by a hypodermic needle, changes in the intraocular pressure, changes in the ocular optics). Second, the dose should be minimized to be selective for the intended synaptic receptors without affecting neurons elsewhere via drug diffusion away from the injection site. Third, for intraocular drug injections, it should be confirmed that the change in retinal output stems from changes at retinal synapses and not from changes in accommodation, pupilloconstriction, or the motility of the extraocular muscles. Fourth, the interpretation should consider all possible pathways through which a change in synaptic transmission could be relayed by the parallel pathways in the visual system in order to affect the oculomotor response.

Such an analysis is often confounded by the closed-loop nature of visually driven oculomotor responses. Changes in visual processing that lead to changes in oculomotor responses will ultimately cause changes in the visual stimulus position as the retinal image is shifted by an eye movement. There is also the possibility of direct oculomotor feedback onto ganglion cells via centrifugal inputs to the retina (Marchiafava, 1976; Martin et al., 1990).

Of all these considerations, perhaps the greatest concern is the specificity of the pharmacological agents. It is for this reason that the analysis of the effects of intravitreal drug injections on eye movements is often complemented by electrophysiology of appropriate neural structures. These electrophysiological and behavioral results, as well as related anatomical and pharmacological evidence, can then be used to postulate the underlying mechanisms of visual system input to oculomotor control.

INTRACRANIAL DRUG ADMINISTRATION

Drug applications to the vertebrate visual system have been either intraocular or intracranial. Intracranial drug applications are often more difficult to interpret because central structures are more complex, less accessible, and located adjacent to other visually related nuclei or fiber tracts. Thus, in addition to the other concerns of drug application, anatomical localization of the site of drug injection is necessary.

Methodology

Physiological concentrations of synaptic drugs are injected into small regions of the brain that had been localized by extracellular recordings. Bolus injections of

0.5–2 μl are slowly made through either hypodermic needles or micropipettes over several minutes. The actual effective drug concentration is not known, only the effective range of the injected solutions (0.1–5 μg/μl). Ideally, the drug effects on oculomotor behavior would begin soon after its injection and would only last for sufficient time to record the effect.

Eye Movements

Spontaneous eye movements

Perhaps the simplest oculomotor behavior to measure is the spontaneous eye movements recorded in awake animals in the dark or in an illuminated laboratory setting. Following intracranial injections of GABAa drugs, spontaneous nystagmus has been observed, consisting of a slow phase eye drift in one direction which is interrupted by a fast resetting jump in the opposite direction.

Receptor antagonists and agonists have caused horizontal nystagmus in opposite directions, such that the direction of the slow phase movement was toward or away from the injected side depending on the drug and site of injection. Bicuculline, a GABAa antagonist, injected into the superior colliculus, evoked a slow phase direction toward the injected side, whereas muscimol, a GABAa agonist, elicited nystagmus slow phases directed away from the injection (Hikosaka and Wurtz, 1985a). In comparison, drug injections into the nearby pretectal nucleus of the optic tract (NOT) had similar effects but occurred more rapidly and transiently (Reisine and Cohen, 1990), suggesting that the NOT may be the primary site of action. Similar spontaneous nystagmus was also observed following muscimol injections into the substantia nigra pars reticulata of the cat and monkey (Boussaoud and Joseph, 1985; Hikosaka and Wurtz, 1985b).

Optokinetic nystagmus (OKN)

Nystagmus during full field (optokinetic) stimuli is normally evoked in all animals exposed to constant velocity stimuli. Boussaoud and Joseph (1985) reported no substantial effect on the cat OKN following muscimol injection of the substantia nigra. Similar drug injections into the NOT produced prominent changes in eye movements during optokinetic stimuli, although most of the changes could be attributed to the dramatic changes in spontaneous eye movements, not the control of optokinetic responses themselves. In monkeys, muscimol injections also affected the velocity storage mechanism which contributes to the slow component of OKN moving toward the contralateral side (Reisine and Cohen, 1990).

Unlike mammals, unilateral pretectal microinjections of a GABA antagonist in frogs had no effect on spontaneous eye movements, yet had substantive effects on OKN (Yücel et al., 1991). The GABAa antagonist SR 95531 increased the nasal-to-temporal response of the contralateral eye without an increase of the temporal-to-nasal response. Yücel and collaborators argued that GABA release in the pretectum serves to create the nasal-to-temporal asymmetry found in lower vertebrates.

VESTIBULAR-OCULAR-REFLEX

Intracranial drug injections have also been used to analyze oculomotor pathways that influence the vestibular-ocular reflex (VOR) (Obata and Highstein, 1970). Whereas retinal image stabilization is influenced by visual and vestibular peripheral inputs, those inputs may share a common velocity storage integrator that in turn feeds the slow phase velocity drive during ocular nystagmus. Consequently, unilateral application of bicuculline and muscimol into the NOT may result in a lateral imbalance of inputs that charges this integrator and thus modifies the VOR (Reisine and Cohen, 1990). Boussaoud and Joseph (1985) also report an asymmetry following muscimol injection of the cat's substantial nigra pars reticulata and a decrease in the VOR gain during rotation toward the ipsilateral side.

One can also modify the optokinetic drive to the oculomotor system by intravitreal injections of GABA drugs (see below). During such drug effects, vestibular stimulation still results in reflexive oculomotor responses, yet the VOR is superimposed on the spontaneous nystagmus elicited by the intravitreal injections (Ariel et al., 1988).

Intramuscular injection of the GABAa agonist baclophen had a selective effect on the monkey's velocity storage system used in the reflexive control of the retinal image by both vestibular and visual inputs (Cohen et al., 1987). Both the VOR time constant and the OKN steady state gain were reduced in a dose-dependent manner. Foveation, nystagmus fast phase movements, and saccades were unaffected by baclophen.

FOVEATION AND SACCADES

In addition to nystagmus, GABAa drugs also affected saccadic eye movements in the monkey following microinjections into the superior colliculus and substantia nigra (Hikosaka and Wurtz, 1985a, 1985b). Following focal injections of muscimol in the superior colliculus, saccades made toward targets flashed in the visual field region that corresponded to the injection site had slower velocities, longer latencies, and smaller amplitudes (Hikosaka and Wurtz, 1985a). Bicuculline injections, on the other hand, resulted in spontaneous saccades toward the visual field region represented by cells at the injection site, even in the absence of a visual target. These drug effects do not appear to be a direct consequence of altered visual sensation because similar saccadic abnormalities were observed when a monkey was trained to make saccades to remembered visual targets.

INTRAOCULAR DRUG ADMINISTRATION

The remainder of this chapter will discuss the effects of intraocular drug application on eye movements. The popularity of this approach is due in part to the huge amount of information known about the retina, the physiological properties of its neurons, their intraretinal anatomical connections, and the neurotransmitters used at their chemical synapses. Yet perhaps the most important reason why eye movements are measured instead of single retinal neurons is the inability to record from retinal cells during an eye movement. Even if one could record from a moving eye, an understanding of intraretinal processing would still be difficult to interpret because most retinal neurons communicate at complex synaptic specializations far from their soma and do not discharge action potentials.

Since retinal cell responses have never been recorded during eye movements, one must rely on our limited knowledge of retina processing and the current pharmacopedia to aid in the design of these experiments. The advantage of such experiments is that an intravitreal drug injection limits a drug effect to the retina of the injected eye. The highest drug concentration will be in that eye, and will be substantially diluted as the drug is absorbed into the bloodstream or cerebral spinal fluid throughout the cranium. Effects of neuronal activity in sensory or oculomotor centers within the brain should be secondary to changes that occur in the retina.

Methodology

In most cases, drugs are injected through a hypodermic needle into the vitreous humour. This may be either preceded by the application of drops of a local anesthetic to the cornea and conjunctiva or the administration of a short-acting general anesthetic such as halothane. The drug injection is made through the superior or temporal side of the globe sufficiently behind the limbus to avoid the lens. Often, the needle tip is visualized through the pupil, enabling its placement close to the retina prior to injection. Despite efforts to maximize the drug concentration near the retina, the vitreous body acts as a huge drug reservoir so that the drug diffuses throughout the retina and has an effect that lasts many hours.

Calibration

It is difficult to estimate the retinal drug concentration following an intravitreal injection. One indirect approach is to compute the drug dilution based on an estimate of the vitreal volume derived from ultrasound or calculations based on the schematic eyes of the animals (which range from 142 μM in goldfish [DeMarco and Powers, 1989] to 3 mM in monkey [Knapp and Schiller, 1984]). A more direct calibration procedure is to inject an inert radioactive substance (of size similar to the drug) into animals of similar ocular dimensions and later measure the radioactivity in the vitreous humour and retina to compute a dilution factor (see Ariel, 1989). Ideally, this radioactive measurement should be made at a time point equivalent to the period when the drug has behavioral effects.

Eye Movement Behaviors

Many classes of retinal ganglion cell exist, serving separate and overlapping roles in perception, foveation, and retinal image stabilization. Following a description of intravitreal drug effects on oculomotor responses, the discussion will focus on the role of rod photoreceptors in detecting light onset and the role of direction-sensitive ganglion cells in encoding global visual field motion.

SPONTANEOUS EYE MOVEMENTS

Some animals show little spontaneous eye movement (i.e., turtle), while other animals (i.e., monkey) often make saccades but to slow tracking eye movements in the absence of moving visual stimuli. However, after monocular intravitreal injections of the GABA antagonists picrotoxin and bicuculline, large spontaneous slow eye movements occur in the temporal-to-nasal direction with respect to the injected eye (see Figure 11.1). Because intravitreal deug effects are long-lasting, these slow temporal-to-nasal drifts in eye position lead to nasal-to-temporal fast phase eye movements to produce horizontal ocular nystagmus.

How does a drug-induced nystagmus correspond to that evoked by intracranial drug administration or electrical stimulation? By blocking retinal GABA inhibition, the effect of bicuculline or picrotoxin may be to excite the contralateral NOT since the NOT directly receives a predominantly crossed retinal projection. Intracranial excitation due to bicuculline injection (Reisine and Cohen, 1990) or electrical stimuli (Mustari and Fuchs, 1990; Schiff et al., 1988) cause horizontal slow phases of nystagmus directed toward the side of the stimulated NOT. Consequently, the temporal-to-nasal slow phases following intraocular injection of GABA antagonists produce the same effect as a contralateral NOT stimulation, suggesting that the two phenomena may be interrelated.

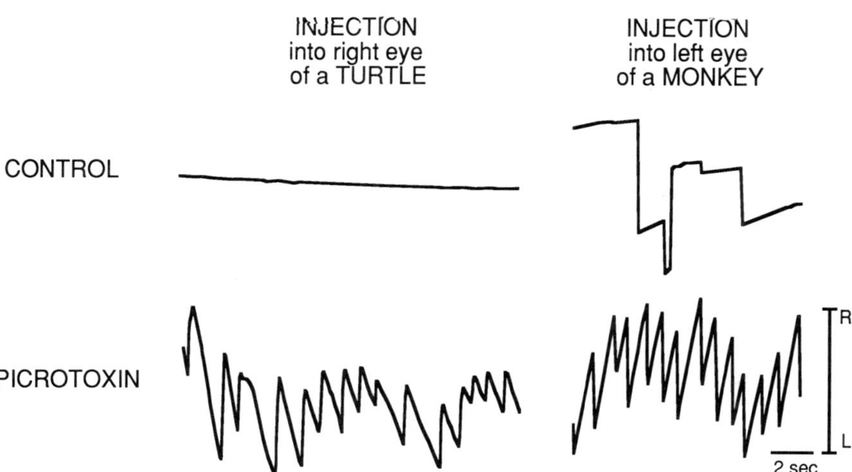

FIGURE 11.1. Effect of intravitreal injection of picrotoxin (3.3 mM) on eye position in the dark for the turtle and for the monkey (Ariel and Tusa, 1992). Note that following the drug application, the slow phase of the nystagmus was temporal-to-nasal with respect to injected eye. The vertical calibration bar represents ± 5° for the turtle and ± 20° for the monkey. R = right; L = left.

This temporal-to-nasal nystagmus after intravitreal picrotoxin injection was observed in turtle (Ariel, 1989), chicken (Bonaventure et al., 1991), rabbit (Ariel et al., 1988), cat (Ariel et al., 1988), and monkey (Ariel and Tusa, 1992), but not in frogs. Since the first five vertebrates show this effect in total darkness, it appears that GABA antagonists increase the spontaneous spike activity of cells that relay information of temporal-to-nasal "retinal slip" to central brain structures. The lack of nystagmus in frog may result from a lack of an increase in the spike rate of frog retinal ganglion cells.

Optokinetic Nystagmus

GABA drugs: The intraocular drug-induced nystagmus appears to be related to an animal's performance of optokinetic nystagmus. First, to various extents, all vertebrates respond better to monocular optokinetic stimuli when moving in the temporal-to-nasal direction. This is the same nystagmus direction evoked by monocular injections of GABA blockers. In higher vertebrates, the nasal/temporal asymmetry is only observed during development or with elaborate stimuli, perhaps due to their frontally placed eyes or their dominant cerebral cortex (Grasse et al., 1984).

However, there are prominent species differences in the response to monocular intravitreal injections of GABA antagonists, which can also be related to differences in optokinetic responses across species. Lower vertebrates (turtle, chicken, and rabbit) respond best to slow speeds (less than 10 °/sec) moving in the temporal-to-nasal direction (see Figure 11.2). In these species, the drug-induced slow phase movements were of slow velocity.

For cats and monkeys, on the other hand, picrotoxin evoked strong nystagmus (slow phase eye velocities from 15 to 50 °/sec). However, in these higher animals, the drug-induced nystagmus was reduced or eliminated when the animal was in a visual environment that was not moving in the temporal-to-nasal direction (Ariel et al., 1988; Ariel and Tusa, 1992). In fact, these higher species still responded to optokinetic stimuli with nystagmus of the appropriate direction and often normal gain (see Figure 11.2).

Whereas GABA blockers elicited a spontaneous nystagmus that overcame the optokinetic drive to the animal's eyes in lower vertebrates, the GABAa agonist muscimol blocked OKN, leaving eyes motionless even in the presence of a rotating visual world (Ariel, 1989). These GABA effects appear limited to drugs that interact with GABAa receptors since GABAb drugs (e.g., baclophen, phaclophen) did not have obvious effects.

APB (2-amino-4-phosphonobutyrate): APB was found to block the retinal ON pathway (Slaughter and Miller, 1981). The ability of APB to block optokinetic nystagmus was first reported in 1984 (Knapp and Ariel, 1984). Since that report in rabbit, other species have been tested (goldfish [DeMarco et al., 1989], frog [Yücel et al., 1989a]; turtle [Ariel and Rosenberg, 1991]; chicken [Bonaventure et al., 1992]; rabbit [Knapp et al., 1988], and cat [Hoffmann, 1986]). Demarco et al.

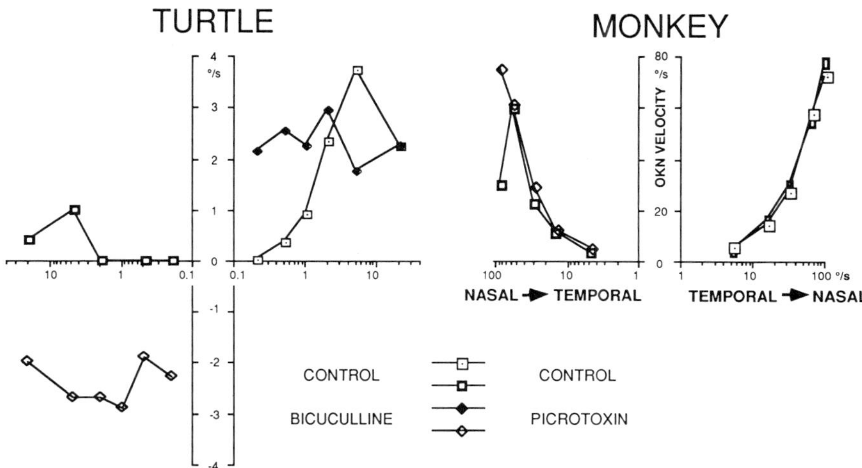

FIGURE 11.2. Graphs of OKN velocity comparing the effects of GABA antagonists on turtle (bicuculline) and monkey (picrotoxin). Data were reformatted from Ariel (1991) and Ariel and Tusa (1992). Note that the control turtle responses show the characteristic temporal/nasal asymmetry and sensitivity to slow speeds. Following the drug injection, monkey OKN was roughly the same as the control, but turtle response to optokinetic stimuli was dominated by the stimulus-independent temporal-to-nasal nystagmus. Note that the negative OKN velocity following bicuculline is a temporal-to-nasal slow phase eye movement during nasal-to-temporal stimulation.

(1989) showed that APB was less effective on goldfish responses to optokinetic stimuli of low spatial frequencies. Otherwise, the APB effect was quite pronounced (see Figure 11.3).

Hoffmann (1986) only found a small effect of APB in cat, and only at high stimulus velocities (see Figure 11.4). If intravitreal APB is decreasing ON-center retinal input to the NOT that encodes slow speeds, Hoffmann suggests that the effect of APB should be greatest at slow speeds, which is not the case (Hoffmann, 1986). However, unlike lower vertebrates, the cat has substantial directional processing in its visual cortex that may be able to compensate for the absence of the direct ON-center retinal input to the NOT. This explanation is consistent with the result that intraocular APB injections were more effective following a bilateral ablation of the visual cortex (Knapp et al., 1988).

Another glutamate analog, cis-2,3-piperidine decarboxilic acid (PDA) has been tested for its effect on OKN (Yücel et al., 1989b). PDA has been purported to block the retinal OFF-system selectively at low doses in the mudpuppy (Slaughter and Miller, 1983). Low doses in frog did not block the slow phase eye movement immediately after the onset of stimulus motion but did eliminate the generation of fast phases of nystagmus (Yücel et al., 1989b). Thus, the frog's ability to stabilize retinal images was eliminated within a few seconds of optokinetic stimulus onset.

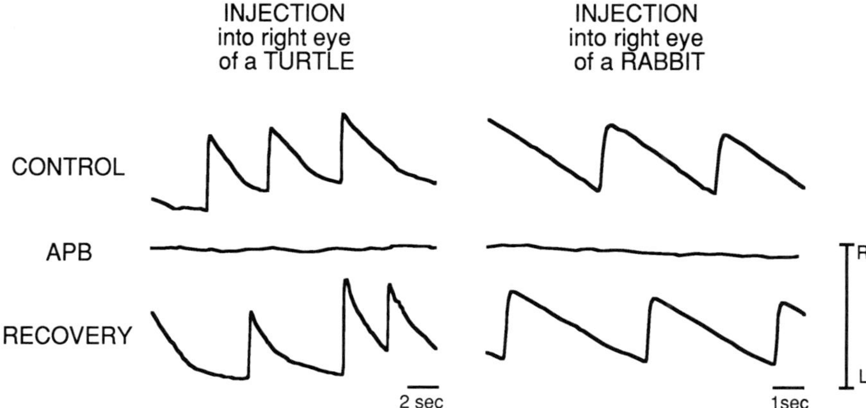

FIGURE 11.3. Effect of intravitreal injection of APB (2.75 mM) on the optokinetic responses to temporal-to-nasal (leftward) stimulation of the injected right eye of the turtle and of the rabbit (Knapp et al., 1988). The vertical calibration is ± 5° for the turtle and ± 20° for the rabbit and the stimulus is moving approximately 3°/sec.

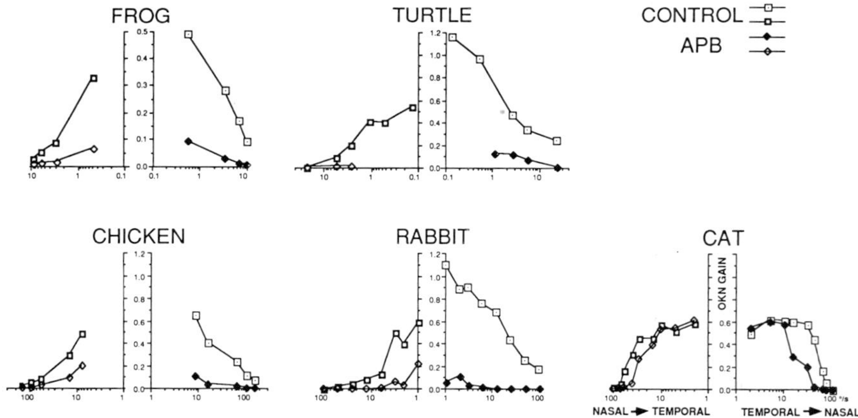

FIGURE 11.4. Graphs of OKN gain for five different species. The top row shows cold-blooded vertebrates and the bottom row shows warm-blooded vertebrates. The data were reformatted from published data as follows: frog (Yücel et al., 1989b), chicken (Bonaventure et al., 1992), turtle (Ariel, 1991), rabbit (Knapp et al., 1988) and cat (Hoffmann, 1986). Note the different vertical scale used in the Frog panel.

Foveation and saccades

The effects of intravitreal injections on foveation and spontaneous saccades have also been examined. Intraocular injections of picrotoxin did not affect a monkey's ability to follow a small target in smooth pursuit, nor was there an impairment in gaze-holding in a illuminated room or stationary optokinetic drum (Ariel and Tusa, 1992).

On the other hand, intravitreal APB appears to affect ocular stability in rabbits (Knapp et al., 1988) but not in monkeys (Dolan and Schiller, 1989). APB also does not appear to reduce saccadic eye velocity or saccade duration in monkeys. Therefore, monkeys can be trained to make saccades to specific targets as part of a psychophysical experiment using intravitreal drug injections. Using this paradigm, Schiller and collaborators showed that following APB injections, a monkey was less able to detect incremental changes in stimulus contrast whereas decremental contrast changes were unaffected in the light-adapted eye (Schiller et al., 1986). All detection was greatly reduced in the dark-adapted eye that was injected with APB (Dolan and Schiller, 1989).

DISCUSSION

A few studies have extended the above findings by recording from central neurons following intravitreal drug injections (frog optic tectum: Jardon et al., 1989; turtle basal optic nucleus: Ariel and Rosenberg, 1991; Schuerger et al., 1990); lateral geniculate nucleus of the rabbit: Knapp and Mistler, 1983; of the cat: Horton and Sherk, 1984; and of the monkey: Schiller, 1984). The correlation of results using electrophysiology and oculomotor behavior lead to a better understanding of the visual processing than either approach alone.

The electrophysiological effects of intravitreal APB were initially described for mudpuppy retinal bipolar cells (Slaughter and Miller, 1981). At that first stage, APB appears selective for the ON-pathway. However, for its effects on oculomotor control to be correctly interpreted, it is important to determine where the ON and OFF pathways project within the brain and whether they remain separate. In the frog tectum, the ganglion cell axon terminals often respond with ON-OFF transients. Intravitreal APB reduced ON responses, enhanced the OFF responses and increased the receptive field center size (Jardon et al., 1989). In the lateral geniculate nucleus, where ON-center and OFF-center geniculate cells were recorded, APB only affected the ON-cells (Horton and Sherk, 1984; Knapp and Mistler, 1983; Schiller, 1984).

Foveation and the Rod Intraretinal Pathway

The result that the dark-adapted monkey is unable to detect dim lights following APB injection suggests that the light responses of rod photoreceptors are transmitted into the inner retina by an APB-sensitive synapse. Electrophysiological studies of retinal cells had shown that the photoreceptor/depolarizing-bipolar-cell is very sensitive to APB (Slaughter and Miller, 1981). Thus, Dolan and Schiller (1989) suggested that scotopic (rod) responses would be found in ON-bipolar cells but not OFF-bipolar cells, a conclusion supported by the intracellular retinal recordings (Dacheux and Raviola, 1986). The behavioral data is also consistent with the finding that dark-adapted ganglion cells are unresponsive following APB (Müller et al., 1988). Thus, results from pharmacological analysis of oculomotor responses demonstrate the utility of this approach in the analysis of retinal processing and the visual pathways leading to perception.

Nystagmus and the Processing of Retinal Slip

The appearance of spontaneous nystagmus following drug application is likely related to the processing of retinal slip. Retinal slip, the average vector of global visual field motion, is encoded in the spike discharges of direction-sensitive cells. Such cells are found both in the accessory optic system and the pretectum and relay their information to the oculomotor system either directly or via the cerebellum or vestibular nuclei. Obviously, then, a pharmacological manipulation could exert its effect at any of these sites or in the visual structures that compute direction-sensitivity of limited parts of the receptive field (the retina or the visual cortex).

The behavioral studies using intact animals cannot readily determine which pathways are being changed by drug application. The nystagmus that results from muscimol injection into the NOT demonstrates that GABAa receptors are present on processes in the NOT, but it is not known whether GABA synapses are essential for null-direction inhibition of visual signals, mutual inhibition by the NOT of each side of the brain, or an inhibitory interaction between the retinal and cortical afferents to the NOT. It is not even known whether there is normally light-modulated GABA release onto these functional receptors.

One approach to reduce the complexity of this system is to reduce its components. In the absence of visual cortex, it has been shown that spontaneous nystagmus still occurred following intravitreal bicuculline injections (Ariel et al., 1988; Schuerger et al., 1990). Thus, a cortical site of directional processing is not necessary for the drug's effect.

A second approach has been to study the effect of synaptic drugs on spike responses recorded in the accessory optic system of the turtle *in vitro* (the basal optic nucleus: [Ariel and Rosenberg, 1991; Schuerger et al., 1990]). In those experiments, the retinal drug concentration was determined either by estimating the dilution or by superfusing the drug (see Figure 11.5). These cells are normally direction-sensitive and give vigorous response to moving stimuli. They respond weakly to diffuse light flashes, predominantly at stimulus offset. During retinal superfusion of 200 μM APB, the light responses were abolished but the off responses remain. This result suggests that the basal optic nucleus receives both ON- and OFF-center retinal input.

Intravitreal injection of GABA antagonists have quite different effects on the cells of the turtle accessory optic system. These drugs, which block direction sensitivity of direction-sensitive (DS) retinal ganglion cells (Ariel and Adolph, 1985), also block the direction sensitivity of brainstem neurons (see Figure 11.5). The simplest explanation of this result is that the direction tuning of turtle brainstem neurons is due to GABA inhibition necessary for direction tuning of retinal DS ganglion cells (Schuerger et al., 1990).

There is also now direct electrophysiological evidence that DS retinal ganglion cells project directly to the turtle accessory optic system (Rosenberg and Ariel, 1991). DS retinal ganglion cells were antidromically driven by microstimulation

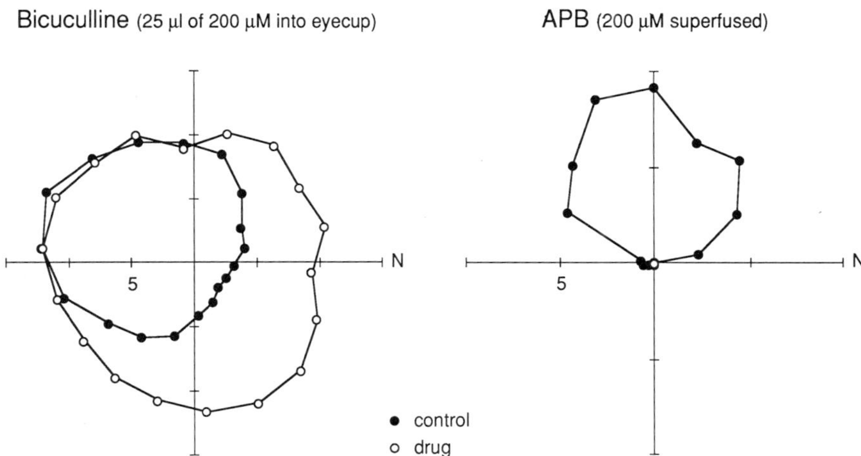

Figure 11.5. Direction-tuning plots of spike responses from two cells recorded in the turtle basal optic nucleus *in vitro*. The polar plots are derived from 3–5 stimulus presentations of nine linear sweeps back and forth. Drugs were applied to the contralateral eyecup. In the left panel, the cell responded best to superior-temporal movement prior to drug application. 25 µl of 200 µM bicuculline was added to the eyecup which contained approximately 100 µl of vitreous. Then the cell responded equally to all directions of motion. In the right panel, the cell responded best to superior motion. The effect of eyecup superfusion of 200 µM APB was to block all visually evoked spike responses.

of the contralateral basal optic nucleus. Moreover, cells of the basal optic nucleus were direction-sensitive, even in the absence of the dorsal midbrain (including the tectum and pretectum) and the entire telencephalon.

Grasse et al. (1990) differentiated the retinal DS processing from the cortical DS processing of the cat by comparing the contralateral and ipsilateral responses of the accessory optic system. Intravitrreal bicuculline injections blocked the contralateral DS responses, but not the ipsilateral DS responses. They argued that intravitreal bicuculline injections do not prevent cortical DS processing. These electrophysiological results in the intact cat and the *in vitro* turtle brain both support the conclusion that intravitreal injection of GABA antagonists may perturb the directional processing within the retina that leads to an anomalous retinal slip signal reaching the brainstem. The presence of cortical processing of retinal slip may then suppress the anomalous signal in higher vertebrates, leading to normal oculomotor behaviors in the light. These results also provide another example of the correlation of oculomotor responses following drug application with the electrophysiological responses of neurons that underly the eye movement behavior. Thus this experimental approach may lead to a better understanding of the parcellation of pathways within the visual and oculomotor systems.

Acknowledgments. My thanks to Alexander F. Rosenberg for providing the electrophysiological data for the turtle basal optic nucleus, and for a critical reading of this manuscript. This work was supported by EY05978 and K02 MH00815.

REFERENCES

Ariel M (1989): Analysis of vertebrate eye movements following intravitreal drug injections: 3. Spontaneous nystagmus is modulated by the GABAa receptor. *J Neurophysiol* 62:469–479

Ariel M (1991): Analysis of vertebrate eye movements following intravitreal drug injections: 4. Drug-induced eye movements are unyoked in the turtle. *J Neurophysiol* 65:1003–1009

Ariel M, Adolph AR (1985): Neurotransmitter inputs to directionally sensitive turtle retinal ganglion cells. *J Neurophysiol* 54:1123–1143

Ariel M, Robinson FR, Knapp AG (1988): Analysis of vertebrate eye movements following intravitreal drug injections: 2. Spontaneous nystagmus induced by picrotoxin is mediated subcortically. *J Neurophysiol* 60:1022–1035

Ariel M, Rosenberg AF (1991): Effects of synaptic drugs on turtle optokinetic nystagmus and the spike responses of basal optic nucleus. *Visual Neurosci* 7:431–440

Ariel M, Tusa RJ (1992): Spontaneous nystagmus and gaze-holding ability in monkeys following intravitreal picrotoxin injections. *J Neurophysiol* 67:1124–1132

Bonaventure N, Kim MS, Jardon B, Yücel H (1992): Pharmacological study of the chicken's monocular optokinetic nystagmus: Involvement of the ON retinal channel evidenced by the glutamatergic separation of ON and OFF pathways. *Vision Res* 32:601–610

Bonaventure N, Wioland N, Bigenwald J (1983): Involvement of GABAergic mechanisms in the optokinetic nystagmus of the frog. *Exp Brain Res* 50:433–441

Boussaoud D, Joseph JP (1985): Role of the cat substantia nigra pars reticulata in eye and head movements: 2. Effects of local pharmacological injections. *Exp Brain Res* 57:297–304

Cohen B, Helwig D, Raphan T (1987): Baclofen and velocity storage: A model of the effects of the drug on the vestibulo-ocular reflex in the rhesus monkey. *J Physiol* 393:703–725

Dacheux RF, Raviola E (1986): The rod pathway in the rabbit retina: A depolarizing bipolar and amacrine cell. *J Neurosci* 6:331–345

DeMarco PJ, Nussdorf JD, Brockman DA, Powers MK (1989): APB selectively reduces visual responses in goldfish to high spatiotemporal frequencies. *Visual Neurosci* 2:15–18

DeMarco PJ, Powers MK (1989): Sensitivity of ERG components from dark-adapted goldfish retinas treated with APB. *Brain Res* 482:317–323

Dolan RP, Schiller PH (1989): Evidence for only depolarizing rod bipolar cells in the primate retina. *Visual Neurosci* 2:421–424

Grasse KL, Ariel M, Smith I (1990): Direction-selective responses of units in the dorsal terminal nucleus of cat following intravitreal injections of bicuculline. *Visual Neurosci* 4:605–617

Grasse KL, Cynader M, Douglas R (1984): Alterations in response properties in the lateral and dorsal terminal nuclei of the cat accessory optic system following visual cortex lesions. *Exp Brain Res* 55:69–80

Hikosaka O, Wurtz RH (1985a): Modification of saccadic eye movements by GABA-related substances: 1. Effect of muscimol and bicuculline in monkey superior colliculus. *J Neurophysiol* 53:266–289

Hikosaka O, Wurtz RH (1985b): Modification of saccadic eye movements by GABA-related substances: 2.Effects of muscimol in monkey substantia nigra pars reticulata. *J Neurophysiol* 53:292–307

Hoffmann KP (1986): Visual inputs relevant for the optokinetic nystagmus in mammals. In: *Progress in Brain Research*, Freund HJ, Buttner U, Cohen B, Noth J, eds. Amsterdam: Elsevier

Horton JC, Sherk H (1984): Receptive field properties in the cat's lateral geniculate nucleus in the absence of ON-center retinal input. *J Neurosci* 4:374–380

Jardon B, Yücel YH, Bonaventure N (1989): Glutamatergic separation of ON and OFF retinal channels: Possible modulation by glycine and acetylcholine. *Euro J Pharmacol* 162:215–224

Knapp AG, Ariel M (1984): Selective blockade of retinal ON channel eliminates horizontal optokinetic nystagmus in rabbits. *Invest Ophthalmol Vis Sci* (Suppl) 25:229

Knapp AG, Ariel M, Robinson FR (1988): Analysis of vertebrate eye movements following intravitreal drug injections: 1. Blockade of retina ON-cells by 2-amino-4-phosphonobutyrate eliminates optokinetic nystagmus. *J Neurophysiol* 60:1010–1021

Knapp AG, Mistler LA (1983): Response properties of cells in rabbit's lateral geniculate nucleus during reversible blockade of retinal ON-center channel. *J Neurophysiol* 50:1236–1245

Knapp AG, Schiller PH (1984): The contribution of ON-bipolar cells to the electroretinogram of rabbits and monkeys: A study using 1-amino-4-phosphonobutyrate (APB). *Vision Res* 24:1841–1846

Marchiafava PL (1976): Centrifugal actions on amacrine and ganglion cells in the retina of the turtle. *J Physiol* 255:137–155

Martin G, Letelier JC, Wallman J (1990): Saccade-related responses of centrifugal neurons projecting to the chicken retina. *Exp Brain Res* 82:263–270

Müller F, Wassle H, Voigt T (1988): Pharmacological modulation of the rod pathway in the cat retina. *J Neurophysiol* 59:1657–1672

Mustari MJ, Fuchs AF (1990): Discharge patterns of neurons in the pretectal nucleus of the optic tract (NOT) in the behaving primate. *J Neurophysiol* 64:77–90

Obata K, Highstein SM (1970): Blocking by picrotoxin of both vestibular inhibition and GABA action on rabbit oculomotor neurones. *Brain Res* 18:538–541

Reisine H, Cohen B (1990): Nucleus of the optic tract (NOT): Effect of muscimol on generation of optokinetic nystagmus and suppression and habituation of vestibular nystagmus. *Soc Neurosci Abstr* 16(2) : 968

Rosenberg AF, Ariel M (1991): Electrophysiological evidence for a direct projection of direction-sensitive retinal ganglion cells to the turtle's accessory optic system. *J Neurophysiol* 65:1002–1033

Schiff D, Cohen B, Raphan T (1988): Nystagmus induced by stimulation of the nucleus of the optic tract in the monkey. *Exp Brain Res* 70:1–14

Schiller PH (1984): The connections of retinal ON and OFF channels to the lateral geniculate nucleus of the monkey. *Vision Res* 24:923–932

Schiller P, Sandell J, Maunsell J (1986): Functions of the ON and OFF channels of the visual system. *Nature* 322:824–825

Schuerger RJ, Rosenberg AF, Ariel M (1990): Retinal direction-sensitive input to the accessory optic system: An in vitro approach with behavioral relevance. *Brain Res* 522:161–164

Slaughter MM, Miller RF (1981): 2-amino-4-phosphonobutyric acid: A new pharmacological tool for retina research. *Science* 211:182–185

Slaughter MM, Miller RF (1983): An excitatory amino acid antagonist blocks cone input to sign-conserving second-order retinal neurons. *Science* 219:1230–1232

Yücel YH, Jardon B, Bonaventure N (1989a): Involvement of ON and OFF retinal channels in the eye and head horizontal optokinetic nystagmus of the frog. *Visual Neurosci* 2:357–365

Yücel YH, Jardon B, Bonaventure N (1989b): Is a retinal input involved in the generation of eye resetting fast phases in the frog eye optokinetic nystagmus? *Neurosci Lett* 97:80–84

Yücel YH, Jardon B, Bonaventure N(1991): Unilateral pretectal microinjections of SR 95 531, a GABAa antagonist: Effects on directional asymmetry of frog monocular OKN. *Exp Brain Res* 83:527–532

12

Evidence for Commissural Interactions in the Nuclei of the Optic Tract of the Opossum

ELIANE VOLCHAN, ANTONIO PEREIRA, JR., RAYMUNDO FRANCISCO BERNARDES, AND CARLOS EDUARDO ROCHA-MIRANDA

The nucleus of the optic tract (NOT) is composed of sparse cells dispersed in the brachium of the superior colliculus at the dorsal surface of the pretectum. Several authors have adduced evidence concerning its involvement in the horizontal optokinetic reflex. The evidence has been derived from stimulation (Collewijn, 1975a; Schiff et al., 1988) and lesion studies (Cazin et al., 1980a; Collewijn, 1975a; Kato et al., 1986; Schiff et al., 1990). Electrophysiological recordings in the NOT of mammals also corroborate this idea (rabbit: Collewijn, 1975b; rat: Cazin et al., 1980b; cat: Hoffmann and Schoppmann, 1981; opossum: Volchan et al., 1989; monkey: Hoffmann et al., 1988; ferret: Klauer et al., 1990). Recordings in the NOT of these species have yielded a high percentage of directionally selective units. The common feature of these units is their preference for stimulus movement toward the recording side (ipsiversive) as opposed to the opposite side (contraversive), the nonpreferred direction. Besides being activated through the contralateral eye, the units of the NOT can also be driven by the ipsilateral eye in all the above-cited mammals, except for the rabbit (Collewijn, 1975b).

Volchan and collaborators (1989, 1990, 1991) have been investigating the nucleus of the optic tract in a marsupial, the South American opossum (*Didelphis marsupialis aurita*). The visual system of this animal has been extensively studied. Lent and collaborators (1976) in a study of the retinofugal projections to mesodiencephalic structures in the opossum, have described four nuclei in the pretectum of this species: anterior, posterior, olivary, and nucleus of the optic tract. These authors observed a strong crossed projection to the NOT, but no uncrossed projection to this nucleus. Later, Linden and Rocha-Miranda (1981) described a subdivision of the NOT which they called *subbrachialis*, with binocular retinal projections and striate cortical projections. The nucleus of the optic tract as defined by Lent and collaborators (1976) was called *intrabrachialis* and had no afferents from the striate cortex. More recently, this nucleus has been reanalyzed by observing the retrogradely labeled cells after inferior olive and intraocular peroxidase injections (Bernardes et al., 1987, 1991). These authors showed that the distribution of these cells conformed with the NOT described by Lent and co-workers (1976). In addition, directionally selective cells were recorded from the superficial pretectum and these matched the region where inferior olive projecting neurons were found (Volchan et al., 1989). The dorsal terminal nucleus (DTN) was described in the opossum by Lent and collaborators (1976) as a small nucleus located at the caudolateral aspect of the superior colliculus. Its limits with the NOT are hard to distinguish. This region has not been explored by the electrophysiological studies that shall be presented here.

The optokinetic reflex (OKR) was studied in the opossum both by electro-oculographic (Nasi et al., 1989) and magnetic search coil techniques (Nasi et al., 1990). These authors found that the gain of the reflex under binocular stimulation approaches 1 at 3°/s, falls slightly at 6°/s, and drops to about 50% at 18°/s and less hereafter. Under monocular stimulation, the gain for temporal-to-nasal stimulation is similar to the binocular condition, with somewhat lower values. The nasal-to-temporal stimulation shows a gain that is 70% that for the opposite direction at 3°/s.

The study of the neural circuits involved in the OKR in this species presents some interesting features for comparative analyses. Studies in the rabbit and rat are usually contrasted with those in the cat and monkey: rats (Hess et al., 1985) and rabbits (Collewijn and Holstege, 1984) have an asymmetric monocular nystagmus, while cats (Harris et al., 1980) and monkeys (Precht, 1981) have an almost symmetric one. This has been variously attributed to the degree of frontalization of the eyes, the presence of an important contingent of uncrossed retinofugal projection, the development of the area centralis, and the strength of corticofugal projection to the NOT. The opossum, however, has frontal eyes and as wide a binocular field as cats and monkeys (Oswaldo-Cruz et al., 1979). The density of ganglion cells with uncrossed projections in its temporal retina exceeds that with crossed projections (Cavalcante et al., 1992; Hokoç et al., 1992) and, although smaller than in the cat (Stone, 1966), this uncrossed contingent is much more expressive than in the rat (Cowey and Perry, 1979) and rabbit (Provis and Watson, 1981). Its retina has no visual streak and a crude area centralis (Hokoç and Oswaldo-Cruz, 1979). As in rabbits, the OKR is asymmetrical (Collewijn and Holstege, 1984). The striate cortex does not project to the NOT in the opossum, but does so in the cat (Schoppmann, 1981) and monkey (Hoffmann et al., 1987). Additionally, the NOT of the opossum lacks uncrossed retinal inputs but has a high percentage of units responding to ipsilateral eye stimulation, similarly to the cat (Hoffmann and Schoppmann, 1981).

The above considerations have led us to believe that among mammals, different strategies may have developed to achieve binocularity at the NOT not always involving the visual cortex. Evidence will be presented here to show the importance of subcortical pathways for the connection between the retina and the ipsilateral NOT and a circuit to achieve binocularity in the NOT will be proposed.

RESPONSES TO BINOCULAR AND TO MONOCULAR VISUAL STIMULATION IN THE OPOSSUM'S NOT

Recordings from the opossum's NOT was performed in paralyzed animals both in pretrigeminal sectioned preparations (Volchan et al., 1989) and alfa-chloralose anesthetized preparations (Volchan et al., 1990, 1991, and unpublished results). The differences in the anesthetized preparation were an increase in the range of spontaneous activity to include values less than 1 spike/sec as well as a tendency for slow rhythmic oscillations of the spontaneous rate. The data presented here will be concerned with experiments done under the alfa-chloralose anesthesia. The units were stimulated with a black-and-white random checkerboard pattern covering a $130 \times 110°$ tangent screen. The nucleus is easily reached 1 mm rostral to the representation of the fixation point at the superior colliculus (Volchan et al., 1981). Typical response histograms for binocular stimulation are shown in Figure 12.1. For recordings at the left NOT, a leftward-moving stimulus produces an enhancement of the frequency of spikes when compared to the preceding period during which the stimulus is left stationary. A rightward movement drops the activity of these units to a level below that of the preceding stationary period. The results are reversed for units recorded at the right side. Thus, in both cases there is excitation for ipsiversive movements and inhibition for contraversive ones.

When stimulating monocularly (see Figure 12.2), the response pattern is similar to that described under binocular condition for units recorded at the NOT contralateral to the stimulated eye. This is the response pattern observed in the majority of the recorded units. Some units, however, showed only the excitatory response for ipsiversive movements or only the inhibitory response for contraversive movements.

At the ipsilateral NOT, contraversive movement usually produces strong inhibition but ipsiversive movement evokes weak if any excitation. Also, about half of the units give no response to ipsilateral eye stimulation.

We shall argue below in favor of an inhibitory pathway between both nuclei, so that excitation of one nucleus evoked by temporonasal movement seen by the contralateral eye would lead to inhibition of the other nucleus, while the opposite movement disinhibits the latter through the inhibition of the former. This hypothesis was derived from the analysis of the response pattern of NOT units under monocular stimulation (Volchan et al., 1990, 1991), by comparing velocity and direction tuning curves and receptive field maps for each eye.

VELOCITY TUNING CURVES

To compare the response strength among units, we chose to represent it by an index: $M - S/M + S$, where M is the frequency of spikes during the period that the stimulus is in motion and S is the frequency of spikes during the preceding stationary period. Positive index expresses excitation while negative index expresses inhibition. The stimulus was moved back and forth along the horizontal axis. The velocity tuning curves are depicted in Figure 12.3. The profile of the curves for increasing stimulus speed when the contralateral eye is stimulated resembles that described previously in the pretrigeminal preparation (Volchan et al., 1989). The ipsiversive direction evokes excitation for a wide range of stimulus velocities and there is a peak of maximum inhibition at about 8°/sec for

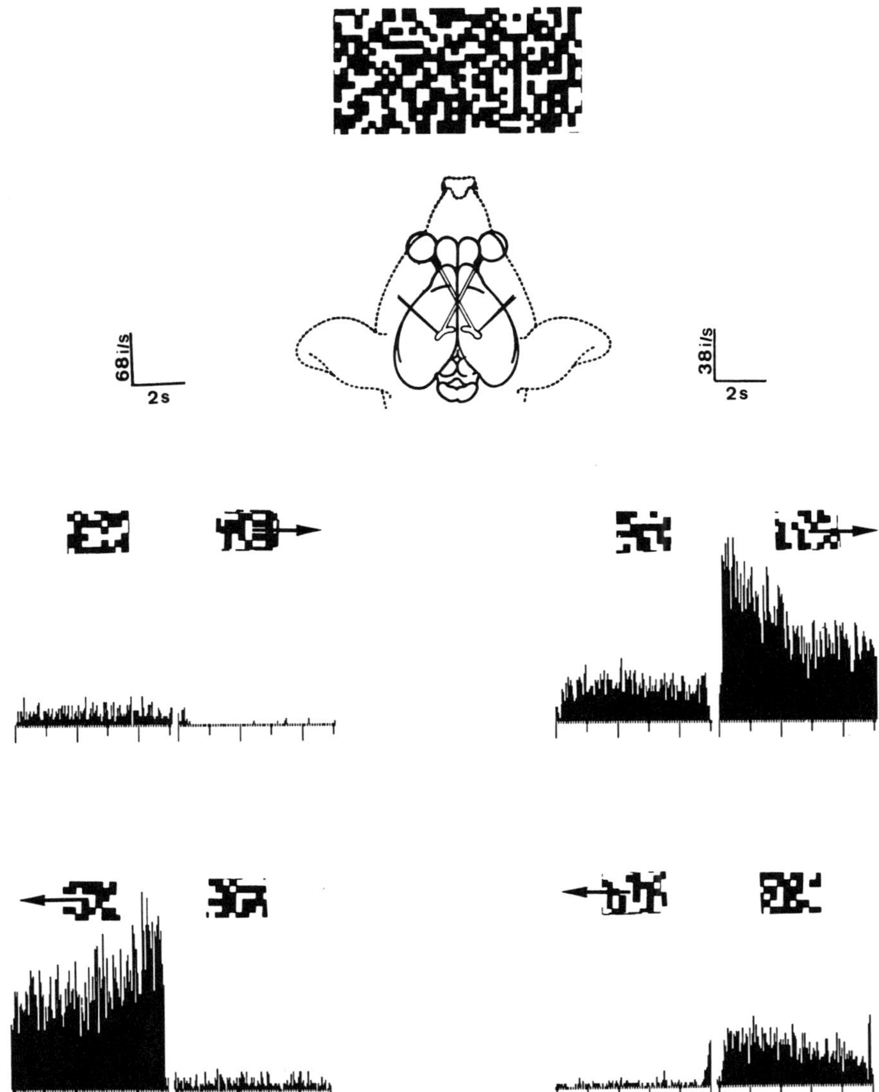

FIGURE 12.1. Responses of NOT units recorded in both hemispheres to wide-field pattern motion in binocular viewing condition. The histograms on the left side correspond to a unit recorded in the left NOT, while those on the right side are from a unit recorded in the right NOT. Before each sweep the spontaneous activity was sampled while the stimulus was kept stationary. Direction of movement is indicated by arrows. i/s: impulses per second. Stimulus velocity was 6.9°/s.

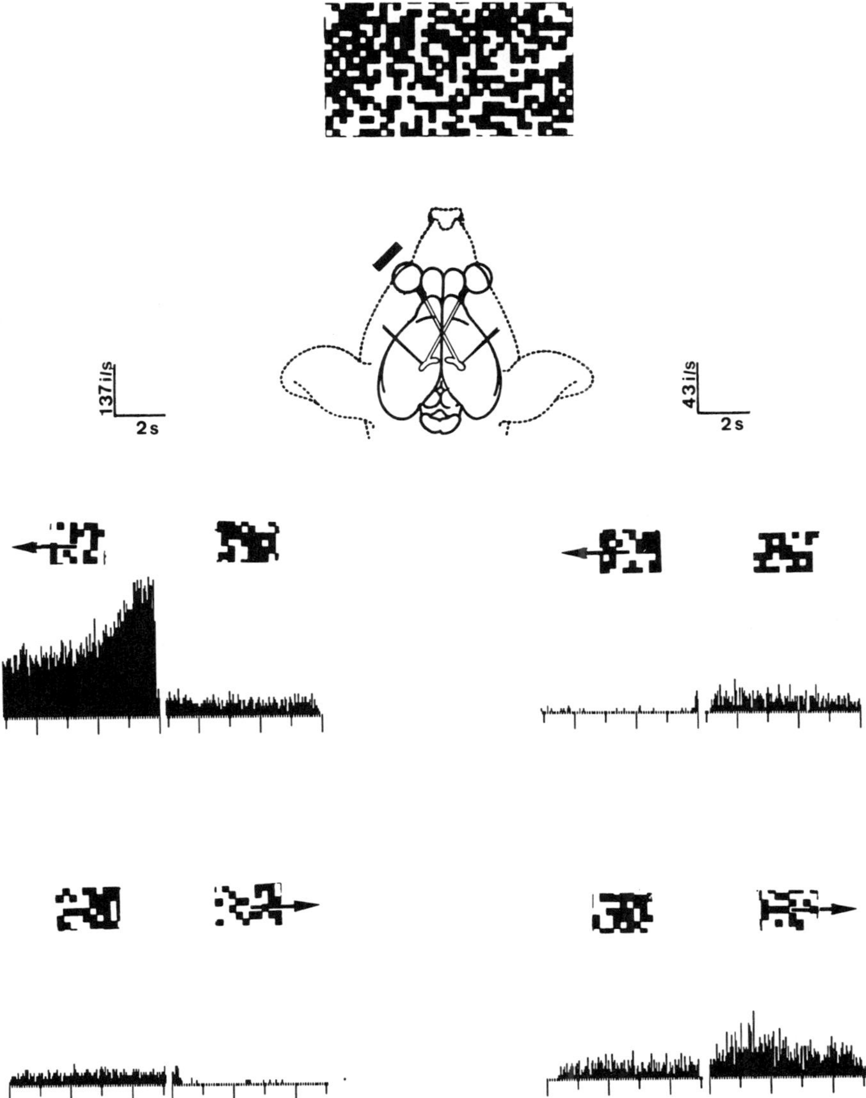

FIGURE 12.2. Responses of NOT units under monocular (right eye) stimulation. The histograms on the left side show the response of a unit recorded in the left NOT which is contralateral to the opened eye. At the right side, the histograms are from a unit recorded at the right NOT, ipsilaterally to the opened eye. Abbrev. as in Figure 12.1.

contraversive stimulation. For ipsilateral eye stimulation, ipsiversive movement produces weak excitation, as mentioned earlier, and contraversive movement shows inhibition as effective as the excitation described for the opposite eye. The inhibitory response curve for the ipsilateral eye approaches that of the excitatory

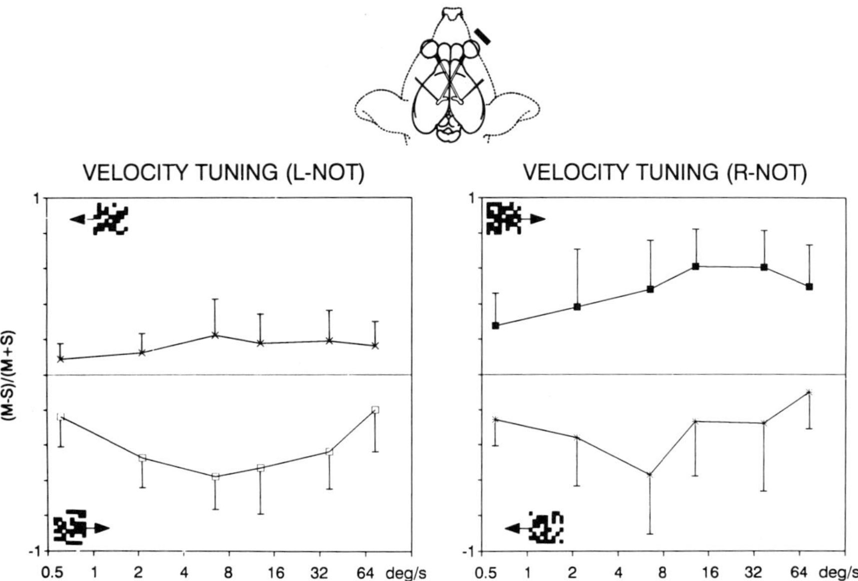

FIGURE 12.3. Average velocity tuning curves for NOT neurons recorded in both hemispheres under monocular viewing condition. Direction of movement is indicated by arrows. *Ordinate*: mean response strength index (see text). *Abscissa*: stimulus velocity (deg/s). Vertical bars represent one standard deviation. L: left; R: right.

curve for the contralateral eye with inverted signs. Conceivably, the former could be caused by the latter, in accordance with our hypothesis.

DIRECTION TUNING CURVES

Figure 12.4 shows examples of polargrams representing the responses to eight different directions of stimulus motion. With only one eye open it is possible to find units on either side of the brain that have polar plot profiles for excitation that approximately match the polar plot profiles for inhibition on the other side. Comparing the performance of each eye, it can be observed that, for contralateral eye stimulation, excitatory vectors tend to be larger than the inhibitory ones and have a wider angular range, and that the reverse is true when the ipsilateral eye is considered. This is confirmed on the average tuning curve in Figure 12.5: the excitation is larger and broader for the contralateral eye and the inhibition, larger and broader for the ipsilateral eye. However, if one supposes that contralateral excitatory vectors are transformed in inhibitory vectors on the other side, one notices that excitation for downward movement in the opposite side lacks its inhibitory counterpart in the ipsilateral side. Two explanations could be raised to explain this discrepancy: the average curve could have a significant amount of noncommissural units with different tuning profiles *or* the nuclei of the accessory optic system could exert a modulatory influence on the NOT, particularly for

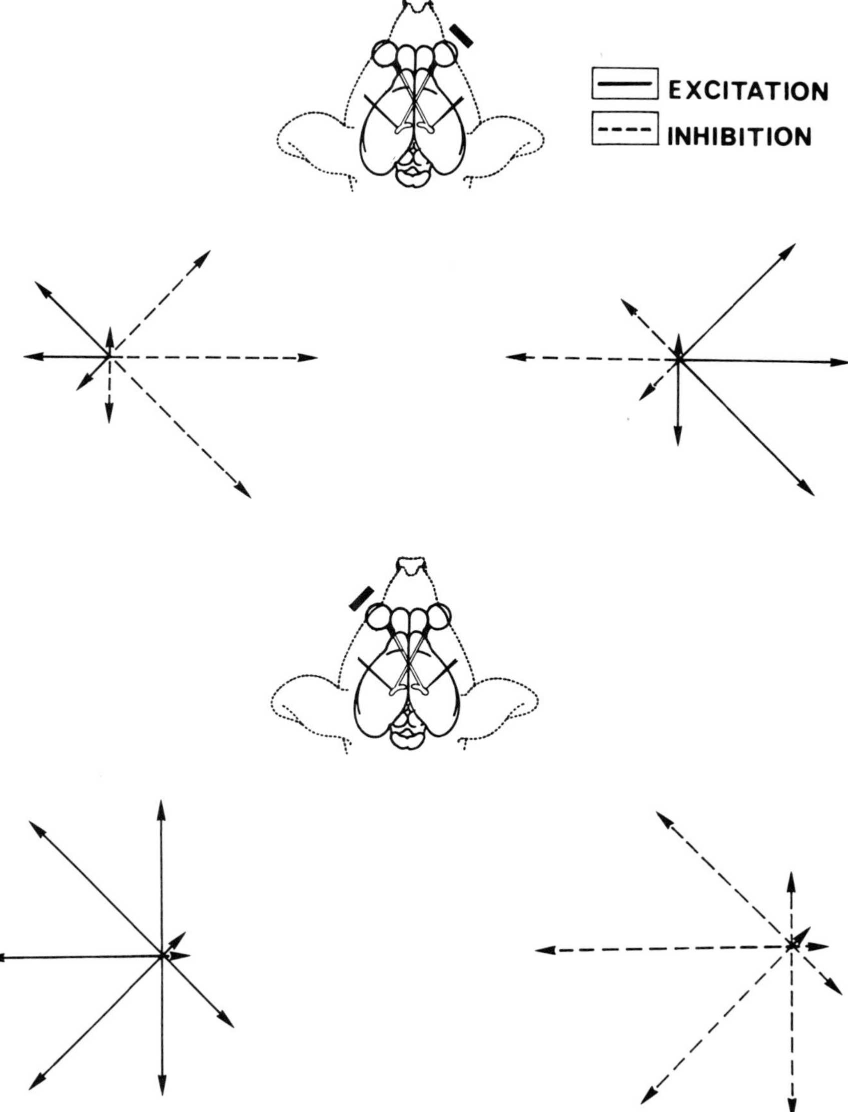

FIGURE 12.4. Representative examples of polargrams of neurons recorded in the NOT of either hemisphere under monocular viewing condition. Eight directions of movement were sampled for each unit. *Top row*; right eye occluded: *Lower row*: left eye occluded. Excitation is indicated by solid lines and inhibition by dashed lines. The vectors length is proportional to the response strength index.

vertical stimulus movement. Vargas and collaborators (1991) have shown a strong projection from the ipsilateral medial terminal nucleus and from both the lateral terminal nuclei to the NOT, and Nasi, Bernardes, Volchan, and Rocha-Miranda (ms. in preparation) have shown bilateral retinal input to the MTN.

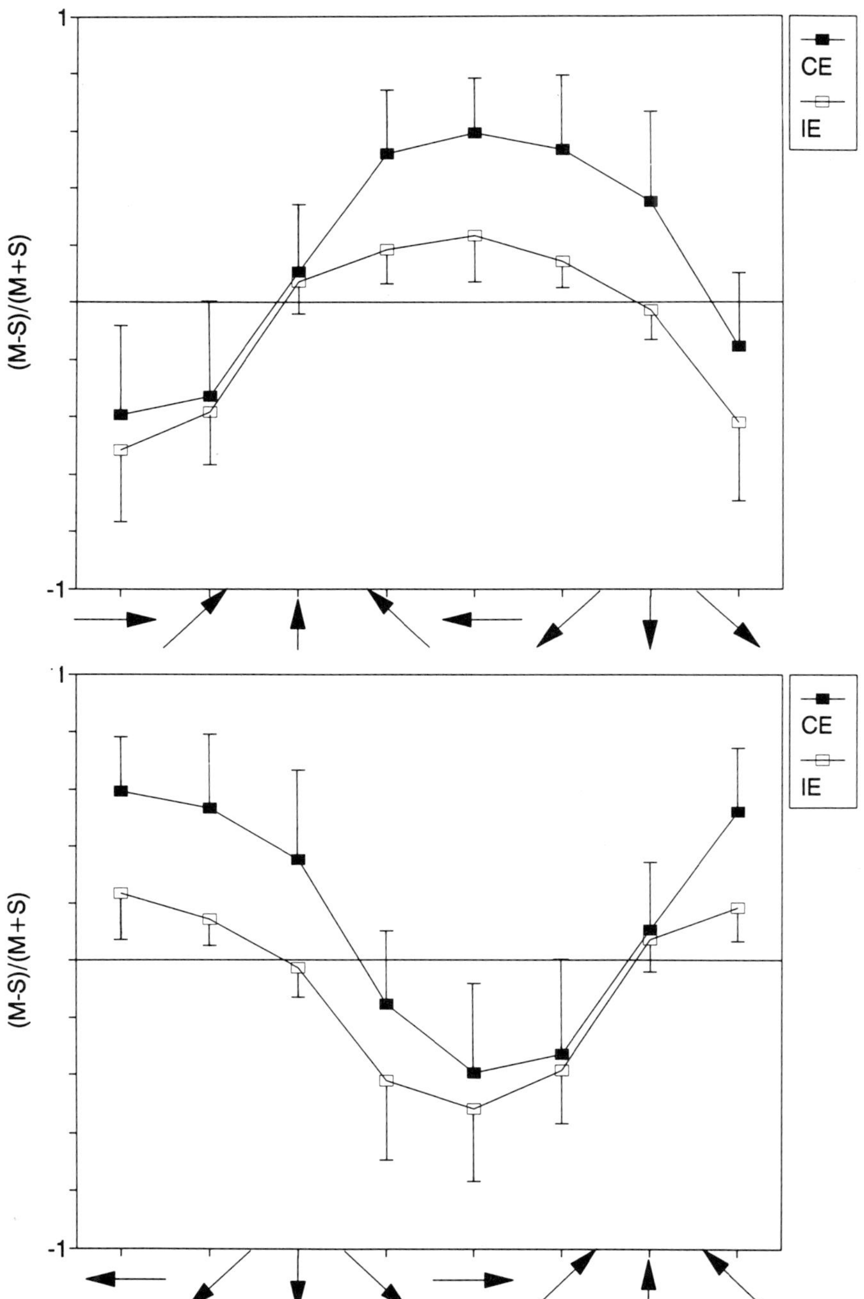

FIGURE 12.5. Average directional tuning curves for NOT neurons under monocular viewing condition. Contralateral eye responses are indicated by filled squares and ipsilateral eye responses by empty squares. Eight directions of movement were used. The same data are represented in the lower graph shifted 180°. *Ordinate*: mean response strength index. *Abscissa*: stimulus motion direction. Vertical bars represent one standard deviation.

LATERALITY AND RECEPTIVE FIELDS

Volchan and collaborators (1990) have shown that while the receptive fields for the contralateral eye lie mostly in the contralateral field, those for the ipsilateral eye lie mostly in the ipsilateral field. If the horizontal extension of receptive fields from Figure 1 of the aforementioned paper is schematically represented as a projection onto the corresponding retina (see Figure 12.6), it becomes clear that a great part of the receptive fields for the ipsilateral eye is mapped onto the nasal retina which is known to project almost exclusively contralaterally (Hokoç et al., 1992). From these data it was concluded that the activity elicited by ipsilateral eye stimulation reaches the NOT by an across the midline pathway.

BLOCKADE BY LIDOCAINE INJECTION

Connections between the NOTs were shown both electrophysiologically, by recording antidromic impulses in one side after electrical stimulation of the other (Volchan et al., 1990), and anatomically, by injecting tracers in one side and labeling cells and terminals in the other (Vargas et al., 1991). Recently, we have shown that the inactivation of one NOT disrupts the response to the ipsilateral eye stimulation in the other NOT. As Figure 12.7 indicates, the inhibition to contraversive stimulus movement is greatly reduced in the ipsilateral NOT while the opposite nucleus is blocked by lidocaine injection.

CONCLUSIONS

These results strongly suggest that: (1) a double decussation is present in the pathway between the retina and the ipsilateral nucleus of the optic tract; (2) the opposite NOT is included in this pathway; and (3) the interaction between the nuclei of the optic tract is inhibitory.

As described for other mammals (rodents: Giolli et al., 1985; rabbit: Van der Want and Nunes-Cardozo, 1988; cat: Horn and Hoffmann, 1987), but unexpected after the observations of Penny and collaborators (1984) in the Virginian opossum, the opossum's NOT was shown (Vargas et al., 1990) to have a large number of GABAergic cells and terminals. These neurons could be responsible for transference of commissural information. Nevertheless, on a preliminary survey, Vargas and collaborators (unpublished results) did not find GABAergic commissural projecting neurons in the NOT using a double labeling technique. Disregarding other possible inhibitory transmitter for the across-the-midline projecting neurons, this transmission could be excitatory onto GABAergic neurons at the opposite NOT. By intrinsic connections, these neurons could then in turn transform excitation into inhibition, generating the response to ipsilateral eye stimulation. In the cat, the visual cortex has a clear role in the response to the ipsilateral eye in the

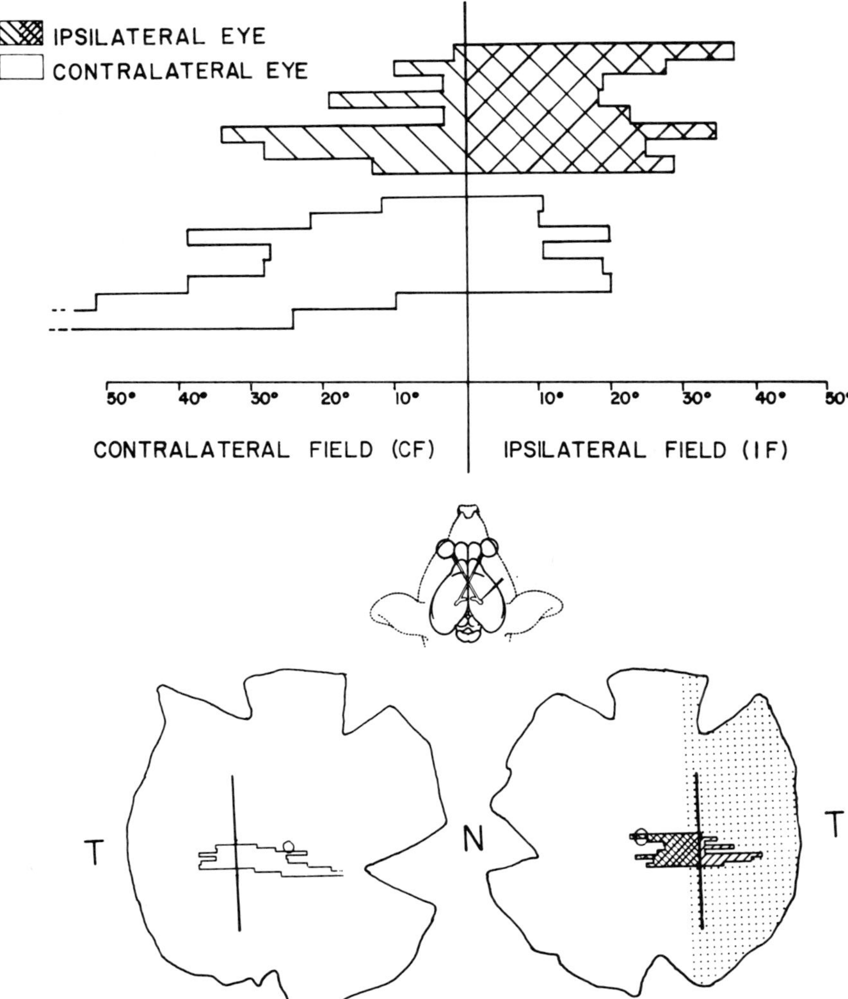

FIGURE 12.6. Information coming from the ipsilateral retina reaches the NOT through a double decussation. *Upper*: Horizontal extension of receptive fields of eight units recorded in the right NOT. *Abscissa*: azimuth in degrees. *Lower*: schematic representation showing the nasotemporal extension on the corresponding retina of the two sets of receptive units depicted above. The dotted region in the right retina represents the population of ipsilaterally projecting ganglion cells (Hokoç et al., 1992). *T*: temporal; *N*: nasal.

NOT (Hoffman, 1981) and DTN (Grasse et al., 1990). In the opossum, it remains to be seen what role if any the visual cortex plays in the activity of the NOT.

Interactions among the subcortical sensory nuclei involved in the optokinetic reflexes were shown in electrophysiological experiments in the rat (Natal and Britto, 1987) and pigeon (Hamassaki et al., 1988). It has been proposed that in

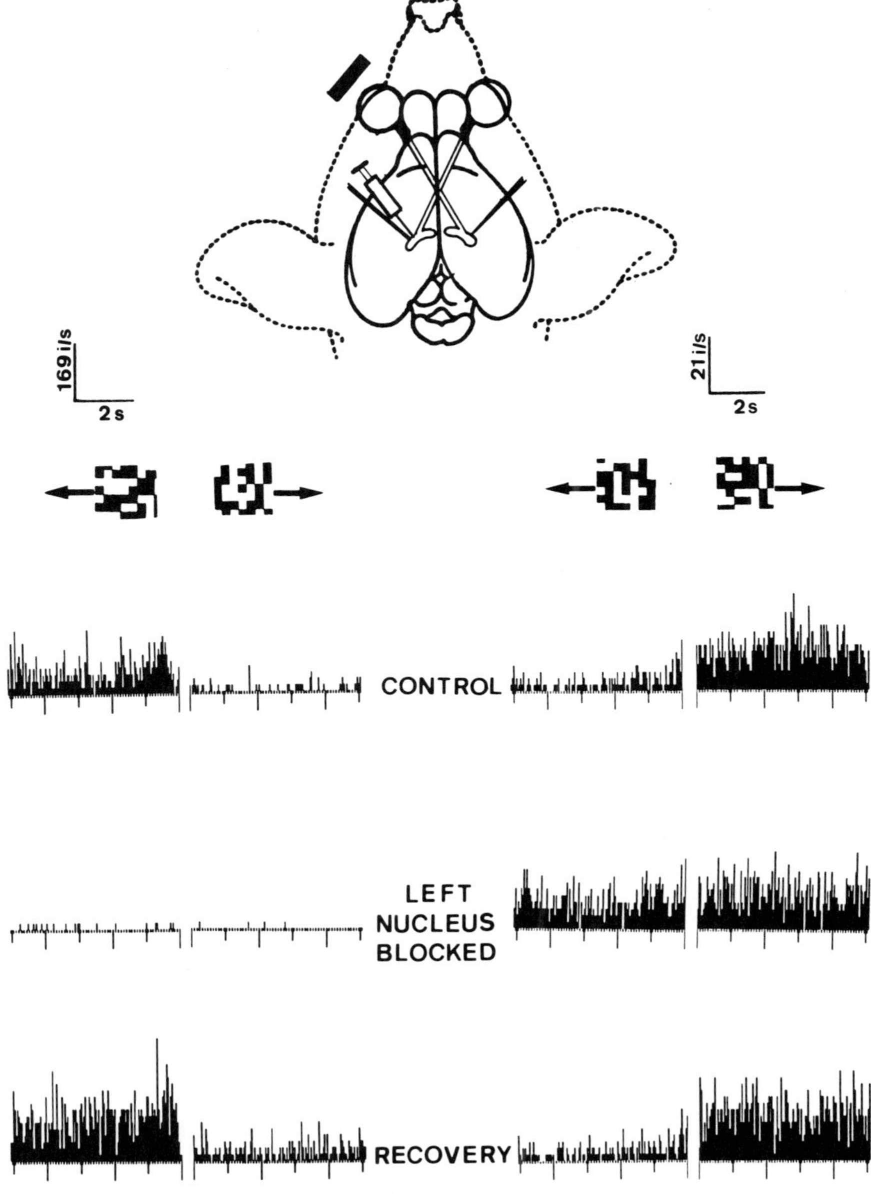

FIGURE 12.7. Unilateral inactivation of the left NOT reduces the strength of the inhibitory response of a neuron in the opposite nucleus when stimulated by the ipsilateral eye. The histograms on the left side depict the multiunit activity recorded in the left NOT, while those on the right side are from a single unit recorded in the right NOT. Notice that the inhibition due to contraversive stimulus movement almost disappears in the right NOT after the opposite nucleus has been blocked. Recovery of the response is illustrated in the lower histograms. Direction of movement is indicated by arrows. i/s: impulses per second. Stimulus velocity was 6.9°/s.

frogs (Yücel et al., 1991) and pigeons (Gioanni et al., 1983a, 1983b) subcortical interactions may play a role in the asymmetric performance of the monocular horizontal optokinetic reflex. In the pigeon (Gioanni et al., 1983a), an across-the-midline interaction between the pretectal nuclei may be involved. A suggestion of modulatory influences between the nuclei of the optic tract in the rat was pointed to by Reber and collaborators (1989). In this species, as well as in the opossum and the monkey, there is evidence for a commissural connection between the nuclei of the optic tract (Mustari et al., 1990; Terasawa et al., 1979; Vargas et al, 1991).

Acknowledgments. The hardware and software support for this research was developed by Mr. Mauricio Thurler Tecles. Financial support was obtained from Conselho Nacional de Defenvolvimento Cientifico e Technológico and Financiadora de Estudos e Projetos.

REFERENCES

Bernardes RF, Nasi JP, Volchan E, Rocha-Miranda CE (1991): Retino-pretectal projections in normal and early unilaterally enucleated opossums. In: *Abstracts of the Second Congress of the Institute of Biophysics Carlos Chagas Filho* (in portuguese) Universidade Federal do Rio de Janeiro, Brazil

Bernardes RF, Volchan E, Rocha-Miranda CE, Nasi JP, Hoffmann K-P (1987): Cells in the optic tract of the opossum with efferent projections to the inferior olive. In: *Extrageniculate Striate Visual Mechanisms: Satellite Symposium of the International Brain Research Organization, Second World Congress of Neuroscience, Budapest*

Cavalcante LA, Allodi S, Reese BE (1992): *Fiber order in the opossum's optic tract.* Submitted for publication

Cazin L, Precht W, Lannou J (1980a): Pathways mediating optokinetic responses of vestibular nucleus neurons in the rat. *Pflugers Arch* 384:19–29

Cazin L, Precht W, Lannou J (1980b): Firing characteristics of neurons mediating optokinetic responses to rat's vestibular neurons. *Pflugers Arch* 386:221– 230

Collewijn H (1975a): Oculomotor areas in the rabbit's midbrain and pretectum. *J Neurobiol* 6:3–22

Collewijn H (1975b): Direction-selective units in the rabbit's nucleus of the optic tract. *Brain Res* 100:489–508

Collewijn H, Holstege G (1984): Effects of neonatal and late unilateral enucleation on optokinetic responses and optic nerve projections in the rabbit. *Exp Brain Res* 57:138–150

Cowey A, Perry VH (1979): The projection of the temporal retina in rats, studied by retrograde transport of horseradish peroxidase. *Exp Brain Res* 35:457–464

Gioanni H, Rey J, Villalobos J, Richard D, Dalbera A (1983a): Optokynetic nystagmus in the pigeon (*Columbia livia*): 2. Role of the pretectal nucleus of the accessory optic system (AOS). *Exp Brain Res* 50:237–247

Gioanni H, Rey J, Villalobos J, Richard D, Dalbera A (1983b): Optokynetic nystagmus in the pigeon (*Columbia livia*): 3. Role of the nucleus ectomamillaris (nEM): Interactions in the accessory optic system (AOS). *Exp Brain Res* 50:248–258

Giolli RA, Peterson GM, Ribak CE, McDonald HM, Blanks RHI, Fallon JH (1985): GABAergic neurons comprise a major cell type in rodent visual relay nuclei: An immunocytochemical study of pretectal and accessory optic nuclei. *Exp Brain Res* 61:194–203

Grasse KL, Ariel M, Smith ID (1990): Direction-selective responses of units in the dorsal terminal nucleus of cat following intravitreal injections of bicuculline. *Visual Neurosci* 4:605–617

Hamassaki DE, Gasparotto OC, Nogueira MI, Britto LRG (1988): Telencephalic and pretectal modulation of the directional selectivity of accessory optic neurons in the pigeon. *Braz J Med Biol Res* 21:649–652

Harris LR, Leporé F, Guillemot J-P, Cynader M (1980): Abolition of optokinetic nystagmus in the cat. *Science* 210:91–92

Hess BJM, Precht W, Reber A, Cazin L (1985): Horizontal optokinetic ocular nystagmus in the pigmented rat. *Neuroscience* 15:97–107

Hoffmann K-P (1981): Neuronal responses related to optokinetic nystagmus in the cat's nucleus of the optic tract. In: *Progress in Oculomotor Research, Dev Neurosci*, Fuchs A, Becker W, eds. New York: Elsevier

Hoffmann K-P, Distler C, Erickson RG, Mader W (1988): Physiological and anatomical identification of the nucleus of the optic tract and dorsal terminal nucleus of the accessory optic tract in monkeys. *Exp Brain Res* 69:635–644

Hoffmann K-P, Erickson R, Distler C (1987): Retinal and cortical projections to the nucleus of the optic tract and dorsal terminal nucleus in macaque monkey. *Soc Neurosci Abstr* 13:1436

Hoffmann K-P, Schoppmann A (1981): A quantitative analysis of the direction-specific response of neurons in the cat's nucleus of the optic tract. *Exp Brain Res* 42:146–157

Hokoç JN, Gawryzewski LG, Volchan E, Rocha-Miranda CE (1992): The retinal distribution of ganglion cells with crossed and uncrossed projections and the visual field representation in the opossum. Anais da Academia Brasileira de Ciências, (in press)

Hokoç JN, Oswaldo-Cruz E (1979): A regional specialization in the opossum's retina: Quantitative analysis of the ganglion cell layer. *J Comp Neurol* 183:385–396

Horn AKE, Hoffmann K-P (1987): Combined GABA-immunocytochemistry and TMB-HRP histochemistry of pretectal nuclei projecting to the inferior olive in rats, cats and monkeys. *Brain Res* 409:133–138

Kato I, Harada K, Hasegawa I, Igarashi T, Koike Y, Kawasaki T (1986): Role of the nucleus of the optic tract in monkeys in relation to optokinetic nystagmus. *Brain Res* 364:12–22

Klauer S, Sengpiel F, Hoffmann K-P (1990): Visual response properties and afferents of nucleus of the optic tract in the ferret. *Exp Brain Res* 83:178–189

Lent R, Cavalcante LA, Rocha-Miranda CE (1976): Retinofugal projections in the opossum: An anterograde degeneration and radioautographic study. *Brain Res* 107:9–26

Linden R, Rocha-Miranda CE (1981): The pretectal complex in the opossum: Projections from the striate cortex and correlation with retinal terminal fields. *Brain Res* 207:267–277

Mustari MJ, Fuchs AF, Kaneko CRS (1990): Descending connections of the macaque nucleus of the optic tract. *Soc Neurosci Abstr* 16:904

Nasi JP, Bernardes RF, Volchan E, Rocha-Miranda CE, Tecles M (1989): Horizontal optokinetic reflex in the opossum. *Braz J Med Biol Res* 22:773–774

Nasi JP, Tecles MJT, Bernardes RF, Volchan E, Rocha-Miranda CE (1990): Frequency and gain of the horizontal optokinetic reflex of normal and early unilaterally enucleated opossums. In: *Abstracts of the 5th Congress of the Federation of Societies of Experimental Biology* (in portuguese), Caxambu, Brazil.

Natal CL, Britto LRG (1987): The pretectal nucleus of the optic tract modulates the direction selectivity of accessory optic neurons in rats. *Brain Res* 419:320–323

Oswaldo-Cruz E, Hokoç JN, Souza APB (1979): A schematic eye for the opossum. *Vision Res* 19:263–278

Penny GR, Conley M, Schhmechel DE, Diamond IT (1984): The distribution of glutamic acid decarboxylase immunoreactivity in the diencephalon of the opossum and rabbit. *J Comp Neurol* 228:38–56

Precht W (1981): Functional organization of optokinetic pathways in mammals. In: *Progress in Oculomotor Research, Dev Neurosci*, Fuchs AF, Becker W, eds. New York: Elsevier

Provis JM, Watson CRR (1981): The distribution of ipsilaterally and contralaterally projecting ganglion cells in the retina of the pigmented rabbit. *Exp Brain Res* 44:82–92

Reber A, Lannou J, Hess BJM (1989): Development of optokinetic neuronal responses in the pretectum and horizontal optokinetic nystagmus in unilaterally enucleated rats. *Arch Ital Biol* 127:225–242

Schiff D, Cohen B, Büttner-Ennever J, Matsuo V (1990): Effects of lesions of the nucleus of the optic tract on optokinetic nystagmus and after-nystagmus in the monkey. *Exp Brain Res* 79:225–239

Schiff D, Cohen B, Raphan T (1988): Nystagmus induced by stimulation of the nucleus of the optic tract in the monkey. *Exp Brain Res* 70:1–14

Schoppmann A (1981): Projections from area 17 and 18 of the visual cortex to the nucleus of the optic tract. *Brain Res* 223:1–17

Stone J (1966): The naso-temporal division of the cat's retina. *J Comp Neurol* 126:585–600

Terasawa K, Otani K, Yamada J (1979): Descending pathways of the nucleus of the optic tract in the rat. *Brain Res* 137:405–417

Van der Want JJL, Nunes Cardozo JJ (1988): GABA immuno-electron microscopic study of the nucleus of the optic tract in the rabbit. *J Comp Neurol* 271:229–242

Vargas CD, Volchan E, Bernardes RF, Rocha-Miranda CE (1991): Subcortical afferents and efferents of the nucleus of the optic tract of the opossum. In: *Abstracts of the Sixth Congress of the Federation of Societies of Experimental Biology* (in portuguese), Caxambú, Brazil

Vargas CD, Volchan E, Hokoç JN (1990): Distribution of GABAergic cells and terminals in the nucleus of the optic tract and accessory optic system of the opossum. In: *Abstracts of the Fifth Congress of the Federation of the Societies of Experimental Biology* (in portuguese), Caxambu, Brazil

Volchan E, Gawryszewski LG, Rocha-Miranda CE (1981): Visuotopic organization of the superior colliculus of the opossum. *Exp Brain Res* 46:263–268

Volchan E, Pereira Jr., Franca JG, Bernardes RF, Assad AR, Rocha-Miranda CE (1991): Ipsi- and contralateral eye response properties at the nucleus of the optic tract (NOT). In: *Third World Congress of Neuroscience Abstracts*, International Brain Research Organization

Volchan E, Rocha-Miranda CE, Franca JG, Bernardes RF, Marrocos MA, Assad AR (1990): Evidence for recrossing of the pathway from the retina to the ipsilateral nucleus of the optic tract. *Braz J Med Biol Res* 23:1057–1060

Volchan E, Rocha-Miranda CE, Picanço-Diniz CW, Zinsmeisser B, Bernardes RF, Franca JG (1989): Visual response properties of pretectal units in the nucleus of the optic tract of the opossum. *Exp Brain Res* 78:380–386

Yücel YH, Jardon B, Bonaventure N (1991): Unilateral pretectal microinjections of SR 95 531, a GABA A antagonist: Effects on directional asymmetry of frog monocular OKN. *Exp Brain Res* 83:527–532

13

A Comparative Survey of Magnification Factor in V1 and Retinal Ganglion Cell Topography of Lateral-Eyed Mammals

CRISTOVAM W. PICANÇO-DINIZ, LUIZ CARLOS L. SILVEIRA, AND EDUARDO OSWALDO-CRUZ

INTRODUCTION

Quantitative analysis of the geometrical transformations that the visual field undergoes when it is projected on central nuclei has been the object of many studies. From Daniel and Whitteridge (1961) until the present day the extent to which V1 cortical topography is determined by its retinogeniculate afferents (Orban, 1984; Pointer, 1986; Tolhurst, 1989; Whitteridge, 1973) has been a key question.

In order to understand these topographical projections it is first necessary to discover the distribution of the receptors and ganglion cells in the retina. Counting ganglion cells in retinal flat mounts revealed two basic patterns—with some intermediate forms—for the topography of retinal ganglion cell density among mammals (Hughes, 1977; Stone, 1983; Whitteridge, 1965, 1973). One pattern in mammals reveals retinal specialization showing an increase of the ganglion cell density along the representation of the horizon, giving rise to the visual streak (e.g., rabbit: Hughes, 1971; agouti, Silveira et al., 1989a). The second pattern reveals an increase in the ganglion cell density in the vertical and horizontal intersect, giving rise to a *fovea centralis* (e.g., man: Curcio and Allen, 1990; van Buren, 1963; Stone and Johnston, 1981; Van Buren, 1963; macaque monkey: Perry and Cowey, 1985; Stone and Johnston, 1981; capuchin monkey: Silveira et al., 1989b), or an *area centralis* (e.g., cat: Hughes, 1975; Stone, 1965; ferret: Henderson, 1985). Similar results were obtained by counting photoreceptors in retinal flat mounts (Curcio et al., 1990; de Monasterio et al., 1985; Müller and Peichl, 1989; Packer et al., 1989; Perry and Cowey, 1985; Steinberg et al., 1973; Wikler and Rakic, 1990).

Studies on the visual topography of the cerebral cortex have shown multiple visual areas that emphasize relatively different regions of the visual field (Kaas, 1980; van Essen, 1979). Extensive quantitative analyses has been carried out comparing different segments of V1 in several species to verify if their size either reflect the distribution of the retinal ganglion cell density or if retinal output is further modified at the subsequent stages of the visual pathway. Some results indicate a V1 overrepresentation in the retinal regions that coincides with the visual axis, the *fovea centralis* in primates (Malpeli and Baker, 1975; Myerson et al., 1977; Perry and Cowey, 1985; Silveira et al., 1989b; Whitteridge, 1965), and the *area centralis* in carnivores (Tolhurst, 1989). Other studies point to a visual field representation in the primary visual cortex that is proportional to the ganglion cell density in primates (Schein and de Monasterio, 1987; Wässle et al., 1989, 1990; Webb and Kaas, 1976) and cat (Albus and Beckmann, 1980; Tusa et al., 1978). In primates and carnivores the optical and visual axes are close to each other, giving rise to a small alpha angle and extensive binocular overlapping in the visual fields of each eye.

Our previous results concerning the agouti (Oswaldo-Cruz et al., 1990; Picanço-Diniz, 1987, 1991) and those of Daniel Whitteridge and Austin Hughes on the rabbit's visual system (Hughes, 1971; Whitteridge, 1965, 1973), have shown that there is a large mismatch between ganglion cell density and V1 magnification factor. The temporal end of the rabbit and agouti visual streak, which corresponds to the nasal (anterior) visual field, is alotted a larger proportion of the cortical representation than the nasal end, which corresponds to the temporal (posterior) visual field. Rabbit and agouti have in common a panoramic visual field, eyes placed laterally, a well-developed visual streak, and divergent optical and visual axes (large alpha angle). Here, we have attempted to verify if these findings would follow a general rule for other lateral-eyed species.

We start by describing some of our own results in the agouti *Dasyprocta aguti aguti*, and then a data compilation concerning the lateral-eyed species for which detailed topography of the visual field representation in V1 as well as the topography of the retinal ganglion cell density is available, such as the rabbit (Hughes, 1971; Provis, 1979), hamster (Dursteler et al., 1979; Tiao and Blakemore, 1976a, 1976b), grey squirrel (Hall et al., 1971; Hughes, 1977; Lane et al., 1971), guinea pig (Choudhury, 1978; do Nascimento et al., 1991; Hughes, 1977), and mouse (Dräger and Olsen, 1981; Wagor et al., 1980). The results obtained by comparing these two variables show a breach in the proportionality between peripheral innervation density, represented by the ganglion cell density, and central sensorial representation, as reflected by the V1 cortical magnification factor.

METHODS

A number of different data sources have been used in the calculations described here. They are pointed out in each case throughout the text. The original papers present detailed information about the procedures adopted to obtain V1 and ganglion cell density maps, but only information essential for text clarity is described here. Ganglion cell isodensity maps were obtained from Nissl-stained retinas. In some species HRP retrograde labeling of ganglion cells was done to verify ganglion cell counts in Nissl-stained retinas. Most of the electrophysiological data came from multi-unit recordings using tungsten microelectrodes. In such experiments animals were anesthetized and ocular movements paralyzed using neuromuscular blocking agents to standardize geometrical relations between retina and visual field. The optic disk visual field projection has been used as a control for residual ocular movements.

Using isoazimuthal/isoelevation lines of the contralateral visual field in V1, and ganglion cell isodensity lines obtained from retinal flat mounts, we have estimated the areal cortical magnification factor (ACMF) and the average density of ganglion cells for equivalent regions of the visual field along the horizon. To compare cortical and retinal maps, we have used spatial coordinates of the optic disk described in the eletrophysiological experiments as reference points upon the retina to trace meridia and parallels assuming constant retinal magnification factor (RMF). We have used enlarged representations of retinal flat mounts mounted on a hemispherical surface, referred to an oculocentric equatorial azimuthal coordinate system or a Mercator projection. The area of equivalent sectors in both retinal and cortical representations was calculated using a digitizing table. Dividing total ganglion cell number in the chosen sectors by the equivalent retinal area, we obtained the averaged ganglion cell density for each region. Dividing cortical area of the representation of this retinal sector by the corresponding visual field area we obtained the ACMF.

We have not considered variation of the RMF, which decreases in the periphery of the optical system. It should be noted that if we correct the present data for the estimates of RMF values in the periphery of the optical system, the mismatch between retinal ganglion cell density and ACMF will increase.

Also, we have not considered the distortions produced by dorsal or dorsolateral view reconstructions of the cerebral cortex. That is expected to be relatively higher in area 17 lateral margin. In this region the magnification factor will be underestimated due to the perspective distortion shortening the cortical distances (see Figure 13.1).

Moreover, it is possible that in the counts carried out in Nissl-stained retinal flat mounts, some authors have included displaced amacrine cells in their ganglion cell counts (Hughes, 1985). However, in some cases ganglion cell counts were checked with specific retrograde labeling (e.g., Silveira, 1985; Silveira et al., 1989a) (see Figure 13.2A).

Finally, it is difficult to determine if implicit uncertainties in the process of making isodensities, isoazimuthal and isoelevation lines, compound or cancel out each other, but the absence of incongruous results between the different data sets can be considered as evidence that our assumptions are valid.

RESULTS

All V1 and LGN maps studied in the present work were asymmetric, with more cortex or LGN devoted to nasal than to temporal fields. Cortical magnification factor (CMF) (Daniel and Whitteridge, 1961) has been used to quantify this

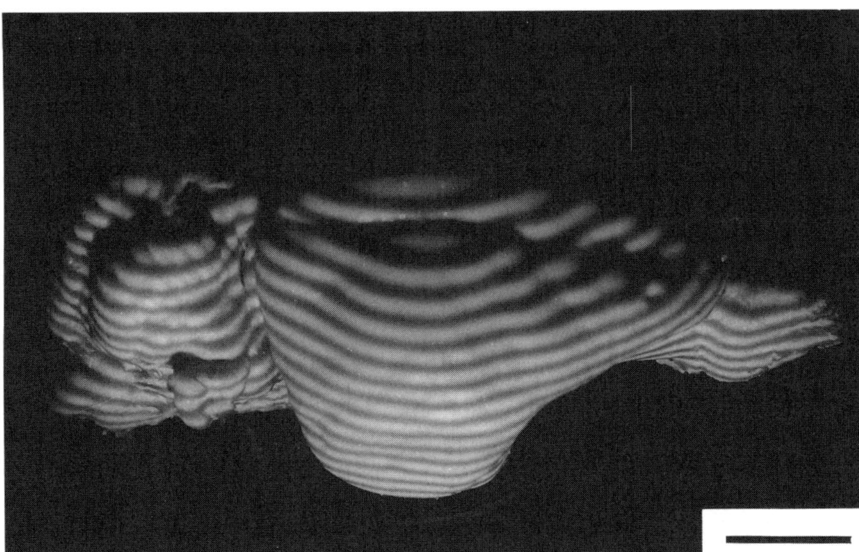

FIGURE 13.1. Dorsolateral view of the agouti brain showing horizontal grating projection on the brain surface to illustrate perspective distortion introduced by dorsal view reconstruction. Note that bars become narrow toward the brain's lateral margin. It should be noted that this distortion reduces the nasotemporal difference along the representation of the horizontal meridian as much as 3.5 times. Thus its effect is to reduce the retinocortical mismatch along this region of the visual field. Scale bar = 10 mm.

asymmetry. The CMF is a local, unidimensional, linear index that depends on the direction of the measurement (Orban, 1984). In order to compare retinal and central representations of the visual field, we have estimated the ACMF, which is a bidimensional, local but areal measurement that is directly comparable to the ganglion cell density (Orban, 1984; Tusa et al., 1978).

Some of the results for the agouti visual system (see Figure 13.2A and 13.2B) have been published before (Picanço-Diniz et al., 1991; Silveira et al., 1989a). Figure 13.2A shows the ganglion cell distribution in an agouti retina retrogradely labeled after multiple, massive HRP injections in the contralateral superior colliculus. In this species, as in other rodents such as the rat (Linden and Perry, 1983), and in lagomorphs such as the rabbit (Vaney et al., 1981), nearly all ganglion cells project to the midbrain. Figure 13.2B shows a comparison between ganglion cell density along the visual streak in the retina and the ACMF along the horizon of the V1 map in the agouti brain. A large mismatch between retinal and cortical representations of the visual field is obvious. Whereas the retinal ganglion cell density is relatively uniform along the visual streak, the ACMF peaks in the cortical representation of the visual axis, in the nasal portion of the visual field.

Data for the rabbit visual system were based on the work of Hughes (1971). They are shown in Figure 13.3. Visual field maps were used to estimate areal

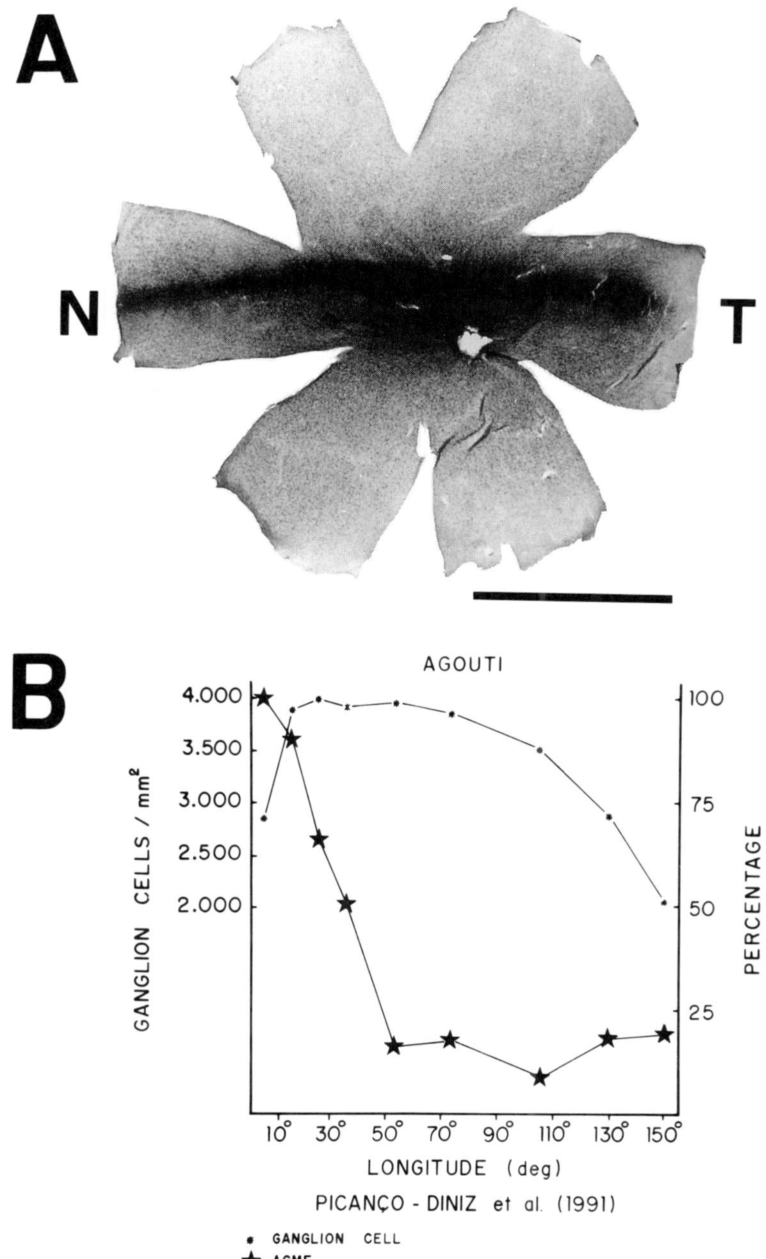

FIGURE 13.2. A. Agouti retinal whole mount retrogradely labeled after massive, multiple HRP injections in the contralateral superior colliculus. In this species almost all retinal ganglion cells project to the midbrain. Compare this figure with ganglion cell isodensity maps from Nissl-stained retinas in Figure 7 of Silveira et al. (1989a). B. Comparison of ganglion cell density and cortical magnification factor in the agouti. Ganglion cell counts were carried in Nissl-stained retinal whole mounts along the visual streak. V1 areal cortical magnification factor was estimated along the cortical representation of the horizontal meridian. Ganglion cell density and ACMF are normalized to the peak value. Scale bar = 10 mm.

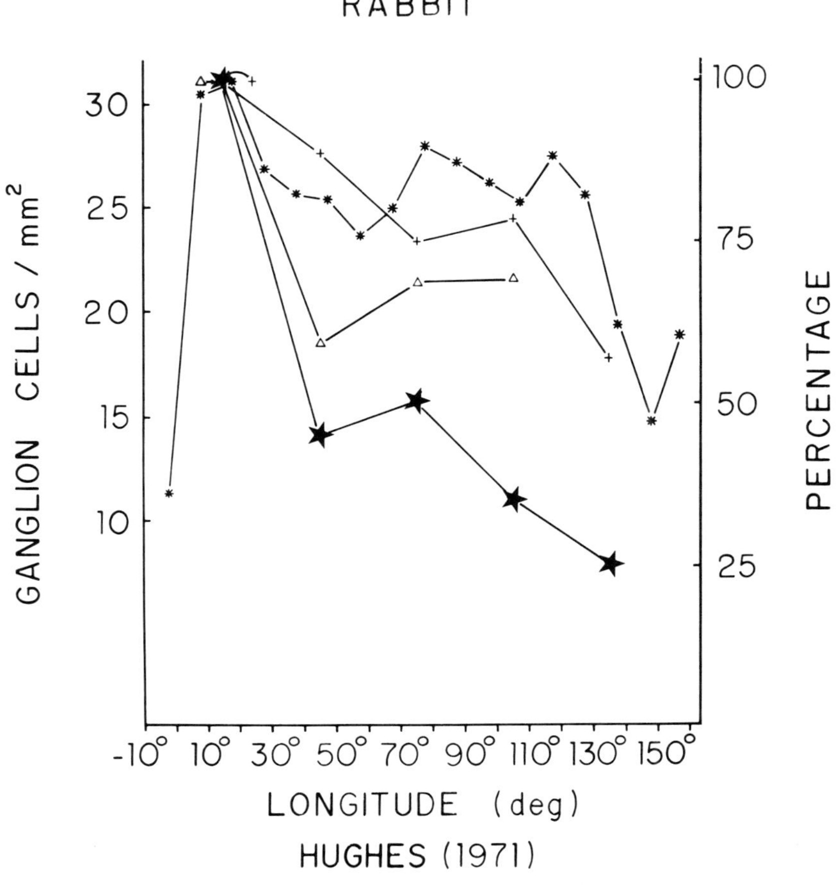

FIGURE 13.3. Rabbit visual-streak ganglion cell counts compared to V1 (ACMF), superior colliculus (ASCMF), and lateral geniculate nucleus (ALGNMF) areal magnification factors along the representation of the horizontal meridian. Ganglion cell density and areal magnification factors have been normalized to the peak value. Measurements were based on Hughes (1971) data.

magnification factor for V1 (ACMF), LGN (ALGNMF), and superior colliculus (ASCMF) along the horizontal meridian. These values were correlated to retinal ganglion cell density along the rabbit visual streak. Note that these two parameters change differently along the horizontal meridian: ganglion cell density shows a plateau along the visual streak which decreases in both extremities of the retina,

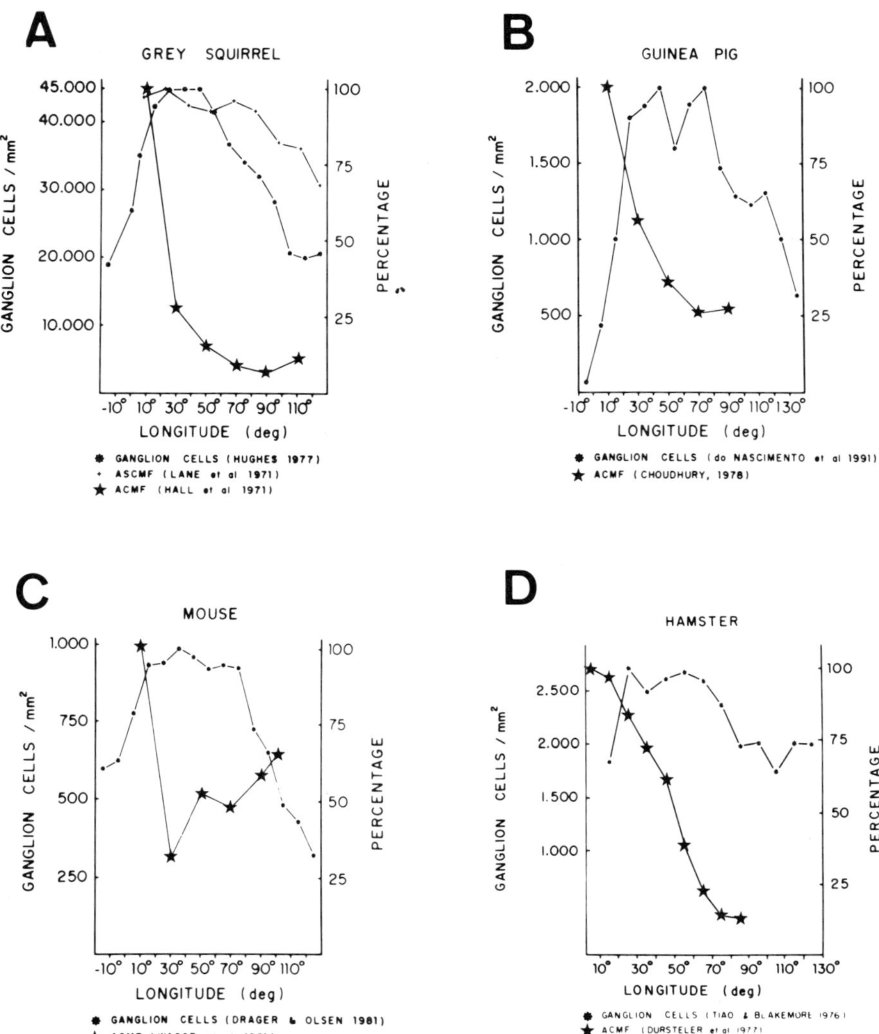

FIGURE 13.4. Ganglion cell counts versus ACMF, ASCMF, and ALGNMF for other lateral-eyed species. Estimatives for grey squirrel (A), guinea pig (B), mouse (C), and hamster (D) were based on the reports cited in the figure. In all the species there is a mismatch between ganglion cell counts and areal magnification factors for the LGN and V1 along the horizon. This mismatch is absent or less obvious for the superior colliculus.

whereas the LGN and V1 areal magnification factor reach their peak values around the intersection of the vertical and horizontal meridians, decreasing toward the temporal periphery of the visual field.

Similar results were obtained for other lateral-eyed mammals (see Figure 13.4). Data shown in Figure 13.4 are derived from published results for grey squirrel

(Hall et al., 1971; Hughes, 1977), guinea pig (Choudhoury, 1978; do Nascimento et al., 1991), mouse (Dräger and Olsen, 1981; Wagor et al., 1980), and hamster (Dursteller et al., 1977; Tiao and Blakemore, 1976a). They show that retinal ganglion cell density are not matched by areal magnification factors for LGN and V1. In summary, in all species analyzed, a large proportion of LGN and V1 is emphasizing nasal visual field. This emphasis is higher than would be expected from the ganglion cell density topography. On the other hand, this mismatch seems to be absent or less obvious in the retinotectal pathway.

DISCUSSION

The results presented here indicate that there are at least two different strategies emphasizing different regions of the visual field in the lateral-eyed visual systems: the tectal projection, roughly reflecting the topography of the retinal ganglion cell density, and the retino-geniculate-striate pathway, emphasizing the representation of the temporal end of the visual streak that projects to the binocular region of the horizon, in the vicinity of the visual axis, in the nasal visual field.

These results are representative for species with different life-styles. Some of them, such as the agouti and rabbit, are diurnal or crepuscular, having retinas with high centroperipheral density gradient and a sharp visual streak (Hughes, 1971; Silveira, 1985; Silveira et al., 1989a). The grey squirrel has high retinal ganglion cell density, low centroperipheral density gradient, and a broad visual streak (Hughes, 1977). The hamster (Tiao and Blakemore, 1976a), mouse (Dräger and Olsen, 1981), and other nocturnal myomorphs have circular or eliptical ganglion cell isodensity contours and low centroperipheral density gradient. The guinea pig (Choudhury, 1978; do Nascimento et al., 1991) has low centroperipheral density gradient and a rudimentary visual streak. Despite these different life-styles and retinal organizations, all these lateral-eyed mammals show a mismatch between retinal and cortical representations of the visual field. This mismatch is larger in those species that have narrow, sharp visual streaks, such as the rabbit and agouti. In this sense Whitteridge's (1973) statement, "All areas of great density of ganglion cells whether they form fovea, areae centrales or horizontal streaks, have large areas of representation on the visual cortex, but this large representative is not always proportional to ganglion cell density", could now be extended to several lateral-eyed species.

These findings might be explained if different ganglion cell classes with different retinal distributions actually contribute in different proportions to the retino-geniculo-cortical and retinotectal pathways. It could be possible that cortical mapping is determined by a ganglion cell class that predominates in the retino-geniculo-striate pathway but is overshadowed by other classes when one evaluates overall ganglion cell density. The large alphalike ganglion cells (Peichl et al., 1987), that peak in the temporal end of the visual streak, around the visual axis, are likely candidates for this role (Provis, 1979; Reese and Cowey, 1986; E.S. Yamada, L.C.L. Silveira, and C.W. Picanço-Diniz, in preparation).

Another possibility is that the overrepresentation of the visual axis is due to intrinsic cortical mechanisms. It has been noted concerning the rabbit primary visual cortex that there is a nasotemporal density gradient of the layer IV nonpyramidal neurons that matches the nasotemporal variation of the cortical magnification factor (Choudhury, 1988). In order to verify if these findings could be extended to the agouti brain, we have estimated the density variation of the *Vicia villosa* positive cells in V1. *Vicia villosa* is a lectin that reacts with membrane receptors from a subgroup of GABAergic neurons (Mulligan et al., 1989). We have found that the density of these cells in the supragranular layers increases toward the lateral margin of V1 where the vertical meridian is represented in the agouti brain. Our preliminary results suggest that there is an increase of inhibitory cells around the visual axis representation. These data could indicate an important role for intrinsic cortical microcircuitry in the genesis of cortical magnification of different segments of the visual field.

Acknowledgments. This work was supported by research grants from Financiadora de Estudos e Projetos (No. 4.3.87.0308.00), Conselho Nacional de Desenvolvimento Científico e Tecnológico, and Pro-Reitoria de Pesquisa e Pós-Graduaçáo da Universidade Federal do Pará. C.W.P.D. and L.C.L.S. have CNPq research fellowships (Nos. 302213/82 and 302216/82). We wish to express our appreciation to M. S. Pompeu de Carvalho, M. Silva Filho, and Z.Z.P. Abnader for research assistance, and to A. Jeronimo Botelho and J.C. Souza for technical assistance.

REFERENCES

Albus K, Beckmann R (1980): Second and third visual areas of the cat: Interindividual variability in retinotopic arrangement and cortical location. *J Physiol (Lond)* 299:247–276

Choudhury BP (1978): Retinotopic organization of the guinea pig's visual cortex. *Brain Res* 144:19–29

Choudhury BP (1988): Distribution of neurons in the pigmented rabbit's visual cortex. *Exp Neurol* 101:458–463

Curcio CA, Allen KA (1990): Topography of ganglion cells in human retina. *J Comp Neurol* 300:5–25

Curcio CA, Sloan KR, Kalina RE, Hendrickson AE (1990): Human photoreceptor topography. *J Comp Neurol* 292:497–523

Daniel PM, Whitteridge D (1961): The representation of the visual field on the cerebral cortex in monkey. *J Physiol (Lond)* 159:203–221

de Monasterio FM, McCrane EP, Newlander JK, Schein SJ (1985): Density profile of blue-sensitive cones along the horizontal meridian of macaque retina. *Invest Ophthalmol Vis Sci* 26:289–302

do Nascimento JLM, do Nascimento RSV, Damasceno BA, Silveira LCL (1991): The neurons of the retinal ganglion cell layer of the guinea pig: Quantitative analysis of their distribution and size. *Braz J Med Biol Res* 24:199–214

Dräger UC, Olsen JF (1981): Ganglion cell distribution in the retina of the mouse. *Invest Ophthalmol Vis Sci* 20:285–293

Dursteler MR, Blakemore C, Garey LJ (1979): Projections to the visual cortex in the golden hamster. *J Comp Neurol* 183:185–204

Hall WC, Kaas JH, Killackey H, Diamond IT (1971): Cortical visual areas in the grey squirrel (*Sciurus carolinensis*): A correlation between evoked potential maps and architectonic subdivisions. *J Neurophysiol* 34:437–452

Henderson Z (1985): Distribution of ganglion cells in the retina of adult pigmented ferret. *Brain Res* 358:221–228

Hughes A (1971): Topographical relationships between the anatomy and physiology of the rabbit visual system. *Doc Ophthalmol* 30:33–159

Hughes A (1975): A quantitative analysis of the cat retinal ganglion cell topography. *J Comp Neurol* 163:107–128

Hughes A (1977): The topography of vision in mammals of contrasting life style: Comparative optics and retinal organization. In: *Handbook of Sensory Physiology: Vol. 7, Part 5: The Visual System in Vertebrates*, Crescitelli F, ed. Berlin: Springer-Verlag

Hughes A (1985) New perspectives in retinal organisation. In: *Progress in Retinal Research*, vol. 4, Osborne N, Chader G, eds. Oxford: Pergamon Press

Kaas JH (1980): A comparative survey of visual cortex organization in mammals. In: *Comparative Neurology of the Telencephalon*, Ebbesson SOE, ed. New York: Plenum Press

Lane RH, Allman JM, Kaas JH, Miezin FM (1971): Representation of the visual field in the superior colliculus of the grey squirrel (*Sciurus carolinensis*) and the tree shrew (*Tupaia glis*). *Brain Res* 26:277–292

Linden R, Perry VH (1983): Massive retinotectal projection in rats. *Brain Res* 272:145–149

Malpeli JG, Baker FH (1975): The representation of the visual field in the lateral geniculate nucleus of *Macaca mulatta*. *J Comp Neurol* 161:569–594

Müller B, Peichl L (1989): Topography of cones and rods in the tree shrew retina. *J Comp Neurol* 282:581–594

Mulligan KA, van Brederode JFM, Hendrickson AE (1989): The lectin *Vicia villosa* labels a distinct subset of GABAergic cells in macaque visual cortex. *Visual Neurosci* 2:63–72

Myerson J, Manis PB, Miezin FM, Allman JM (1977): Magnification in striate cortex and retinal ganglion cell layer of owl monkey: A quantitative comparison. *Science* 198:85–857

Orban GA (1984): *Neuronal Operations in the Visual Cortex*. Berlin: Springer-Verlag

Oswaldo-Cruz E, Picanço-Diniz CW, Pompeu MS, Silveira LCL (1990): Asymmetries of the representation of the contralateral visual field on area 17 of the cortex of the anaesthetized agouti do not match the retinal ganglion-cell regional densities. *J Physiol (Lond)* 422:8P

Packer O, Hendrickson AE, Curcio CA (1989): Photoreceptor topography of the adult pigtail macaque (*Macaca nemestrina*) retina. *J Comp Neurol* 288:165–183

Peichl L, Ott H, Boycott BB (1987): Alpha ganglion cells in mammalian retinae. *Proc R Soc Lond, (Biol)* 231:169–197

Perry VH, Cowey A (1985): The ganglion cell and cone distributions in the monkey's retina: Implications for central magnification factors. *Vision Res* 25:1795–1810

Picanço-Diniz CW (1987): *Organização do sistema visual de roedores da Amazônia: Topografia das áreas visuais da cutia. "Dasyprocta aguti."* Ph.D. Thesis, Instituto de Biofísica Carlos Chagas Filho, Universidade Federal do Rio de Janeiro, Brazil

Picanço-Diniz CW, Silveira LCL, de Carvalho MSP, Oswaldo-Cruz E (1991): Contralateral visual field representation in area 17 of the cerebral cortex of the agouti: A comparison between the cortical magnification factor and retinal ganglion cell distribution. *Neuroscience* 44:325–333

Pointer JS (1986): The cortical magnification factor and photopic vision. *Biol Rev* 61:97–119

Provis JM (1979): The distribution and size of ganglion cells in the retina of the pigmented rabbit: A quantitative analysis. *J Comp Neurol* 185:121–138

Reese BE, Cowey A (1986): Large retinal ganglion cells in the rat: Their distribution and laterality of projection. *Exp Brain Res* 61:375–385

Schein SJ, de Monasterio FM (1987): Mapping of retinal and geniculate neurons onto striate cortex of macaque. *J Neurosci* 7:996–1009

Silveira LCL (1985): *Organização do sistema visual de roedores da Amazônia: Óptica ocular e distribuição das células ganglionares retinianas*. Ph.D. Thesis, Instituto de Biofísica, Universidade Federal do Rio de Janeiro, Brazil

Silveira LCL, Picanço-Diniz CW, Oswaldo-Cruz E (1989a): Distribution and size of ganglion cells in the retinae of large Amazon rodents. *Visual Neurosci* 2:221–235

Silveira LCL, Picanço-Diniz CW, Sampaio LFS, Oswaldo-Cruz E (1989b): Retinal ganglion cell distribution in the *Cebus* monkey: A comparison with the cortical magnification factors. *Vision Res* 29:1471–1483

Steinberg RH, Reid M, Lacy PL (1973): The distribution of rods and cones in the retina of the cat (*Felis domesticus*). *J Comp Neurol* 148:229–248

Stone J (1965): A quantitative analysis of the distribution of ganglion cells in the cat's retina. *J Comp Neurol* 124:337–352

Stone J (1983): *Parallel Processing in the Visual System: The Classification of Retinal Ganglion Cells and Its Impact on the Neurobiology of Vision*. New York: Plenum Press

Stone J, Johnston E (1981): The topography of primate retina: A study of the human, bushbaby, and New- and Old-World monkeys. *J Comp Neurol* 196:205–223

Tiao YC, Blakemore C (1976a): Regional specialization in the golden hamster's retina. *J Comp Neurol* 168:439–458

Tiao YC, Blakemore C (1976b): Functional organization in the visual cortex of the golden hamster. *J Comp Neurol* 168:459–482

Tolhurst DJ (1989): Cortical magnification and peripheral innervation density. In: *Seeing Contour and Colour*, Kulikowski JJ, Dickinson CM, Murray IJ, eds. Oxford: Pergamon Press

Tusa RJ, Palmer LA, Rosenquist AC (1978): The retinotopic organization of area 17 (striate cortex) in the cat. *J Comp Neurol* 177:213–236

van Buren JM (1963): *The Retinal Ganglion Cell Layer*. Springfield, IL: Thomas

van Essen DC (1979): Visual areas of the mammalian cerebral cortex. *Ann Rev Neurosci* 2:227–264

Vaney DI, Peichl L, Wassle H, Illing RB (1981): Almost all ganglion cells in the rabbit retina project to the superior colliculus. *Brain Res* 212:447–453

Wagor E, Mangini NJ, Pearlman AL (1980): Retinotopic organization of striate and extrastriate visual cortex in the mouse. *J Comp Neurol* 193:187–202

Wässle H, Grünert U, Röhrenbeck J, Boycott BB (1989): Cortical magnification factor and the ganglion cell density of the primate retina. *Nature* 341:643–646

Wässle H, Grünert U, Röhrenbeck J, Boycott BB (1990): Retinal ganglion cell density and cortical magnification factor in the primate retina. *Vision Res* 11:1897–1911

Webb SV, Kaas JH (1976): The sizes and distribution of ganglion cells in the retina of the owl monkey, *Aotus trivirgatus*. *Vision Res* 16:1247–1254

Whitteridge D (1965): Geometrical relations between the retina and the visual cortex. In: *Mathematics and Computer Science in Biology and Medicine*, Medical Research Council, ed. Longdon: H.M.S.O.

Whitteridge D (1973): Projection of optic pathways to the visual cortex. In: *Handbook of Sensory Physiology: Vol. 7, Part 3: Central Processing of Visual Information. Part B. Visual Centers in the Brain*, Jung R, ed. Berlin: Springer-Verlag

Wikler KC, Rakic P (1990): Distribution of photoreceptor subtypes in the retina of diurnal and nocturnal primates. *J Neurosci* 10:3390–3401

14

Columnar Organization of Primate Visual Cortex

ROGER B.H. TOOTELL AND RICHARD T. BORN

Most developmental studies of visual system functional organization have concentrated, quite sensibly, on areas in which the adult organization was already well understood. In vertebrates, these include the ocular dominance, orientation and retinotopic maps in mammalian visual cortex, and the retinotopy (and in experimental preparations, the ocular segregation) in frog optic tectum. Developmental aspects of these topics are covered elsewhere in this volume. Here we review the wealth of new data on the functional architecture of monkey visual cortex, a system particularly ripe for future developmental enquiry.

Such studies are particularly promising because the functional architecture is so highly ordered in this region of the brain (e.g., DeYoe and Van Essen, 1988; Felleman and Van Essen, 1991; Livingstone and Hubel, 1988; Zeki and Shipp, 1988), and also because the architecture is likely to be similar to that in man. This new data also makes it possible to test whether developmental mechanisms already described in striate cortex and the tectum (e.g., NMDA receptors, dendritic exuberance followed by retraction, etc.) can be generalized to other cortical areas. Because the recently-described functional architectural features are extensions of, and often connected with, known striate cortical architectures, it also becomes possible to ask developmental questions about the relative control and timing of intra- and intercortical connections between specific columns in each area. For instance, does column A in one area develop before or after column B in a higher-tier area?; does column A induce column B?; and so on.

TELEOLOGY OF COLUMNS AND GENERAL COMMENTS

Cortical laminae were obvious from the very earliest cytoarchitectonic studies (e.g., Brodman, 1905). It is becoming increasingly apparent that functional columns (cells sharing some common functional property, grouped radially, perpendicular to the cortical surface) are an equally profound and common feature of the cortical architecture. Since functional columns are not usually revealed by cytoarchitectonic stains, their widespread demonstration has had to wait for the advent of current functional labeling techniques, such as deoxyglucose (DG) labeling, optical recording with and without voltage-sensitive dyes, stains for metabolic enzymes, and others.

It is not clear why columns should occur in so many cortical areas. One could conceivably design a cortex in which functionally similar cells are grouped together in different layers, or at oblique angles to each other, or in alternating compartments in different layers, or in numerous other geometric arrangements. One could even distribute functionally similar cells randomly but retain systematic connections between them.

Two general theories have been offered for the existence of cortical columns (Barlow, 1986; Cowley, 1979). First, they might greatly simplify the developmental coding necessary for cortical organization. Whether cells in a given column or module are initially equipotential with maps imposed on them, or whether they migrate to a predetermined map position, it should take much less information to code a cell's position or connections relative to a simple two-dimensional map, rather than to each of the millions of cells surrounding it.

Second, cells that are functionally similar may share more connections than cells that are dissimilar (Gilbert and Wiesel, 1989; Livingstone and Hubel, 1984a,b). If so, then collections of similar cells within columns (or any other arrangement, for that manner) will reduce the average length of neural connection in that region of the brain. This model works only if functionally similar cells share more connections than functionally dissimilar cells. Furthermore, it only justifies a grouping of like cells; it does not favor one specific grouping arrangement over another (e.g., columns vs. separate laminae).

Teleology is a tricky business at best, and here we muddy the waters further with two additional possibilities. First, the small blood vessels in the cerebral cortex are oriented radially, parallel to the long axis of the columns. Since the metabolic demand of active neurons is higher than that of inactive neurons, it may be that functionally similar cells are grouped together parallel with the microvasculature (i.e., in columns) so that blood flow can be increased *specifically* to active neurons. The tight correlation between neural columnar activity and microvascular blood flow is the basis for optical recording of "intrinsic" images, which have been shown to correlate well with neural maps. This becomes something of a chicken-and-egg problem (which came first, the radial microvasculature or the parallel neural columns), but it is certainly a useful arrangement.

Finally, there is a more subtle aspect to the cortical columnar maps. In primary visual cortex, cells are anatomically arranged to preserve the original neighborhood relations seen in the peripheral receptors, the "retinotopic" organization. Homologous architectures exist in primary auditory and somatosensory cortex, in which cells are arranged into cochleotopic and somatotopic maps, respectively. The retinotopic organization can be quite precise: in the macaque, map borders as measured by DG uptake can be resolved more precisely than the acuity limits of the animal at the same eccentricity (Tootell et al., 1988b) (see Figure 14.1).

In progressively higher tier cortical visual areas, the receptive fields become progressively larger and the stimulus selectivity becomes progressively more abstract, until all traces of the initial precise retinotopy are lost. Two-dimensional

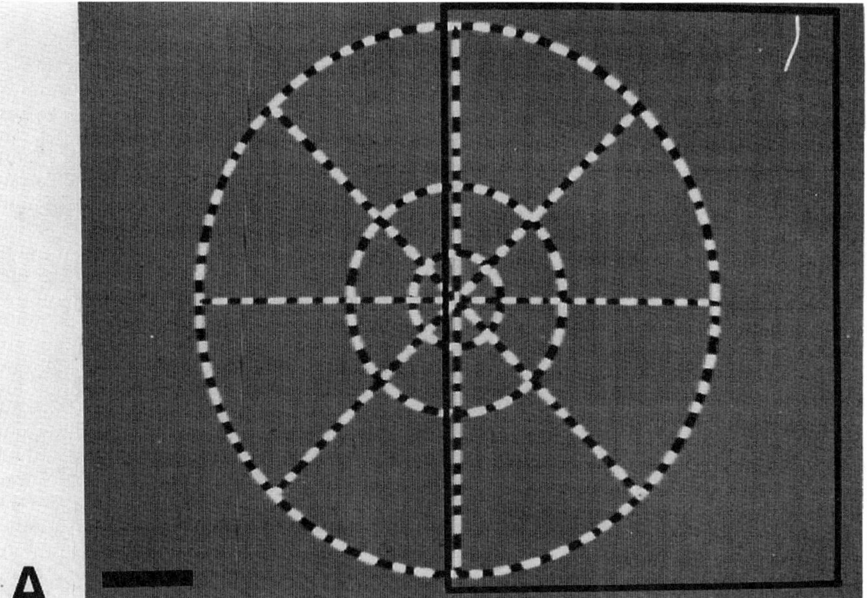

FIGURE 14.1. Deoxyglucose labeling of retinotopic organization in macaque striate cortex. A. The visual stimulus viewed monocularly by an anesthetized, paralyzed monkey. The black-and-white checks are flickered in couterphase at 3 Hz on a uniform grey background. Calibration bar = 2°. B. An autoradiograph taken from a flattened tissue section, including the central half of striate cortex. The areas of striate cortex that were activated by the retinotopically restricted stimulus are clearly resolvable in the DG image. The sharpest retinotopic borders occur in layer 4Cb, which show a half-amplitude fall-off of 140 μm. Calibration bar = 1 cm. Reprinted with permission of AAAS from (1982): Deoxyglucose analysis of retinotopic organization in primate striate cortex. *Science* 218: © 1982 AAAS.

information in the primary retinotopic map is thus converted to a *nonretinotopic* code in the same two dimensions (that is, a folded plane parallel with the cortical layers). The retinotopic "columns" in early areas like primary visual cortex (and even the LGN) are transformed into orientation columns, direction columns, and other types of columns in higher tier cortical areas.

The division of cortex into separate *areas* is closely related to the division of cortical areas into *columns* and *modules*. Cortical *areas* are based on differences in the topography of: (1) the retinotopy, (2) histological staining patterns, (3) connections, and/or (4) stimulus sensitivity (e.g., Felleman and Van Essen, 1991). Importantly, cortical *columns and modules* have also been defined on the basis of differences in the topography of histological staining (e.g., Horton and Hubel, 1981; Tootell et al., 1983), function (e.g., Livingstone and Hubel, 1984a, 1987), and even the retinotopy (e.g., Zeki and Shipp, 1989). It would not be surprising if mechanisms used in the development of cortical areas were similar to those used in the development of columns and modules.

STRIATE CORTICAL ARCHITECTURE

The most well-studied examples of cortical columns are probably the ocular dominance and orientation columns in area V1. This bias is easy to explain: until recently the evidence for columns in additional cortical visual areas has been quite scarce. Nevertheless, it is ironic that columnar development has been studied almost entirely in primary visual cortex, because this area shows striking laminar variations not present in other visual cortical areas (see below).

Why should the functional architecture of striate cortex be so highly laminated, compared to that in extrastriate visual areas? One reason may be that there is an unusually heavy information load passing through the area into extrastriate cortex. Much of the information distributed throughout extrastriate visual cortex passes first through striate cortex (e.g., Felleman and Van Essen, 1991). Furthermore, two strictly segregated inputs (the magno- and parvocellular LGN inputs) and three segregated outputs (from the blobs, interblobs, and layer 4B) need to be routed and processed through striate cortex in a systematic manner (DeYoe and Van Essen, 1988; Felleman and Van Essen, 1991; Livingstone and Hubel, 1988; Zeki and Shipp, 1988). Unlike the LGN (another throughput area), a great deal of receptive field rearrangement takes place in striate cortex. These processing constraints may also help explain why striate cortex is the largest known area in primate visual cortex (Felleman and Van Essen, 1991; Van Essen and Maunsell, 1980), with a cell density twice as high as in other cortical areas (e.g., Garey, 1971).

Numerous "extra" laminar differences are prominent in anatomical, physiological, and functional labeling studies of V1. Anatomically, 12 layers (1, 2, 3, 4A, 4B, 4Ca, 4Cb, 4D, 5A, 5B, 6A, and 6B) have now been differentiated in the area (e.g., Lund and Boothe, 1975; Lund et al., 1975), compared to the nominal 6 layers in most cortical areas. Even the name and early identification of "striate" cortex are based on an unusual laminar feature, the striking band of high myelination in layer 4B (Gennari, 1782; Vicq d'Azyr, 1786).

Figure 14.2 shows a diagram of all the single unit properties not appearing at prior subcortical nuclei, together with the striate cortical layers in which they first appear. Because single units have been more thoroughly investigated in striate cortex than in other cortical areas, it is difficult to say with certainty that there are *more* laminar differences in striate cortical receptive fields than in higher cortical areas. However, the elaborate laminar differences in physiology certainly appear related to the elaborately laminated anatomy.

Some functional labeling techniques, like the optical recording procedure, give no information about possible laminar variations. Another procedure, the deoxyglucose (DG) labeling method, has confirmed and clarified the wealth of laminar variations in V1 suggested by the earlier anatomical and physiological studies. In addition, the DG studies suggested several new functional distinctions, such as differences in peak spatial frequency tuning between blob and interblob cells (Born and Tootell, 1991a; Tootell et al., 1988c), and blob/interblob differences of function in layers 4B, 5, and 6 (Tootell et al., 1988c, 1988d, 1988e).

Several years ago, single unit data suggested that cells in layer 4Ca are oriented, rather than nonoriented like cells in the companion geniculorecipient layer 4Cb (Livingstone and Hubel, 1984a). Early DG studies (Livingstone and Hubel, 1984a) also suggested that layer 4Ca was orientation-specific. However, that study relied on comparisons between sections 70 μm thick; this is a possible source of confusion when attributing DG patterns to layers only slightly thicker than that. In addition, the initial study used only a single DG label, which leaves open the possibility that the periodicities in layer 4Ca were related to a bloblike architecture rather than an orientation architecture in 4Ca.

To resolve these questions, we labeled orientation columns using a double-labeled deoxyglucose technique, and stained the same sections for cytochrome oxidase to reveal laminar boundaries in the same images. An anesthetized, paralyzed macaque monkey viewed a drifting rectangular wave grating at a single orientation, binocularly. The spatial and temporal frequencies of the grating were systematically varied over a range appropriate for striate cortical neurons. Tritiated-deoxyglucose was injected to label the activation produced by a horizontal grating, and ^{14}C-deoxyglucose was injected to label the effects of the grating at a vertical orientation.

Figure 14.3 shows the result as a difference image. The orientation columns appear as black-and-white patches, produced by the horizontal versus vertical gratings, respectively. Laminar boundaries produced by automated density analysis of the cytochrome oxidase staining patterns from the same section are superimposed on the DG image. It is clear that the orientation-specific "columns" extend through layer 4Ca, but skip layer 4Cb.

Let us consider another example. Some evidence suggests that New World monkeys lack ocular dominance columns, based on an apparent lack of modula-

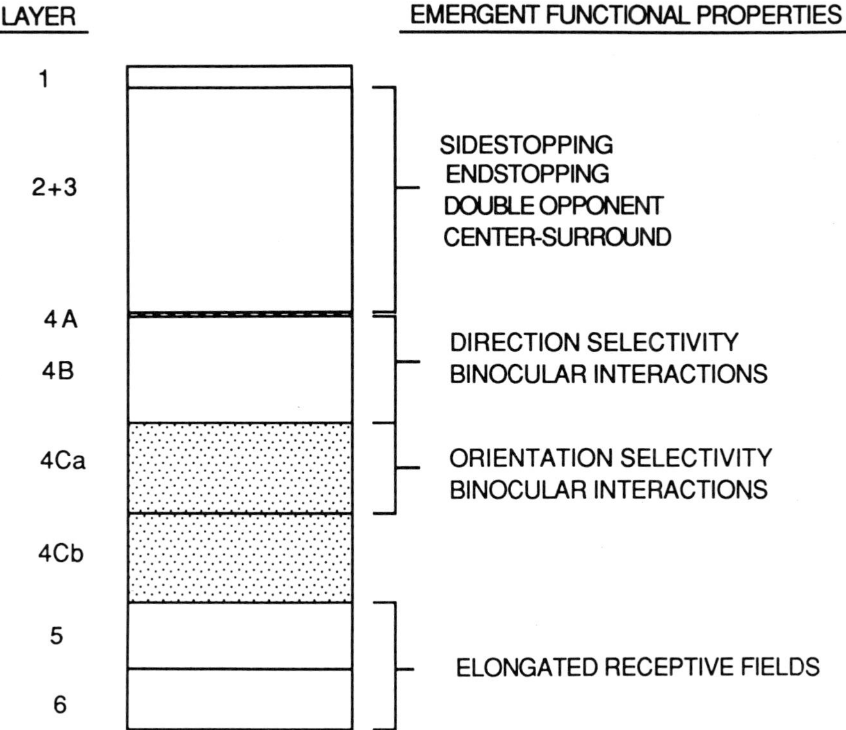

FIGURE 14.2. Emergent receptive field properties in different layers of macaque striate cortex. The diagram shows a cross-section of macaque striate cortex, with the cortical surface uppermost and the white matter at the bottom. Cytochrome oxidase blob- and interblob compartments are not distinguished. Input from the LGN is nonoriented, monocular and center-surround in receptive field shape. Receptive fields in layer 4Cb are similar to those seen in the LGN, but at least some receptive fields in 4Ca are orientation-specific and show binocular input (Blasdel and Fitzpatrick, 1984; Livingstone and Hubel, 1984a). Layer 4B, which gets most of its input from 4Ca, is populated with oriented, mostly simple receptive fields that are often directional (Blasdel and Fitzpatrick, 1984; Dow, 1974; Hawken et al., 1988; Hubel and Weisel, 1968), an attribute not seen at prior levels. Layers 2 + 3 contain double-opponent receptive fields in the cytochrome oxidase blobs (Livingstone and Hubel, 1984; Ts'o and Gilbert, 1988), which are not seen at prior levels. In the layer 2 + 3 interblobs side-stopping (Born and Tootell, 1991b) and end-stopping (Livingstone and Hubel, 1984) appear, apparently for the first time. In the lower layers the receptive fields become greatly elongated (Gilbert, 1977).

tion in the transneuronal transport of tritiated amino acids into layer 4C of New World monkeys (e.g., Kaas et al., 1978). This makes it difficult to rationalize the presence of ocular dominance columns in terms of stereopsis, which is present in New World monkeys (Allman, 1988). To test this possibility, we showed owl monkeys a very general stimulus (a drifting square wave grating shown at all

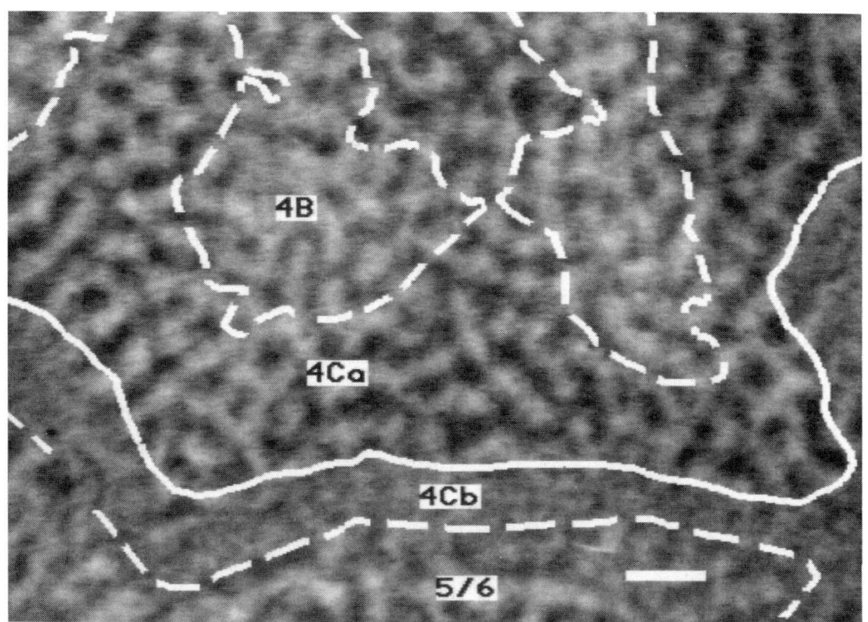

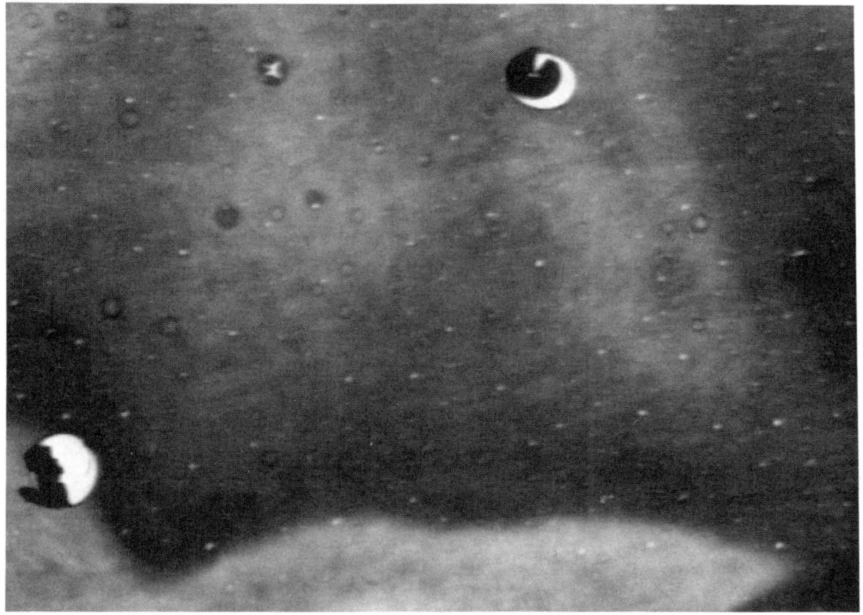

See caption FIGURE 14.3 on next page.

orientations, at systematically varied velocities and spatial frequencies) through one eye. DG was injected to label this monocular activity. Figure 14.4 shows what appears to be ocular dominance columns in layer 4 of owl monkey striate cortex. Control experiments using identical visual stimuli viewed binocularly suggest that the patterns in Figure 14.4 are not due to bloblike biases in the stimulus sensitivity or metabolic activity of the cells. These DG-labeled columns in the New World monkey are admittedly lower in contrast and more restricted to the middle layers than those in the Old World macaque, when tested identically (Tootell et al., 1988a). However, the data does suggest that ocular dominance columns are evolutionarily conserved, and this allows us to reach more general conclusions about the teleology and usefulness of such columns.

EXTRASTRIATE VISUAL CORTICAL FUNCTIONAL ARCHITECTURE

These and many additional deoxyglucose experiments have shed light on the functional architecture of extrastriate cortex as well. With the exception of a uniform grey field, every visual stimulus tested has produced columns of high activation within as many as 10 different visual areas. These areas include V1, V2, V3, V3A, DP, V4, VP, MT, and V4t. In extrastriate areas, the columns have a stereotypic laminar profile, much simpler than that in V1. They are darkest (most active) in layer 4, with monotonically decreasing contrast in super- and subgranular layers.

Double-label deoxyglucose tests allowed us to test whether such columns are specific for stimulus orientation, color, binocular disparity, velocity, and other stimulus parameters. In a typical test, we compared the effects of a stimulus of one orientation to the effects of the same stimulus at an orthogonal orientation, or one color to its color complement, or one direction to its opposite direction, and so forth.

From these experiments we conclude that the columnar organization so obvious in V1 is replicated in kind in extrastriate visual areas. In V2 and VP, there is unambiguous single- and double-label DG evidence for orientation columns (Tootell and Born, 1990; Tootell and Hamilton, 1989). In MT, there is a

FIGURE 14.3. Double-label DG evidence for orientation columns in layer 4Ca of macaque striate cortex. A. A difference image from a flattened section, in which activation produced by a horizontal grating is coded white and activation produced by a vertical grating is coded black. Regions of equal activation thus appear uniformly grey, and the regions of unequal orientation-specific activation appear as alternating white and black bands. B. Cytochrome oxidase staining patterns from corresponding regions of the same tissue section. The laminar boundaries shown in panel A are based on densitometric measurements of the cytochrome oxidase staining patterns in B. It is clear that there is a high-contrast orientation map in layer 4Ca, a medium-contrast orientation map in layers 4B and 5, and no orientation map in 4Cb. Calibration bar = 1 mm.

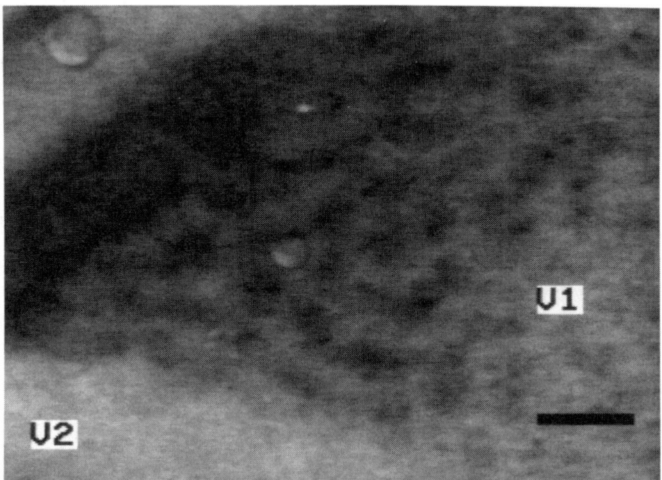

FIGURE 14.4. DG evidence for ocular dominance patches in owl monkey striate cortex. The monkey was stimulated with a very general visual stimulus (a rectangular wave grating of varied spatial frequency, drift speed, orientation, and direction) through only one eye, while DG was injected. The patchy patterns of high activity occur only in layer 4 of striate cortex (V1), and do not occur in other areas (e.g., V2), nor in control animals shown the same visual stimuli binocularly. Calibration bar = 1 mm.

columnar, 360° map of preferred direction (Tootell and Born, 1991). In the thin stripes of V2 and in many additional areas, there is evidence for columns of bloblike cells, sensitive to color differences and to low spatial frequencies, when tested appropriately (Tootell and Born, 1990; Tootell and Hamilton, 1989). There are also many systems of columns that we do not yet understand well: these hold great promise for understanding cortical function when they can be tested in detail.

It is also instructive to consider *changes* in the columnar organization between cortical areas that are known to be interconnected. For instance, cells receiving dominant input from each eye are kept segregated in the six geniculate layers and in the ocular dominance columns in striate cortex. The ocular dominance columns extend through all layers of striate cortex, at least in metabolic labeling studies. However, there is no trace of ocular dominance columns nor patches in V2, nor in any other cortical area. This finding corresponds well with the prevalence of ocular biases reported in V1 single units, compared with their apparent absence in units recorded from V2 and other extrastriate areas. Together, these data suggest that input from each eye is kept separate until V1, where it is systematically combined, and the merging of monocular inputs appears complete (or at least transformed into a binocular code) by the time information reaches the next cortical areas. Presumably, this merging of inputs must be done systematically, because it is critical to know which eye the input came in from, in order to compute binocular disparity, vergence movements, and so on.

The progression of orientation-specific information offers an interesting contrast. Orientation-specific cells are common in many extrastriate visual areas (e.g., Felleman and Van Essen, 1991; Hubel and Wiesel, 1970; Zeki, 1983), and evidence for orientation-specific columnar maps has now been shown in at least areas V1 (Blasdel and Salama, 1986; Hubel and Wiesel, 1968; Hubel et al., 1978), V2 (Hubel and Livingstone, 1987; Tootell and Born, 1990; Tootell and Hamilton, 1989), and VP (Tootell and Born, 1990). These areas form a serial chain three areas long, in which orientation specificity is retained and mapped. Interestingly, the intercolumnar distance becomes progressively wider as one ascends the chain. Thus, the orientation maps are conserved across at least three areas, while ocular dominance maps appear to serve a more transitory purpose, apparently complete within primary visual cortex.

The evidence for direction columns in area MT is also interesting. Single unit studies have shown that direction-specific cells can be found in several visual cortical areas, including (layers 4B and 6 of) V1, area V2, V3, V3A, VP, and especially MT and MST (e.g., Zeki, 1978, 1983). Double-label DG tests show that 360° maps of direction are found only in area MT and perhaps MST, but not in areas V1, V2, V3, V3A, or VP (Tootell and Born, 1991). This finding brings up several interesting questions. For instance, why does a direction map appear in MT, but not in the many directional cells of striate cortical layer 4B, which projects directly to MT (Lund et al., 1975; Maunsell and Van Essen, 1983; Rockland and Pandya, 1981)? Also, how are the 180° maps of orientation columns in V1 and V2 related to the 360° map of direction columns in MT? It is easy to imagine how a 180° map of orientation or axis-of-motion can be extracted from a 360° map of direction, by simply combining the signals from directions 180° apart. However, the evidence suggests the reverse, a 360° map in MT extracted from 180° maps in prior areas V1, V2, and so on. This is harder to imagine, based solely on map coordinates.

A final example shows an interesting aspect to the intracortical functional anatomy that bears on receptive field analysis in the "width" axis. In V1, both single unit and deoxyglucose evidence indicates that "side-stopping" is common in cells in the layer 2 and 3 interblob regions (Born and Tootell, 1991b). Since these cells are major projection neurons into subregions of area V2 (e.g., Livingstone and Hubel, 1984), it made sense to look for the presence of side-stopping in area V2 cells.

We designed a double-label deoxyglucose experiment to look for such receptive field properties. In this experiment, an anesthetized, paralyzed macaque viewed a vertical achromatic grating, binocularly. For one stimulation condition, the grating was a square wave grating of 5.0 cycles/degree. For the other stimulation condition, the grating was a rectangular wave grating, of duty cycle 0.2, and 2.0 c/deg spatial frequency. The rectangular wave grating was designed to maximally stimulate the V1 side-stopped cells.

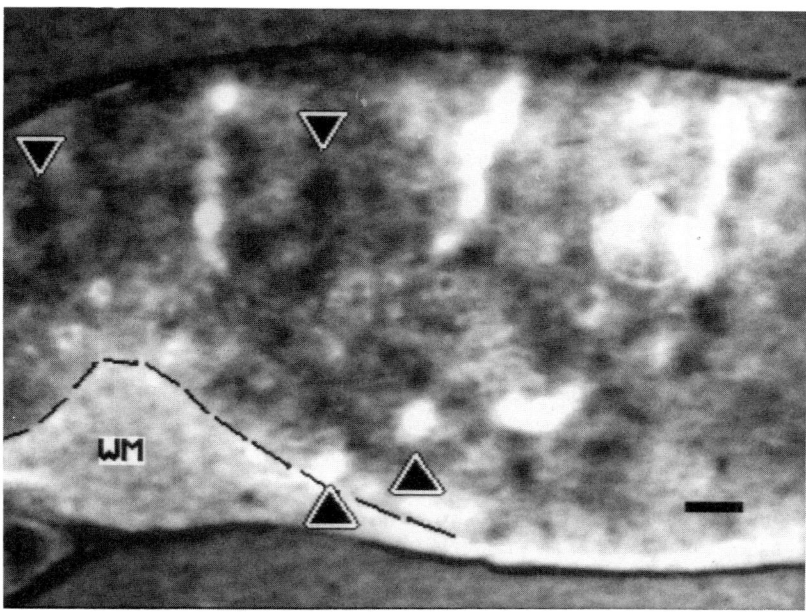

FIGURE 14.5. Double-label deoxyglucose evidence for columns of side-stopped versus non-side-stopped cells in cortical area V2. The figure shows the difference between the autoradiograph produced by a rectangular grating (which activates side-stopped cells) and the autoradiograph produced by a size-matched square wave grating (which activates side-stopped cells less or not at all). Both autoradiographs were produced by the same tissue section, cut parallel with the flattened surface from V2. Activated patches produced by the square wave are shown in white, two of which are indicated by upward-pointing triangles. Activated patches produced by the rectangular wave are coded in black, two of which are indicated by downward-pointing triangles. The black and white patches are columnar in shape when reconstructed in three dimensions. This constitutes evidence for columns of cells that are preferentially activated by the rectangular wave grating, presumably side-stopped. WM = white matter. Calibration bar = 1 mm.

Figure 14.5 shows the results of this experiment, as a difference image in area V2. Clearly, different regions are activated by the rectangular versus the square wave gratings. Since both gratings were presented at the same orientation, at the same binocular disparity, and at the same contrasts and velocities, this strongly suggests that side-stopped cells are grouped into specific columns within V2, and that cells without side-stopping are arranged with a different topography. Obviously, further single unit work is required to understand the mapping differences, but the DG results are certainly suggestive.

Approaches such as the ones described here will continue to clarify the functional architecture in visual cortex. Already, however, it is clear that columnar functional architectures are the rule rather than the exception throughout visual cortex. By studying them further we should be able to map out the connections between each columnar system, and begin to study the associated developmental mechanisms.

REFERENCES

Allman J (1988): Variations in visual cortex organization in primates. In: *Neurobiology of Neocortex*, Rakic P, Singer W, eds.

Barlow HB (1986): Why have multiple cortical areas? *Vision Res* 26:81–90

Blasdel GG, Fitzpatrick D (1984): Physiological organization of layer 4 in macaque striate cortex. *J Neurosci* 4:880–895

Blasdel GG, Salama G (1986): Voltage-sensitive dyes reveal a modular organization in monkey striate cortex. *Nature* 321:579–585

Born RT, Tootell RBH (1991a): Spatial frequency tuning of single units in macaque supragranular striate cortex. *Proc Natl Acad Sci U S A* 88:7066–7070

Born RT, Tootell RBH (1991b): Single-unit and 2-deoxyglucose studies of side inhibition in macaque striate cortex. *Proc Natl Acad Sci U S A* 88:7071–7075

Brodman K (1905): Beitrage zur histologischen lokalisation der Grosshirnrinde. *J Psychol Neurol* 6:275–400

Cowey A (1979): Cortical maps and visual perception: The Grindley Memorial Lecture. *Q J Exp Psychol* 31:1–17

DeYoe EA, Van Essen DC (1988): Concurrent processing streams in monkey visual cortex. *Trends Neurosci* 11:219–226

Dow BM (1974): Functional classes of cells and their laminar distribution in monkey visual cortex. *J Neurosphsiol* 37:927–946

Felleman DJ, Van Essen DC (1991): Distributed hierarchical processing in primate cerebral cortex. *Cerebral Cortex* 1:1

Garey LJ (1971): A light and electron microscopic study of the visual cortex of the cat and monkey. *Proc R Soc Lond (Biol)* 21–40

Gennari F (1782): *De peculiari structura cerebri nonnullisque eius morbis. Paucae aliae anatom. observat. accendunt.* Parma, Italy

Gilbert CD (1977): Laminar differences in receptive field properties of cells in cat primary visual cortex. *J Physiol (Lond)* 288:391–421

Gilbert CD, Wiesel TN (1989): Columnar specificity of intrinsic horizontal and corticocortical connections in cat visual cortex. *J Neurosci* 9:2432–2442

Hawken MJ, Parker AJ, Lund JS (1988): Laminar organization and contrast sensitivity of direction-selective cells in the striate cortex of the Old World monkey. *J Neurosci* 8:3541–3548

Horton JC, Hubel DH (1981): Regular patchy distribution of cytochrome oxidase staining in primary visual cortex of macaque monkey. *Nature* 292:762–764

Hubel DH, Livingstone ML (1987): Segregation of form, color and stereopsis in primate area 18. *J Neurosci* 7:3378–3315

Hubel DH, Wiesel TN (1968): Receptive fields and functional architecture of monkey striate cortex. *J Physiol (Lond)* 195:215–243

Hubel DH, Wiesel TN (1970): Cells sensitive to binocular depth in area 18 of the macaque monkey visual cortex. *Nature* 225:41–42

Hubel DH, Wiesel TN, Stryker MP (1978): Anatomical demonstration of orientation columns in macaque monkey. *J Comp Neurol* 177:361–380

Kaas J, Heurta MF, Weber JT, Harting JK (1978): Pattern of retinal terminations and laminar organization of the lateral geniculate nucleus of primates. *J Comp Neurol* 182:517–554

Livingstone MS, Hubel DH (1984a): Anatomy and physiology of a color system in the primate visual cortex. *J Neurosci* 4:309–356

Livingstone MS, Hubel DH (1984b): Specificity of intrinsic connections in primary visual cortex. *J Neurosci* 4:2830–2835

Livingstone MS, Hubel DH (1987): Connections between layer 4B of area 17 and thick cytochrome oxidase stripes of area 18 in the squirrel monkey. *J Neurosci* 7:3371–3377

Livingstone M, Hubel DH (1988): Segregation of form, color, movement and depth: Anatomy, physiology and perception. *Science* 240:740–749

Lund JS, Boothe RG (1975): Interlaminar connections and pyramidal neuron organization in the visual cortex, area 17, of the macaque monkey. *J Comp Neurol* 159:305–334

Lund JS, Lund RD, Hendrickson AE, Bunt AH, Fuchs AF (1975): The origin of efferent pathways from the primary visual cortex, area 17, of the macaque monkey as shown by retrograde transport of horseradish peroxidase. *J Comp Neurol* 164:287–304

Maunsell JHR, Van Essen DC (1983): The connections of the middle temporal visual area (MT) and their relationship to a cortical hierarchy in the macaque monkey. *J Neurosci* 3:2563–2586

Rockland KS, Pandya DN (1981): Cortical connections of the occipital lobe in the rhesus monkey: Interactions between areas 17, 18, 19, and the superior temporal sulcus. *Brain Res* 212:249–270

Tootell RBH, Born RT (1990): Columns beyond V1 and V2 in macaque visual cortex: A double-label deoxyglucose study. *Soc Neurosci Abstr* 16:292

Tootell RBH, Born RT (1991): Architecture of primate area MT. *Soc Neurosci Abstr* 17:524

Tootell RBH, Hamilton SL (1989): Functional anatomy of the second cortical visual area (V2) in the macaque. *J Neurosci* 9:2620–2644

Tootell RBH, Hamilton SL, Silverman MS, Switkes E (1988a): Functional anatomy of macaque striate cortex: 1. Ocular dominance, binocular interactions and baseline conditions. *J Neurosci* 8:1500–1530

Tootell RBH, Switkes E, Silverman MS, Hamilton SL (1988b): Functional anatomy of macaque striate cortex: 2. Retinotopic organization. *J Neurosci* 8:1531–1568

Tootell RBH, Silverman MS, Hamilton SL, DeValois RL, Switkes E (1988c): Functional anatomy of macaque striate cortex: 3. Color. *J Neurosci* 8:1569–1593

Tootell RBH, Hamilton SL, Switkes E (1988d): Functional anatomy of macaque striate cortex: 4. Contrast and magno-parvo streams. *J Neurosci* 8:1594–1609

Tootell RBH, Silverman MS, Hamilton SL, Switkes E, DeValois RL (1988e): Functional anatomy of macaque striate cortex: 5. Spatial frequency. *J Neurosci* 8:1610–1624

Tootell RBH, Silverman MS, DeValois RL, Jacobs GH (1983): Functional organization of the second cortical visual area of primates. *Science* 220:737–739

Ts'o DY, Gilbert CD (1988): The organization of chromatic and spatial interactions in the primate striate cortex. *J Neurosci* 8:1712–1727

Van Essen DC, Maunsell JH (1980): Two-dimensional maps of the cerebral cortex. *J Comp Neurol* 191:255–281

Vicq-d'Azyr R (1786): *Traité d'anatomie et de physiologie*. Paris: F.A-Didot

Zeki SM (1978): Uniformity and diversity of structure and function in rhesus monkey prestriate visual cortex. *J Physiol* 277:273–290

Zeki SM (1983): The distribution of wavelength and orientation selective cells in different areas of monkey visual cortex. *Proc R Soc Lond (Biol)* 217:449–470

Zeki SM, Shipp S (1988): The functional logic of cortical connections. *Nature* 335:311–317

Zeki SM, Shipp S (1989): Modular connections between areas V2 and V4 of macaque monkey visual cortex. *Eur J Neurosci* 1:494–506

15

In Search of the Canonical Microcircuits of Neocortex

RODNEY J. DOUGLAS AND KEVAN A.C. MARTIN

> "The machinery may be roughly uniform over the whole striate cortex, the difference being in the inputs. A given region of the cortex simply digests what is brought to it, and the process is the same everywhere. It may be that there is a great developmental advantage in designing such a machinery once only, and repeating it over and over monotonously, like a crystal"—Hubel and Wiesel, 1974

EXPLORING THE CEREBRAL JUNGLE

The circuitry of the neocortex is the lodestone that has attracted generations of neuroscientists. Few have felt the pull more strongly than Ramon y Cajal, who viewed the cerebral cortex as an "enigma of enigmas," a "cerebral jungle" with "impenetrable thickets in which so many explorers had lost themselves" (Ramon y Cajal, 1937). His success in discovering the basic circuit of the cerebellum led him to anticipate, with misplaced optimism as he later admitted, that he could repeat the performance with the neocortex. Contemporary neuroscientists are still battling through the same jungle. At times they do not seem much closer to understanding the basic organization of the neocortical circuits than did Ramon y Cajal (Martin, 1988a,b). Perhaps this is an indication that there are no basic circuits to be found.

In developing our own direction we have made one central assumption that separates us from the historical stream that has Ramon y Cajal as its source: that the cortical circuitry is essentially simple. His view, that the circuits are extremely complex, probably arises from the tertiary structure of the circuits, much as the folding of a protein adds a complex dimension to a basically simple linear sequence of amino acids, allowing a protein to function as a receptor, enzyme, or piece of specialized tissue. Similarly, although they may be overlapped or interleaved with each other to form apparent complexity, we assume that the basic or "canonical" circuits will be relatively simple (Douglas et al., 1989).

From silicon technology we are already familiar with the concept of a simple processing unit, the transistor, that can be repeated over and over on a silicon chip to provide more processing power. Whether or not Ramon y Cajal's dream of discovering the basic principles of organization of the microcircuits is realized by the current generation of neuroscientists, there are many good reasons for making the attempt, and it is the purpose of this chapter to provide some of the reasons why. Neuroanatomists studying the microcircuits of neocortex have produced widely contrasting views of the nature of the circuits, ranging from high degrees of specificity between individual neurones (Palay, 1967) through "(quasi)-randomness" (Szentágothai, 1978), to the view that "no theory, mathematical or other, relying on specific circuits can be maintained" (Sholl, 1956). But the real reason that 100 years after Ramon y Cajal we still have no consensus on the form of a basic cortical circuit, or even agreement that such circuits exist, may simply be because the available techniques have not been well-matched to the magnitude of the problem. Our own view is that Ramon y Cajal's original intent of discovering the general principles of cortical organization remains a valid and important route of exploration and we have begun the search anew for the canonical microcircuits of the visual cortex (Douglas and Martin, 1991b; Douglas et al., 1989).

Canonical microcircuits exist in the brain. They have been described in the olfactory bulb (Shepherd, 1988) and in the cerebellum (Ramon y Cajal, 1904). All these circuits form cortical (barklike) structures. It is tempting to think that cortical structures are ubiquitous because they solve basic problems of organizing neurons into processing units or canonical microcircuits that are efficient in terms of wiring, relatively simple and easy to build, and modular in design. The implication is that where there is a cortex, there will be a basic microcircuit. Repeated modular-like structures are used throughout nature, perhaps most prominently and effectively in the design of the DNA molecule. Thus it might be expected that the rapid expansion of the neocortex has been made possible by a modular design and a relative ease in the replication of standard modules. As in the case of DNA, many of the organizing principles and functions of the neocortical microcircuits will only become apparent when we understand the basic form of the canonical microcircuits.

Already we have made some progress in developing a canonical microcircuit based on the circuits in cat visual cortex. The details, justification, and logic of the development of our first version of a canonical microcircuit have been published and we will not cover the same ground here (Douglas and Martin, 1991b, 1992; Douglas et al., 1989). However, to give some flavor of the kind of concept we will be discussing here, we illustrate the microcircuit in Figure 15.1.

Our version of a canonical microcircuit for neocortex consists of two groups of pyramidal cells interconnected with a single group of smooth neurons. All three groups receive direct excitatory input from the thalamus. The pyramidal cells are excitatory, the smooth cells are inhibitory. The pattern of connections and the statistics of the synaptic contributions of the different groups were obtained from extensive anatomical studies. The function of the components has been established through a combination of computer simulation and intracellular electrophysiology (Douglas et al., 1989). The circuit contains the minimum number of components that are consistent with the experimental data. Clearly, much more elaborate

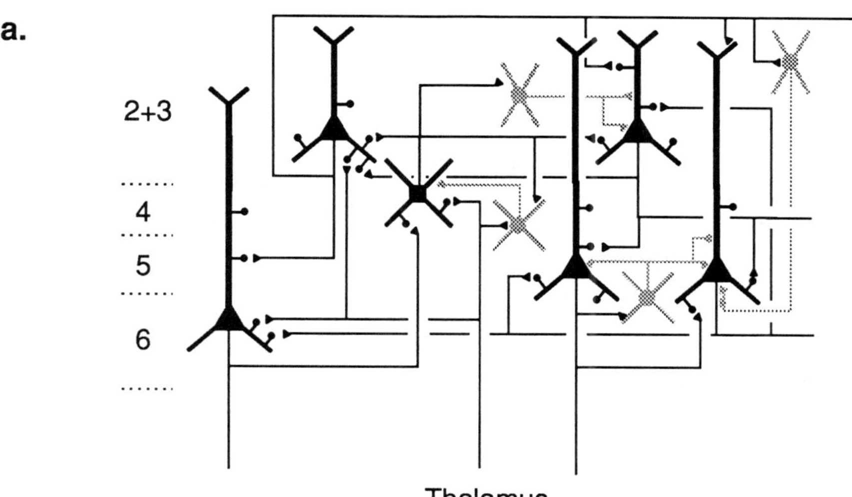

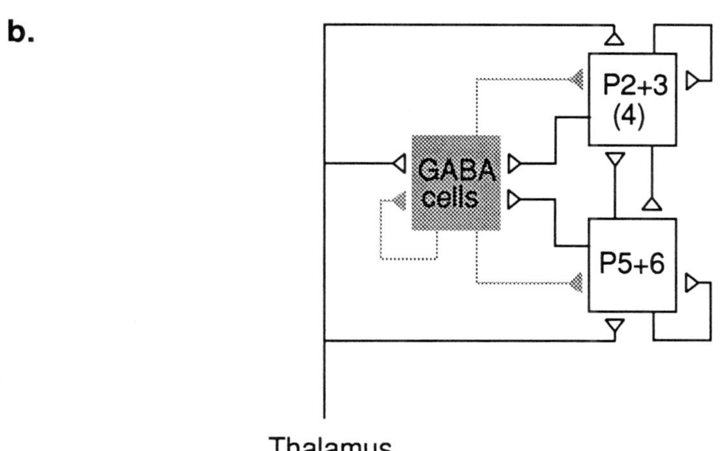

FIGURE 15.1. A. Anatomical characteristics of representative neurons in the six layers of primary visual cortex. Spiny (excitatory) cells shown in black, smooth (inhibitory) cells in grey. B. Canonical microcircuit that successfully predicts the intracellular responses of cortical neurons to pulse stimulation of thalamic afferents (Douglas et al., 1989). Three populations of neurons interact with one another: one is inhibitory (GABA cells, grey); two are excitatory (open), representing superficial (P2 + 3, 4) and deep (P5 + 6) layer pyramidal neurons. Model neurons are similar to those in Figure 15.2. Each population received excitatory input from thalamus. The inhibitory inputs activate both GABAa and GABAb receptors on pyramidal cells. The GABAa mediated inhibitory input to deep pyramidal population is relatively stronger than that to superficial population.

circuits can and have been devised, but in specifying a basic circuit for the neocortex we cannot allow Occam's beard to grow too long.

FROM SINGLE CELLS TO CANONICAL MICROCIRCUITS

For the past 40 years, the single unit has been the sole currency of discourse about cortical function (see Douglas and Martin, 1991a). The advantage of the single unit approach is that it is technically and conceptually straightforward and it links directly to psychophysical phenomena. Ramon y Cajal's early view of the pyramidal neuron as the "psychic cell" is not far from the modern view that single units code for the features of objects in the external world (Barlow 1953, 1972; Hubel and Wiesel, 1962; Lettvin et al., 1959). This thesis continues to dominate ideas about the relation between the brain and the perceptions and behavior of the organism.

It is now thought that neurons performing similar functions are grouped together to form a cortical area, with many different cortical areas subserving the different functions (e.g., Livingstone and Hubel, 1988; Zeki and Shipp, 1989; but see Martin 1988a). In the visual system these areas are thought to be arranged in hierarchies of increasingly complex function, for example, "early vision" is carried out by the microcircuits of the primary visual cortex and is concerned with extracting features of objects, like edges, orientations, or directions of movement (De Yoe and Van Essen, 1988). The output of the primary visual cortex is then shipped to other cortical areas where it is used for further "extraction of information."

The process of "extraction" often involves a synthesis of elements formed at a prior stage of processing, for example, the nonoriented center-surround receptive fields of the visual thalamus collectively form the orientated and directional units of the primary visual cortex, which in new combinations are thought to form the more complex receptive fields found in the visual areas of the prestriate cortex (Barlow, 1972; Livingstone and Hubel, 1988). Inevitably, the focus has been on the connections between these areas, because they determine the relations of members of the hierarchy, rather than on the connections within an area (DeYoe and Van Essen, 1988; Zeki and Shipp, 1989).

The concept of a hierarchy keeps faith with the dogma that the activity of single neurons represents the necessary information (Barlow, 1972). The central tenet of Balow's "neuron doctrine" is that the information required by the system is visible at the level of the single neuron, that is, there is no "hidden" information contained within the combinatorial possibilities of the microcircuit. "All the news that's fit to print," as the motto of the *New York Times* goes, is printed by single neurons and so peristimulus time histograms (PSTHs) jostle for space on the pages of every neuroscience journal. It is incontrovertible that the activity of recording the spike output of single neurons has been extraordinarily valuable. Nevertheless, the PSTHs may only be footnotes to the main messages, which are encoded in the activity of collectives of neurons and not in single units. It is the implicit messages of the collective that we now have to learn to read.

The debate between single neurons and collectives has a long history, succinctly put by Sherrington (1941) as "one pontifical nerve-cell . . . the climax of the whole system of integration," versus a "million-fold democracy, whose each unit is a cell." The former view has dominated, because for the past 30 years physiologists, not anatomists, have led the development of hypothetical microcircuits for the neocortex. The size of the electrophysiologists' tool, the microelectrode, has meant that their concepts have been based willy-nilly on the output of single neurons. The response properties of any single neuron was explained by the action of just a few presynaptic neurons (Douglas and Martin, 1991a; Martin, 1988b). Modern anatomy shows that this view must now change.

The problem with the Golgi technique, which was used to such effect by Ramon y Cajal, is that it does not reveal myelinated axons. For this reason most of the best examples of the morphology of single neurons come from neonatal brains. Only with the introduction of the technique of intracellularly injecting neurons with the enzyme horseradish peroxidase (HRP), were neurons in the adult neocortex visualized in their entirety for the first time (Gilbert and Wiesel, 1979; Martin and Whitteridge, 1984). For those familiar with the Golgi material, the new view down the microscope was baroque. The familiar dendritic tree had sprouted a filligreed axon consisting of an extensive network of local axon collaterals beaded with boutons *terminaux* and *en passant*.

It was obvious simply by looking, that each neuron was forming synapses with hundreds if not thousands of other neurons. More detailed studies of the HRP-filled material revealed that even when a bouton-laden collateral was in the vicinity of a target dendrite, it rarely made more than one synapse on the dendrite (see Martin, 1988b). Electron microscopic studies had already shown that each neuron receives many thousand synapses (see White, 1989). Simple division leads to the conclusion that each neuron receives convergent excitation from thousands of neurons. This conclusion was further strengthened by theoretical predictions (Douglas et al., 1987), confirmed by subsequent experimental studies (Mason et al., 1991), which showed that the functional connection between pairs of neurons was weak.

Although these observations should have focused our attention on the importance of thinking in terms of circuits rather than single neurons, instead they concentrated our attention on the function of the different cell types (Martin 1984, 1988b; Somogyi and Martin, 1985). Our concern was to understand what basket cells, or what chandelier cells "do." This in itself could be a valuable activity (Douglas and Martin, 1990), but the speculation as to the function of these components remains couched in terms of single cells. In reality the function of any one cell is determined by the combined effects of hundreds or thousands of cells, as the connectivity clearly showed.

Similarly, the effect that any one neuron could have on its many targets depends on the functional state of each target, varying in effect from moment to moment. Whatever the general mechanisms of cortical processing might be, they appear to depend on the averaging over a broad range of inputs rather than on selective responses to discrete inputs. The problem is that it is impossible to conceptualize the rich interactions that take place in even 1 sq. mm of neocortex. To get around this problem, cortical physiologists have conventionally resorted to the unreality of stick-and-ball models of the cortical circuits. What we need in order to develop our concepts is a model of the cortical circuits that can incorporate the details and richness of the observed biology, but at the same time has all the conceptual accessibility of the simple models produced by physiologists. The solution we have provided is the canonical microcircuit (see Figure 15.1).

A COPERNICAN REVOLUTION FOR NEOCORTEX?

There is no general theory of neocortex. Barlow (1978) has referred to this state of neuroscience as being "pre-Harvey-an," analogous to the understanding of the structure and function of the heart before William Harvey discovered the circulation of the blood, when the heart was thought to warm the blood or be the source of romantic love. It is true that virtually all models of cortical function are purpose-built. They have been developed to explain a particular phenomena, for example, the orientation tuning of neurons in the primary visual cortex has been explained in one model by the geometric arrangement of the excitatory afferents (Hubel and Wiesel, 1962). The problem with purpose-built models is that each different aspect of neuronal selectivity requires a different model.

The failure to develop global theories means that each observed phenomena remains separate and unrelated to other phenomena. To use Barlow's analogy, it is as if there is one theory for the function of heart valves, another for the ventricles, and yet another for the baroreceptors. These theories may have nothing in common with each other, except that they all concern the heart. Although much can be discovered about the structure and function of heart valves, ventricles, and baroreceptors in isolation from one another, the failure to assemble this knowledge into a coherent global picture of a heart pumping blood ultimately makes nonsense of hypotheses that are specialized to the degree of explaining the function of only one component in isolation from others. In a similar way, the often ingenious explanations of individual phenomena of neocortex, do not add up to a coherent theory of the neocortex. On the contrary, the accumulation of such ad hoc hypotheses actually may obstruct the emergence of a coherent theory. In such a situation the temptation is to carry on collecting interesting data and contribute further to what Rushton (1965) called "the ever thickening forest of fact."

The solution to such an impasse, whether it be in cosmology, particle physics, or molecular biology, has been a revolution. The discovery of the structure of DNA by Watson and Crick, for example, completely changed our understanding of the coding and replication of genetic information. Hubel (1979) may well be correct in his prediction that a revolution of Copernican proportions may never reach neurobiology. The reason is that we already know too much about the brain and the function of its component parts, especially neocortex. We believe that the main purpose of the brain is to process, store, and use information to ensure the survival of its transport system, the body. The questions we now ask are less about what the brain does, but how, given the constraints imposed by biology on speed and accuracy, do the brain circuits achieve such a high level of processing? What cortical neurobiology lacks is a framework within which we can develop a coherent answer to this question (Hubel, 1979).

ARCHES OVER NEOCORTEX

In science, the "what" and the "how" are not easily separated. Global theories of gravitation, quantum theory, origin of species, and molecular genetics have their power in providing a framework within which they can explain a wide range of seemingly disparate and unlinked phenomena. The collection of data in neurobiology has been more or less haphazard, depending on the fashion of technique and the whims of the individual investigator. Concepts like Barlow's "neuron doctrine," or Hubel and Wiesel's hypothesis of cortical columns, provided a framework for research, but they have also ended up reinforcing the status quo rather than providing the impetus for the development of new concepts. Mastering the technique of single unit recording, for example, ensured a season ticket for lunch. Any part of the brain could and was probed by microelectrodes. The doctrinal views of cortical organization and function did not touch many levels of cortical research. For example, they did not promote the study of microcircuitry, or of biophysics, or of neurochemistry. Wide-ranging and influential as they were, they cannot be considered to be global.

This lack of an overarching theory for the neocortex means that related pieces of data remain unrelated and levels that should be connected remain apart. This fragmentation is as good a reason as any for developing concepts of neocortical function at the level of canonical microcircuits, because we are then propelled into the arena of global explanation. If such concentration on theory appears to direct attention away from experiment, we need only point to areas of science where rapid advance in the experimental domain has occurred only because the difficult terrain ahead has been intensively scouted by theorists. Experimental work in neurobiology is largely driven by data or led by technique. Seldom have neuroscientists asked themselves, "What does it all mean?" The pursuit of data has its own compulsions.

It is here that the canonical concept can make an enormous contribution in weaving together the many different threads of experimental data and eliminating from consideration areas of investigation that are irrelevant. This latter process may seem dogmatic, but it is a natural part of science. Because of the particular theoretical frameworks we choose in science we do not now investigate the properties of phlogiston, or the ether, or the role of the heart in romantic love. Similarly, by adopting a framework of the canonical circuit, we provide a focus and direction for future research, which eliminates the random walk that is a feature of science where there are many competing hypotheses, but no overarching theory.

If we consider the array of techniques now applied to understanding the function of the neocortex, it is evident that there are many different levels at which the problem can be attacked: the techniques of anatomy, pharmacology, neurochemistry, *in vivo* and *in vitro* physiology, and computational neuroscience all have a part to play, *but only a part*, and therein lies the rub. It was Barlow's hope that his neuron doctrine would provide a framework in which the necessary experimental observations of cortical function could be made at the level of single neurons. It now seems clear that this is not possible; no single technique can provide the necessary depth of analysis that is required to understand cortical function.

In order to assemble the blueprint of the microcircuit (see Figure 15.1) into a working model, many different pieces of data have to be joined together. Thus, anatomy, biophysics, physiology of circuits, and computer modeling all contribute to the larger picture of cortical function, but together in a single coherent framework for the first time. This joining together does, however, raise an important technical issue of how the translations are to be brought about between the different levels. How do observations of the activity of single channels get translated into the action potential output of single cells? How do the ultrastructural observations of the organization of synapses get translated into their biophysical actions? How do the different input-output relations of single cells scale into the performance of a circuit of such neurons? Such problems have not had to be addressed before because a comprehensive synthesis of data drawn from different levels has not been attempted previously.

Two problems then manifest themselves. The first is that we need to link observations within a level, for example, a horizontal linkage between different aspects of anatomical investigation. The second problem is to link different levels, from anatomy to physiology, from single cells to local circuits, and so on. The solution to these problems may lie in another facet that is intrinsic to the development of a canonical microcircuit, that of minimalism.

The process of achieving a minimalist representation of a given structure or phenomenon is one that has preoccupied artists for centuries. This process is not the same as the familiar one of reductionism in science. What we are attempting to achieve is a reduction in the number of factors that have to be considered, not a reduction from a higher level to a lower one. In neuroscience, the best example of such a minimalist representation is that of Rall's reduction of a branched dendritic tree to an equivalent cylinder using a formal mathematical algorithm (Rall, 1977). This reduction is not one of level but one of number: many dendrites are reduced to one. Rall's minimalist representation of the dendritic tree has provided a powerful tool for the biophysical analysis of electric current flow in neurons. In a somewhat analogous way we can reduce the number of components we have to deal with in the canonical microcircuit. However, it is rarely possible to achieve a formal correspondence between the detailed and the simplified model. In many cases the correspondence can only be demonstrated empirically, not analytically.

Through a process of simplification we can begin to approach the problem of translation in a rational, logical way. The simplified representation at any level can be used to generate the data for the next level of analysis. We have used this device to produce a minimalist representation of different morphological types of cortical neurons (see Figure 15.2). In simulations (see Figure 15.3) both the subthreshold

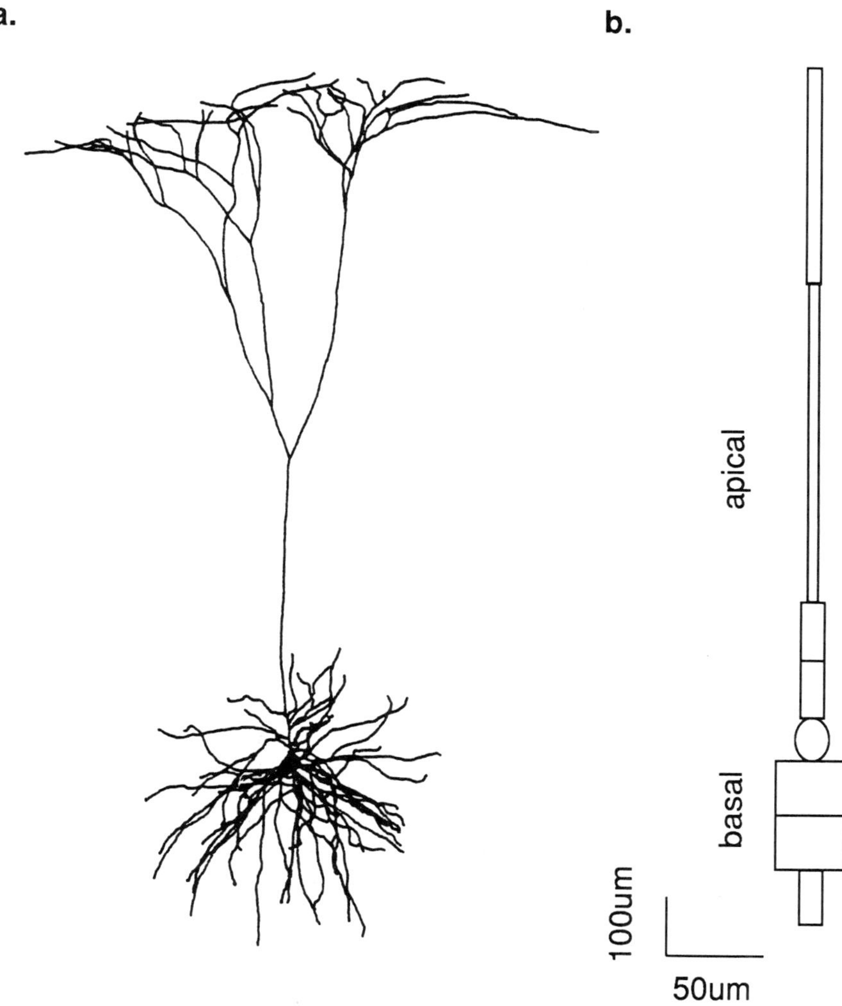

FIGURE 15.2. A. Reconstructed layer 5 pyramidal neuron recorded in cat visual cortex (area V1) and intracellularly labeled with horseradish peroxidase. B. Simple compartmental model used to represent the detailed three-dimensional structure of pyramidal neuron. Lower three cylinders represent basal dendritic arborization, upper four cylinders the apical dendrite. The dimensions of cylinders preserve the dendritic surface area of the original neuron. Each cylinder, and ellipsoidal soma, is allocated a particular profile of passive, active, and ligand-gated conductances.

and suprathreshold responses of these minimalist neurons are remarkably similar to those of neurons recorded intracellularly in the visual cortex.

Single examples of such simplified neurons can be used to represent the average behavior of a group of neurons of the same morphological type. Thus in Figure 15.1, the pyramidal neurons in the superficial layers are represented by one

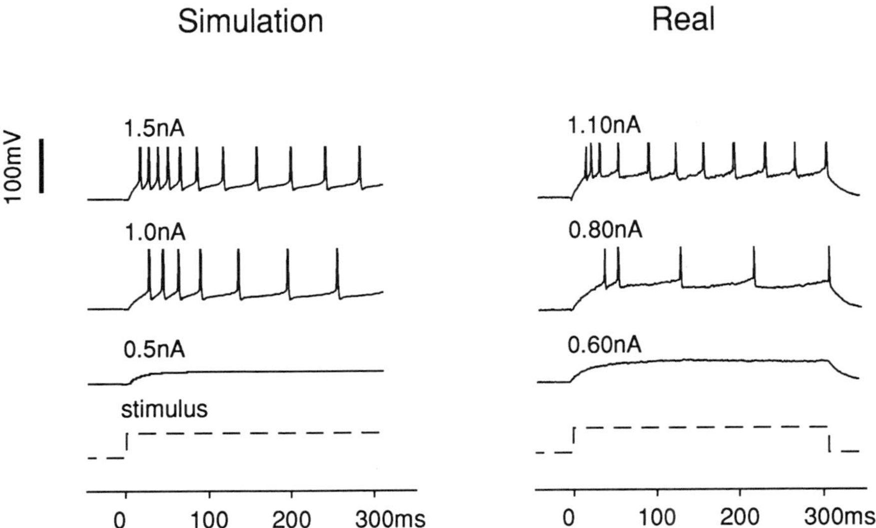

FIGURE 15.3. A. Digital computer simulation of action potential discharge in response to sustained intrasomatic current injection using compartmental model outlined in Figure 15.2B. B. Response of neuron recorded *in vitro* in cat visual cortex to intrasomatic current injection. Action potentials in illustrations have been truncated (Berman, Douglas and Martin, unpublished).

minimalist pyramidal cell, those of the deep layers by another, and the GABAergic neurons by a third. The average activity of thousands of neurons can be translated into the activity of a single representative. In an analogous way, channels can be represented by synapses, synapses by average conductances of dendrites, details of individual dendrites can be simplified to create a generic dendrite, and so forth. The power of this approach is that all details can be taken into the account, asssessed, and incorporated in as detailed or simplified a form as required. If nothing else, the canonical microcircuit acts as a repository for the collection, assimilation, cataloguing, and accessing of a very wide range of information about cortex.

VISUAL CORTEX: THE MODEL FOR NEOCORTEX

It is generally accepted and frequently reiterated in grant proposals that the visual cortex is the model for neocortex. This myth propagates despite the fact that the primate primary visual cortex is self-evidently the most specialized piece of neocortex we know (see Martin, 1988a). This should not present much of a problem if we believe that canonical microcircuits form the basic or organizational principle of neocortex, because they will be found in the most specialized cortical regions and in the least specialized. The cat visual cortex is ideal in this respect. It does not have the additional complications of the primate, but it has all the advantages that have made visual cortex such a dominant area of research. It is the best intermediate between highly specialized and more general neocortical areas.

In practice, the visual cortex has not been exploited as a model system. Instead, the experimental and theoretical analysis of visual cortex has been concerned with the specific properties of visual processing. Explorations of the properties of single neurons have led to a focus on isolated properties like orientation or direction selectivity. The many models that have been devised to explain the formation of visual receptive fields confine their accounts to specific functions and often the neural basis for these models has not been considered in any depth (see Martin, 1988b). Models that explain orientation selectivity do not also explain direction selectivity, or velocity sensitivity. Thus visual cortex is often viewed as a collection of specialist circuits, each designed to perform a different function. Where more general theories have been advanced, most attention has been devoted to the form of coding of the visual input. These models range from specific feature detectors (e.g., Barlow, 1953) to various Fourier-like filter systems for spatial and temporal frequency (e.g., Shapley and Lennie, 1985). The form in which they have been configured makes it difficult for them to be generalized to the processing requirements of other cortical areas.

A further problem arises with the sophistication of the stimuli that are used in the visual system. Because the structure of visual receptive fields has been mapped in such minutiae, stimuli can be tailored by changing their size, texture, contrast color, position in space, velocity, direction of motion, and so on. Even superficially simple stimuli, like bars or edges, become complex when all these factors are added together. Because we do not understand the receptive fields of any other modality in as much depth as those of primary visual cortex, it is virtually impossible to make the transformation from the mechanisms occurring in visual processing to those occurring in auditory processing or motor output. Yet, implicit in the notion of a canonical microcircuit for the neocortex is the idea that form follows function: basic commonalities in structure reflect a commonality in functional mechanisms. Thus we expect that there will be canonical functions as well as canonical structures.

The problem is to find a way of testing this hypothesis. Structural similarities are relatively straightforward to analyze. The techniques, from Golgi-staining inward, already exist; it is only a shortage of manpower that prevents their being widely applied to different cortical areas. The situation is different with the function of cortical areas. In only a few cases do we have more than an approximate idea of the stimulus requirements of cortical neurons. Even in the visual system, we understand the stimulus requirements of the neurons in only one or two areas in any detail. At the present rate of progress we cannot afford to wait until the function of the many other cortical areas have been explored before we discover whether they have a common functional denominator. We need a way of analyzing the basic functional properties of areas even when the details of the stimulus requirements are not known.

The solution we have arrived at is to use a pulse stimulus (Douglas et al., 1989). There are good reasons for this. Traditionally, pulse stimuli have been used for the analysis of dynamical systems and in analyzing the dynamics of the nervous system, an electrical pulse stimulus has long been a part of the armamentarium of the physiologist. In the visual system, however, pulse stimuli have been used largely to examine the conduction velocities of nerve fibers using extracellular recording (e.g., Martin and Whitteridge, 1984). However, it can be used equally well in all cortical areas and it obviates the requirement to know the response properties of the neurons to natural stimuli. Necessarily, the information the pulse stimulus can provide is limited, but it can provide essential information about the dynamics and response characteristics of the component neurons in the cortical microcircuits. The same stimulus can then be used in all cortical areas to obtain comparative data regarding the equivalence or differences of the circuits in the different cortical areas.

The pulse stimulus has been a powerful tool for the development of our version of the canonical microcircuit for visual cortex (Douglas and Martin, 1991b; Douglas et al., 1989). By recording intracellularly, we were able to analyze the subthreshold membrane voltage events that are invisible to extracellular microelectrodes (Douglas and Martin, 1991b; Douglas et al., 1989). Figure 15.4 shows the basic sequence of events following a single electrical pulse applied to the white matter underlying the cortex: a short-duration (30 msec) depolarization is followed by a long-lasting hyperpolarization (200–300 msec). The hyperpolarization is generated by the combined response of the GABAa and the GABAb receptor, which have different reversal potentials and time constants. The depolarization which may be sufficiently large to reach spike threshold, arises as a complex of excitatory inputs, some from thalamic afferents themselves, but mainly from intracortical excitatory pathways. The pulse provides for a temporal separation of the different components that cannot be readily achieved with natural stimuli.

The pulse response was also able to differentiate between deep and superficial layer pyramidal cells. The deep cells showed apparently stronger GABAa

FIGURE 15.4. Intracellular responses of real and model cortical neurons to brief electrical pulse stimulation of the thalamocortical afferents. The morphology of the neurons was identified by intracellular injection of horseradish peroxidase. Intracellular traces of real neurons are averaged, and so individual action potentials are attenuated unless they have a constant latency; the "intracellular" traces were obtained from model neurons in the canonical microcircuit (see Figure 15.1B). They are single sweeps, but action potentials have been cropped at -30 mV. A. Response of real layer 3 pyramidal neuron. Early depolarization evoked an action potential. The depolarization was followed by sustained hyperpolarization that reached maximum amplitude at a latency of about 100 msec. In this, and some other neurons, a rebound depolarization (often accompanied by a burst of action potentials) occurred during relaxation of the hyperpolarization. B. Response of model superficial pyramidal cell was similar to real cell. No attempt was made to model the rebound excitation. C. Response of model smooth cell consists of multiple spikes because adaptation currents were not included. No smooth cells were HRP-filled in these experiments. D. Response of actual layer 6 pyramidal cell. Hyperpolarization occurred rapidly, and reached a maximum at a latency of about 25 msec. E. Response of model deep pyramid was similar to the actual cell (Douglas and Martin, unpublished).

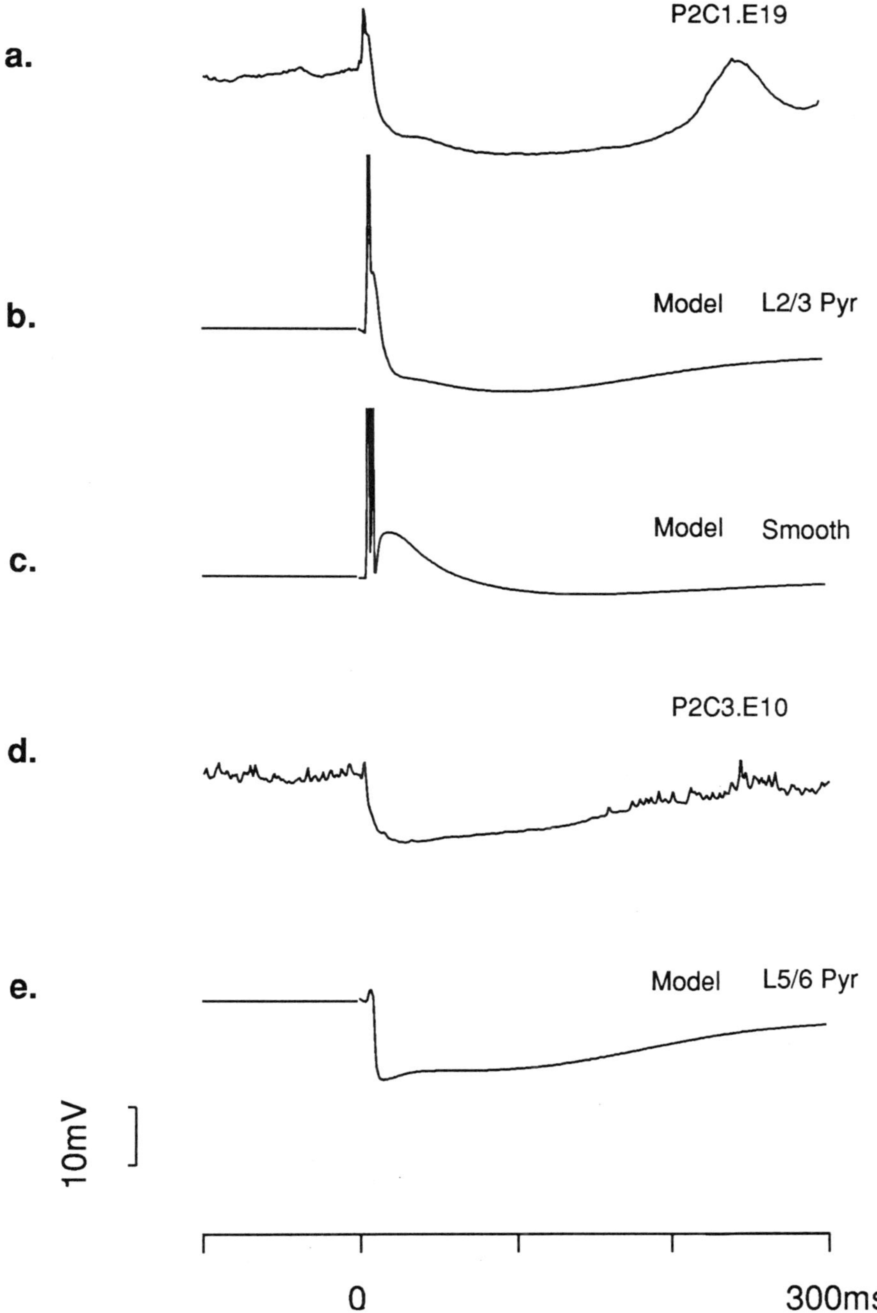

a.
P2C1.E19
b.
Model L2/3 Pyr
c.
Model Smooth
d.
P2C3.E10
e.
Model L5/6 Pyr
10mV
0
300ms

inhibition, which results in a truncated depolarization, and a shorter latency to maximum hyperpolarization (see Figure 15.4). All these phenomena can be simulated with remarkable fidelity by the model microcircuit using a minimum of assumptions. The success of the approach for visual cortex suggests that similar tactics will be equally successful in the analysis of the microcircuits in other cortical areas.

CANONICAL DEVELOPMENT AND PLASTICITY OF CORTICAL CIRCUITS

The mechanisms of epigenesis of the cortical microcircuits are not understood. As Hubel and Wiesel (1974) pointed out, there may be considerable advantages in having a set of circuits that are repeated over and over like a crystal. However, the only repeated system that has been extensively studied physiologically and anatomically is the ocular dominance system. Most of the anatomical work here has concentrated on the distribution of the terminals of the thalamic afferents, not on the cortical circuits themselves (see Hubel and Wiesel, 1977). Conversely, most of the physiological recordings are from cortical cells, not from the thalamic terminals. These differences in the source of the data have led to some conflicting interpretations as to the mechanisms underlying the formation and plasticity of this system (Burchfiel and Duffy, 1981; Hubel et al., 1977). The canonical microcircuit potentially has an important role to play in providing a rigorous model of the adult structure and function against which developing and abnormal circuits can be evaluated. By understanding the form and functional relevance of adult microcircuits we can focus more easily on the critical elements of the developing circuits and detect the changes occurring during normal development and during rearrangements produced by experience.

The canonical microcircuit also offers new interpretations of old experiments made using the paradigm of monocular deprivation. This paradigm is not ideal for the study of cortical plasticity because it confounds the processes of normal development with those of abnormal visual experience. Nevertheless, it is the standard model for plasticity in the central nervous system, which makes it a challenge for models of cortical structure and function, like the canonical microcircuit. The effects of monocular deprivation on the physiology of single cells in the primary visual cortex have been known for some time. These results show that if one eye is placed at some disadvantage, usually by suturing the eyelids together during the critical period, most neurons in the visual cortex come to be functionally dominated by the advantaged eye. Structurally, the thalamic afferents that are driven by the advantaged eye come to occupy more territory in cortex at the expense of the disadvantaged eye (see Hubel and Wiesel, 1977).

The general assumption has been that the ocular dominance assessed physiologically provides a direct reflection of the underlying distribution of thalamocortical synapses. The shift in ocular dominance has been explained by a competitive mechanism: the synaptic sites initially occupied by the deprived axons are thought to be displaced by the nondeprived axons (Hubel et al., 1977). However, in a recent study of monocular deprivation in cats we were able to study the detailed morphology and connectivity of single physiologically characterized axons. We found that the extent of the deprived thalamocortical axon arbors was far larger than would be predicted from the ocular dominance assessed physiologically and that the nondeprived axons made synapses on ectopic sites like the dendritic shaft (Friedlander et al., 1991). Unlike most models of cortical plasticity, the simplified neurons of the canonical model offer a means of assessing theoretically the functional consequences of siting synapses on dendritic shafts rather than spine heads.

The canonical microcircuit emphasizes that the response of individual cortical neurons is determined mainly by excitation derived from the neurons of the local cortical circuit. Even in layer 4, the main termination zone of the thalamic afferents, less than 30% of synapses are derived from the thalamus (see White, 1989). In the canonical model, the role of the local cortical circuits is to amplify this primary excitation arriving from the thalamus. Figure 15.5 demonstrates how this amplification mechanism operates in response to a natural stimulus. In this view the ocular dominance of any neuron will be critically dependent on the cortical circuits, not on the excitation of the thalamic afferents, as current ideas hold. The introduction of the canonical microcircuit into discussions of the familiar model of cortical plasticity provides for a more global explanation of the mechanisms underlying ocular dominance changes during development and monocular deprivation.

We need not consider merely the relative weighting of the left and right eye input to single neurons. Instead, we can include many factors that are clearly important, but that have not been considered in theories of ocular dominance. These include the biophysical properties of the geniculocortical synapse and the postsynaptic receptors, the cable properties of the target neuron, the role of the inhibitor and of intracortical excitation. All these can now be incorporated and analyzed in whatever detail required in the canonical microcircuit.

NEW TOOLS FOR SIMULATING CORTICAL FUNCTION

If nothing else, the neocortex is a marvelous machine whose mode of operation continues to evade our understanding, despite a rapidly rising flood of data. For decades the dream has been that we will be able to build machines whose performance would compare with that of the brain. So far we have failed. Our visual system remains a more powerful image-processor than all the world's supercomputers put together (see Mead, 1989). Comparison of the energy consumption of the brain and current computers also makes sobering reading: the most efficient digital VLSI circuits consume about 10^{-9} joule per operation (an operation is defined as a multiply-and-add). Neurons consume only 10^{-16} joule per operation. If the brain were to operate as inefficiently as the digital VLSI circuits, it would require 10 megawatt of power, that is, it would melt! (Mahowald and Mead, 1991).

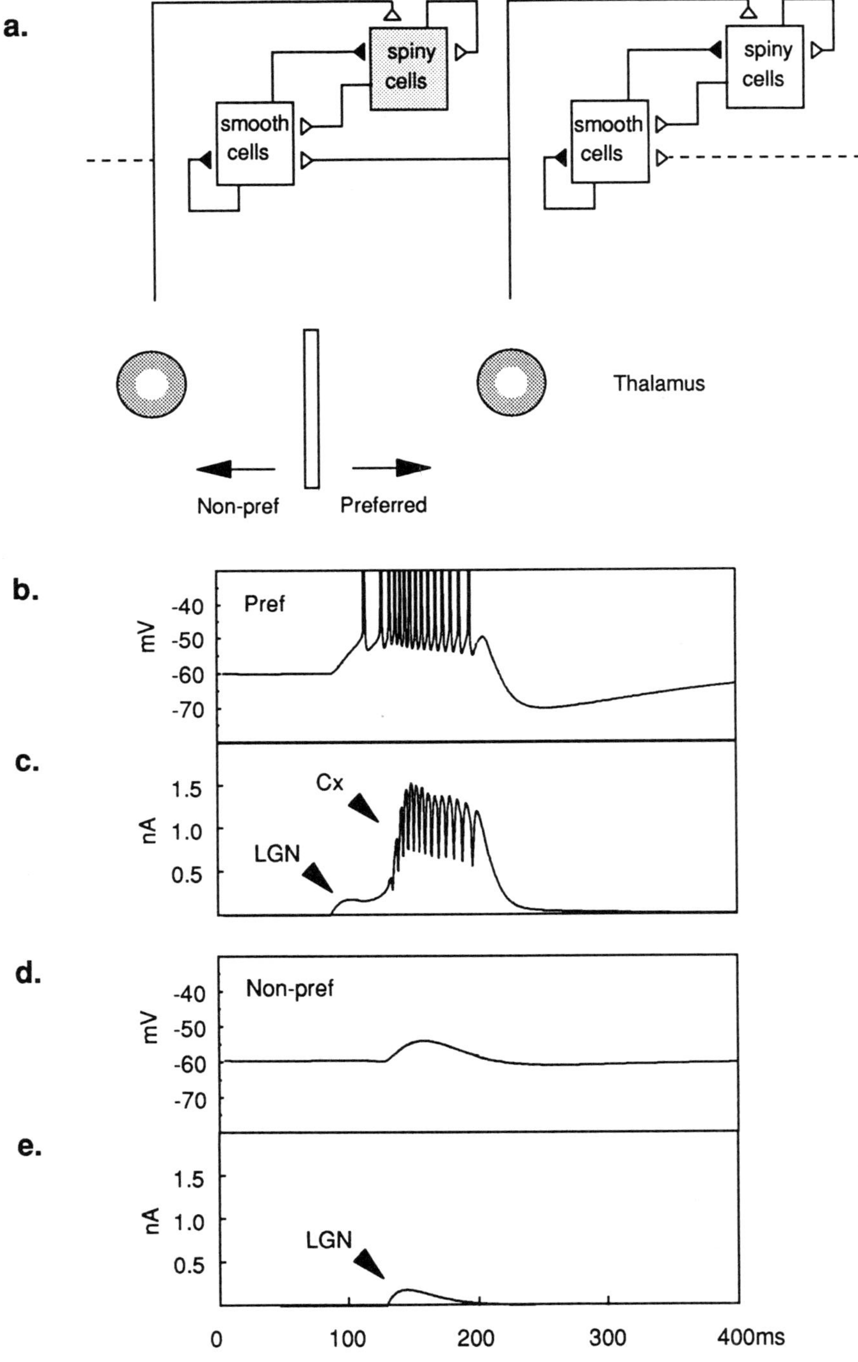
a.
spiny cells
smooth cells
spiny cells
smooth cells
Thalamus
Non-pref
Preferred
b.
Pref
mV
-40
-50
-60
-70
c.
nA
1.5
1.0
0.5
Cx
LGN
d.
Non-pref
mV
-40
-50
-60
-70
e.
nA
1.5
1.0
0.5
LGN
0
100
200
300
400ms

We have begun to realize that biological computation is very different from digital computation and that the neocortex may indeed be the source of yet-undiscovered strategies of processing. Thus the blueprint of the cortical circuits remains a powerful attraction. However, in recent decades most effort has gone into mapping the connections and neuronal responses of different areas of visual cortex. This has resulted in a macrocircuit diagram of epic proportions showing how the different visual areas are interrelated (Felleman and Van Essen, 1991). The attack on the microcircuitry has been rather less coherent and much more data-driven, to the extent of appearing directionless as to the long-term goal. As we have discussed above, introduction of a simplifying "minimalist" approach means that we can renew the search for a cortical blueprint with some hope of realizing Ramon y Cajal's dream.

Of course, even if we could map out the synaptic connections of the cortical neurons in sufficient detail, we would still not be able to build a replica neocortex. The example of the cerebellar cortex, whose circuitry was discovered by Ramon y Cajal, but whose function remains uncertain, is warning enough that the connectivity is only one of many steps that have to be taken. The major problem, however, is that even if we had all the data at hand we could not assemble it in a coherent model that we could easily validate by computer simulation. To simulate even a simple response from model neurons of the type we have described here takes many minutes on a fast personal computer. Simulating the behavior of small assemblages of realistic model neurons in reasonable times (minutes rather than hours) could only be contemplated using supercomputers. For larger assemblages processing more complex inputs, the task would simply take too long to be practicable. If we are to explore effectively the operation of cortical circuits we require a new set of tools on which simulations can be performed at speeds that are closer to those of the real brain. One avenue we are exploring is that of *analogue* very large scale integration (VLSI; Mead, 1989).

FIGURE 15.5. Hypothetical mechanism whereby the canonical microcircuit explains direction selectivity (Douglas and Martin, 1991b, 1991c). A. Thalamic afferents stimulate two circuits that each consist of the superficial half of the canonical model shown in Figure 15.1B. The traces shown in b–e are "recordings" from a single spiny cell located in the left hand (shaded) group. In the preferred direction of motion the spiny cell is excited both by the thalamic afferents and by recurrent excitation from other spiny cells, restricted only by feedback inhibition from the associated smooth cells. In the nonpreferred direction of motion, feedforward inhibition by the smooth cells suppresses the spiny cell's excitation and prevents recurrent excitation. The major component of excitation that is observed in spiny cells during the preferred response is due to cortical excitation. The thalamic input provides a relatively small initial component that is easily controlled by inhibition in the nonpreferred direction. B, C. Somatic potential, and total synaptic current in the spiny cell during preferred stimulation. The major fraction of synaptic current is derived from intracortical reexcitation. D, E. Somatic potential, and total synaptic current, in the spiny cell during nonpreferred stimulation. Small inhibition is sufficient to hold small geniculate input below threshold, avoiding cortical reexcitation.

Analogue VLSI uses the same components (transistors, resistors, and capacitors) and the same well-established methods for fabricating the CMOS (complementary metal-oxide-semiconductor) chips that are found and used in digital computers. The difference is that for analogue computations the transistors are made to operate in their subthreshold mode, whereas in typical digital circuits they operate well above threshold. In the analogue circuits the mode of operation is by smoothly varying electric currents rather than by discrete bits representing ones and zeros. The physics of the analogue silicon circuits is very similar to that of neurons. For example, as in neurons, charge is conserved, which means that electric currents add and subtract. The current flowing into a neuron is an exponential function of the voltage across the membrane; similarly the current-voltage relationship of a transistor in subthreshold operation is an exponential function.

Silicon "neurons" have now been built (see Mahowald and Douglas, 1992; Mahowald and Mead, 1991; Mead, 1989). Their performance is remarkably like the actual neurons on which they are based. This may not seem remarkable until we point out that such model neurons *operate in real time*. What this means is that a circuit of silicon neurons would also operate in real time, regardless of how many neurons were used. As a demonstration of the validity of this claim, a CMOS circuit for the outer plexiform layer of the mud puppy retina has been made (Mahowald and Mead, 1991). It contains a 50×50 array of photoreceptors and the associated circuitry of horizontal and bipolar cells. In tests it simulates the performance of the biological retina with remarkable fidelity, *in real time*. The technology of analogue VLSI thus solves the major problem of simulating the behavior of large numbers of realistic model neurons. The advantages of using such a method for exploring the functions of the canonical microcircuits of neocortex are immense.

CONCLUSION

The canonical microcircuit (Douglas et al., 1989) was a child born of necessity. At the very least we needed a means of drawing together many different strands of investigation into a unified theoretical framework. It has proved its worth in that role. The canonical microcircuit is placed at a level of analysis that enables it to link even high-level computational models of cortical function with the actual machinery of cortex. The form of the canonical microcircuit was designed to be universal so that it could be configured for any cortical area. Applied to the visual cortex, the circuit simulates aspects of known visual physiology that no other model can (Douglas and Martin, 1991b, 1992).

The close relationship between the abstracted canonical microcircuit and its specific implementation in a particular cortical area provides a means of testing the microcircuit in a number of different ways. The data so obtained can then be used to elaborate the basic microcircuit, or to configure it more precisely for specific functions in particular cortical areas. We are at the threshold of a new revolution in which the single neuron will be replaced by the microcircuit, in which digital computational methods are replaced by analogue, in which we increasingly adopt hardware rather than software for testing models, and in which there is a coherent framework in which the wide-ranging research on cortical microcircuits can be brought together under a single roof. The canonical microcircuit may be the map to guide us through the thickets of the cerebral jungle.

Acknowledgments. We thank the M.R.C., the EC, Wellcome Trust, the Royal Society for support. Misha Mahowald gave lucid explanations of analogue VLSI. K.A.C.M. is the Henry Head Research Fellow of the Royal Society.

REFERENCES

Barlow HB (1953): Summation and inhibition in the frog's retina. *J Physiol* 119:69–88

Barlow HB (1953): Single units and sensation: A neuron doctrine for perceptual psychology. *Perception* 1:371–394

Barlow HB (1978): Three theories of cortical function: In: *Developmental Neurobiology of Vision*, Freeman RD, ed. New York: Plenum Press

Burchfiel JL, Duffy FH (1981): Role of intracortical inhibition in deprivation amblyopia: Reversal by microintophoretic bicuculline.*Brain Res* 206:479–484

De Yoe EA, Van Essen DC (1988): *Trends Neurosci* 11:219–226

Douglas RJ, Martin KAC (1990): Control of neuronal output by inhibition at the axon initial segment. *Neural Computation* 2:283–292

Douglas RJ, Martin KAC (1991a): Opening the grey box. *Trends Neurosc* 14:286–293

Douglas RJ, Martin KAC (1991b): A functional microcircuit for cat visual cortex. *J Physiol* 440:735–768

Douglas RJ, Martin KAC (1992): Exploring cortical microcircuits: A combined anatomical, physiological, and computational approach. In *Single Neuron Computation*, Zornctzer S, Davis J, McKenna T, eds. Boston: Academic Press

Douglas RJ, Martin KAC, Whitteridge D (1987): Estimation of the amplitudes of theoretical unitary excitatory post-synaptic potentials in neurones of the cat striate cortex by a combination of biophysical and anatomical measurements. *J Physiol* 394:110P

Douglas RJ, Martin KAC, Whitteridge D (1989): A canonical microcircuit for neocortex. *Neural Computation* 1:480–488

Felleman DJ, Van Essen DC (1991): Distributed hierarchical processing in the primate cerebral cortex. Cerebral Cortex 1:1–47

Friedlander MJ, Martin KAC, Wassenhove-McCarthy D (1991): Effects of monocular visual deprivation on geniculocortical innervation of area 18 in cat. *J Neurosci* 11:3268–3288

Gilbert CD, Wiesel TN (1979): Morphology and intracortical projections of functionally characterized neurons in the cat visual cortex. *Nature* 280:120–125

Hubel DH (1979): The brain. *Sci Am* 241:38–47

Hubel DH, Wiesel TN (1962): Receptive fields, binocular interaction and functional architecture in the cat's visual cortex. *J Physiol* 160:106–154

Hubel DH, Wiesel TN (1974): Uniformity of monkey striate cortex: A parallel relationship between field size, scatter, and magnification factor. *J Comp Neurol* 158:295–306

Hubel DH, Wiesel TN (1977): The functional architecture of the macaque monkey visual cortex: The Ferrier Lecture. *Proc R Soc Lond (Biol)* 198:1–59

Hubel DH, Wiesel TN, Le Vay S (1977): Plasticity of ocular dominance columns in monkey striate cortex. *Philos Trans R Soc Lond (Biol)* 278:377–409

Lettvin JY, Maturana HR, Pitts WH, McCulloch WS (1959): What the frog's eye tells the frog's brain. *Proc Inst Radio Eng* (N.Y.) 47:1940–1951

Livingstone MS, Hubel DH (1988): Segregation of form, color, movement, and depth: Anatomy, physiology, and perception. *Science* 240:740–749

Mahowald MA, Douglas RJ (1991): A silicon neuron. *Nature* 354:515–518

Mahowald MA, Mead C (1991): The silicon retina. *Sci Amer* 264:76–82

Martin KAC (1984): Neuronal circuits in cat striate cortex. In *Cerebral Cortex*, vol. 2, Jones EG, Peters A, eds. New York: Plenum Press

Martin KAC (1988a): From enzymes to visual perception: A bridge too far? *Trends Neurosci* 11:380–387

Martin KAC (1988b): From single cells to simple circuits: The Wellcome Prize Lecture. *Q J Exp Physiol* 73:637–702

Martin KAC, Whitteridge D (1984): Form, function, and intracortical projections of spiny neurons in the striate visual cortex of the cat. *J Physiol* 356:463–504

Mason A, Nicoll A, Stratford K (1991): Synaptic transmission between individual pyramidal neurons of the rat visual cortex *in vitro*. *J Neurosci* 11:72–84

Mead C (1989): *Analog VLSI and Neural Systems*. Reading, MA: Addison-Wesley

Palay SL (1967): Principles of cellular organization in the nervous system. In: *The Neurosciences*, Quarton GC, Melnechuk T, Schmitt FO, eds. New York: Rockefeller University Press

Rall W (1977): Core conductor theory and cable properties of neurons. In: *Handbook of Physiology: the Nervous System: Vol. 1. Cellular Biology of Neurons, Part 1*, Kandel ER, ed. Bethesda, MD: American Physiological Society

Ramon y Cajal S (1904): *La Textura del sistema nervioso del hombre y los vertebrados*. Madrid: Moya

Ramon y Cajal S (1937): *Recollections of My Life*, Craigi EH, trans. Philadelphia: American Philosophical Society

Rushton WAH (1965): Visual adaptation: The Ferrier Lecture. *Proc R Soc Lond (Biol)* 162:20–46

Shapley R, Lennie P (1985): Spatial frequency analysis in the visual system. *Ann Rev Neurosci* 8:547–583

Shepherd GM (1988): A basic circuit for cortical organization. In: *Perspectives on Memory Research*, Gazzaniga MC, ed. Cambridge: MIT Press

Sherrington CS (1941): *Man on His Nature*. Cambridge: Cambridge University Press

Sholl DA (1956): *The Organisation of the Cerebral Cortex*. London: Methuen

Somogyi P, Martin KAC (1985): The role of inhibitory interneurons in the function of area 17. In: *Models of the Visual Cortex*, Rose D, Dobson VG, eds. New York: Wiley

Szentágothai J (1978): Specificity versus (quasi-) randomness in cortical connectivity. In: *Architechtonics of the Cerebral Cortex*, Brazier MAB, Petsche H, eds. New York: Raven Press

White EL (1989): *Cortical Circuits: Synaptic Organization of the Cerebral Cortex—Structure, Function and Theory*. Boston: Birkhauser

Zeki S, Shipp S (1989): The functional logic of cortical connections. *Nature* 335:311–317

16

Visual Responses Outside the Classical Receptive Field in Primate Striate Cortex: A Possible Correlate of Perceptual Completion

RICARDO GATTASS, MARIO FIORANI, JR., MARCELLO GONÇALVES PEREIRA ROSA, MARIA CARMEN GIRALDEZ PEREIRA PIÑON, AGLAI PENNA BARBOSA DE SOUSA, AND JULIANA GUIMARÃES MARTINS SOARES

In this chapter we address the dynamic properties of single neurons in primary visual cortex (V1) that can be related to the completion phenomenon. This phenomenon accounts for the perceptual filling in of blind regions in the visual field, such as the optic disk, small retinal lesions (Bender and Teuber, 1946; Sergent, 1988), or lesions in the geniculo-striatal projection system (Pöppel, 1985, 1986). We propose that completion is achieved by dynamic changes in receptive field size of cells in V1.

The dimensions of receptive fields in monkey striate cortex have been considered virtually unaffected by stimulus contingencies. In contrast, responses of neurons in the extrastriate areas V2, V3A, V4, and MT are modified by stimulation of silent peripheries with stimuli differing in texture, wavelength, spatial frequency, and direction of motion (Allman et al., 1985; Desimone and Shein, 1987; Tanaka et al., 1986; Zeki, 1983). In addition, responses in V2 can be induced by illusory contour stimuli (Peterhans and von der Heydt, 1989a) and the boundaries of excitatory receptive fields in V4 can be modulated by spatial attention (Moran and Desimone, 1985). Most of these effects described in prestriate cortex were shown not to occur in V1 neurons, findings that contributed to the notion that receptive fields in V1 have stable boundaries.

The *Cebus* monkey, a diurnal New World primate, has well-defined ocular dominance columns in V1 (Rosa et al., 1988). These columns are easily defined in tangential section through layer 4C stained for cytochrome oxidase in monocularly enucleated monkeys (Rosa et al., 1988). In preparations of the contralateral hemisphere, the cortical regions corresponding to the enucleated eye appear as cytochrome oxidase-poor regions in layer 4C, while those corresponding to the ipsilateral eye appear as cytochrome oxidase-rich regions. However, the region where the blind spot of the contralateral eye is represented in layer 4C in V1 is occupied by projections from the ipsilateral eye. Therefore, the representation of the blind spot of the contralateral, enucleated eye appears as a cytochrome oxidase-rich region in layer 4C in V1 and that of the monocular crescent appears as a cytochrome oxidase-poor sector (Rosa et al., 1988).

We studied single neurons in V1 at the representation of the optic disk and in the opercular surface using collinear bars with coherent motion. We found that single neurons in V1 show response properties analogous to the perceptual phenomenon of completion (Fiorani et al., 1990). Following a brief report of our findings (Fiorani et al., 1990), several psychophysical and physiological studies addressed the phenomenon of completion (Ramachandran and Aiken, 1991; Ramachandran and Gregory, 1991; Roger-Ramachandran and Ramachandran, 1991) and the issue of the dynamic properties of V1 neurons (Pettet and Gilbert, 1991).

METHODS

Detailed descriptions of our methods were presented elsewhere (Fiorani et al., 1992; Rosa et al., 1988; Sousa et al., 1991). Briefly, we recorded single- and multiunit activity in the region of representation of the blind spot in the primary visual cortex in anesthetized and paralyzed *Cebus* monkeys. The position of the eye was monitored continuously during the experiments by the image of a laser beam reflected in a mirror attached to the contact lenses. The outline of the blind spot was mapped with the aid of a reversible ophthalmoscope.

Three types of stimulation were used. The interpolated responses in the blind spot were first studied under photopic conditions with long 3D-colored bars moving tangential to the surface of the translucent hemisphere positioned in front of the animal. Subsequently, computer-controlled stimulation was used with either a tangent screen or a computer superVGA monitor for stimulus presentation. Most of the cells were studied with black-and-white bars and edges projected onto a tangent screen, with low contrast in the photopic range. We used relatively low contrast stimuli to avoid the problems of scattered light. The background illumination was 10 cd/m^2 and the stimuli were 0.3 Log units above or below the background. A small percentage of the cells was studied with white or colored isoluminant bars over a gray background presented on the screen of a computer monitor (Nec MultiSync 5D) located 57 cm from the animal.

We studied the completion effect by means of two different experiments. In the first we used the natural blind spot and in the second we added neutral masks to create a sort of "artificial blind spot" in other retinal locations. Masking was achieved by inserting opaque gray cardboard squares of several sizes to cover the excitatory minimum response fields (Barlow et al., 1967). Interpolated responses were studied in cells that responded to long bars, thus end-stopped cells or cells with inhibitory surround were not included. After 3 weeks of recording, the animals were monocularly enucleated under ketamine anesthesia, allowed to survive 18 days and then deeply anesthetized and perfused with a mixture of aldehydes. Alternate sections were stained for cell bodies with cresyl violet or reacted for cytochrome oxidase. Small electrolytic lesions were used to locate the recording sites; the cytochrome oxidase stain allowed us to locate the representation of the optic disk in V1 (see Rosa et al., 1988).

In another group of animals we injected retrograde fluorescent tracers in different portions of V1, after restricted physiological mapping, to study the feedback projections to V1 from extrastriate cortical areas (Sousa et al., 1991).

OCULAR DOMINANCE AND ORIENTATION COLUMNS

Area V1 in the *Cebus* is topographically organized and shows local anisotropies related to the orientation of the ocular dominance columns (Gattass et al., 1987; Rosa et al., 1988). It has a columnar organization for orientation and ocular dominance that is similar to that of the *Macaca* (Rosa et al., 1992). Like in the macaque, penetrations oblique to the cortical surface show gradual changes in ocular dominance and orientation selectivity, with occasional discontinuities. The degree of ocular dominance varies from layer to layer, but it is possible to see contralateral and ipsilateral domains at different depths in oblique penetrations. In contrast, penetrations perpendicular to the cortical surface show less variation in ocular dominance or orientation selectivity (Rosa et al., 1992).

COMPLETION EFFECT

In this section we show that neurons within the cortical representation of the optic disk interpolate receptive field position for the contralateral eye based on the extension of the stimuli beyond the boundaries of the blind spot. We also show that the ability to interpolate receptive field position across large distances is present in neurons in other portions of V1. In these neurons, the masking of their receptive fields uncovers further excitatory regions up to 10 times bigger than the size of the receptive field as defined with moving bars and measured from poststimulus histograms. That is, it uncovers active surrounds with areas up to 100 times bigger than those of the classical receptive fields.

We became interested in the problem of completion when we were studying the visual topography of V1 using paralyzed preparations. In these preparations, monkeys with only one eye open had their multiunit receptive fields mapped with 3D-bars, under photopic conditions. We expected that with one eye open it should be possible to map blind regions corresponding to the optic disk and retinal vessels. Instead, we saw a continuous progression of receptive centers passing through the projection of the optic disk, both with the contralateral and the ipsilateral eyes open (see Figure 16.1). Within the representation of the blind spot in V1 the locations of the interpolated response fields, mapped through the contralateral eye, were comparable to the locations of the receptive fields mapped through the ipsilateral eye.

Figure 16.2 summarizes the "apparent dilemma" and the tests that we used to characterize the interpolated response as well as the extension of the interpolation fields, heretofore referred to as active surround. Sweeping bars longer than the diameter of the blind spot induced responses along restricted regions of space,

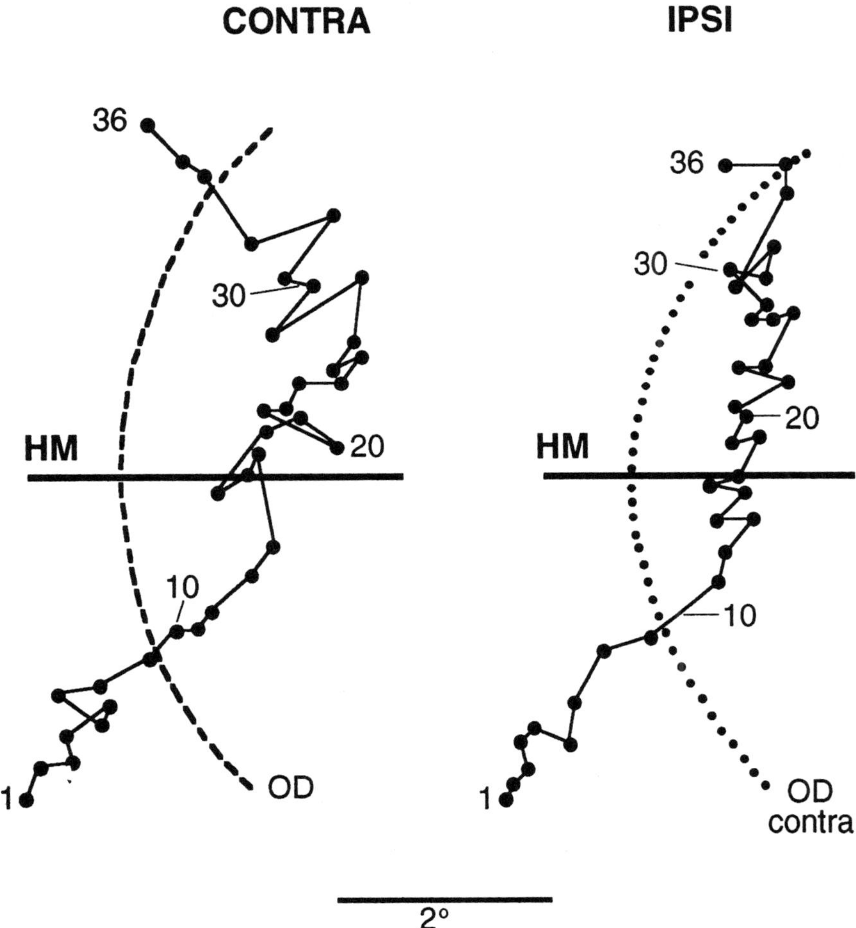

FIGURE 16.1. Location of receptive fields mapped through the contralateral (contra) and ipsilateral (ipsi) eyes corresponding to 36 recording sites in V1 across the representation of the optic disk (OD). Dotted lines represent the temporal border of the optic disk. HM = horizontal meridian.

marked by asterisks in Figure 16.2. In cells that lacked orientation selectivity the location of the excitatory regions could be defined as an interpolated receptive field, shown by a black square in Figure 16.2. Masking the whole projection of the optic disk did not change the location of the interpolated receptive field or the strength of the responses. Sweeping bars restricted to one side of the blind region yielded either poor responses or no responses in some neurons. The neurons that gave no response when the stimulation was restricted to one side were classified as completion neurons, while those that showed considerable responses upon stimulation of either side of the blind spot were considered neurons with discontinuous receptive fields. In both cases a large mask covering the surround of the optic disk,

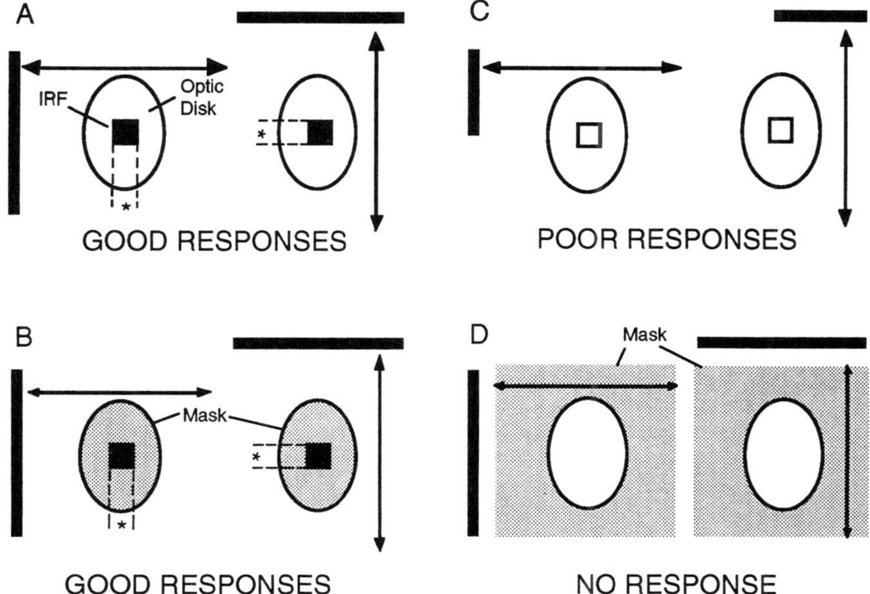

FIGURE 16.2. Schematic representation of the responses of an interpolated response field (IRF) located in the representation of the optic disk to a bar moving perpendicular to its long axis in two orientations with (B and D) and without (A and C) masks. Arrows represent axis of movement. Asterisks mark location of induced responses.

but leaving the disk itself uncovered, abolished the responses. This procedure argues against stray light and eye movement as a source of artifact.

Figure 16.3 illustrates the properties of a neuron with a discontinuous receptive field located at the representation of the optic disk. The recording site of this neuron was located in layer 4B, above the area of rich and continuous cytochrome oxidase stain in layer 4C. Poststimulus histograms after bilateral and unilateral stimulation of regions surrounding the optic disk showed that bilateral stimulation elicited vigorous responses (see Figure 16.3A), while unilateral stimulation of regions surrounding the optic disk elicited small responses (not shown in the figure). The stimulation with a small bar through the center of the blind spot elicited no response (see Figure 16.3B). Figure 16.3C shows that there was a statistically significant potentiation of the cell response in the condition of bilateral stimulation. The response to bars A and B together is larger than the algebraic sum of the response to A plus the response to B. Figure 16.3D shows the response to the ipsilateral (open triangles) and contralateral (filled squares) stimulation as a function of orientation. It shows that the orientation selectivity of directly driven and interpolated responses is similar. That is, the orientation selectivity is the same whether we test directly the "classical" ipsilateral field or the "phantom," interpolated, contralateral field. In the mask-free condition, the ipsilateral eye response is more vigorous than that of the contralateral eye. But for tests with

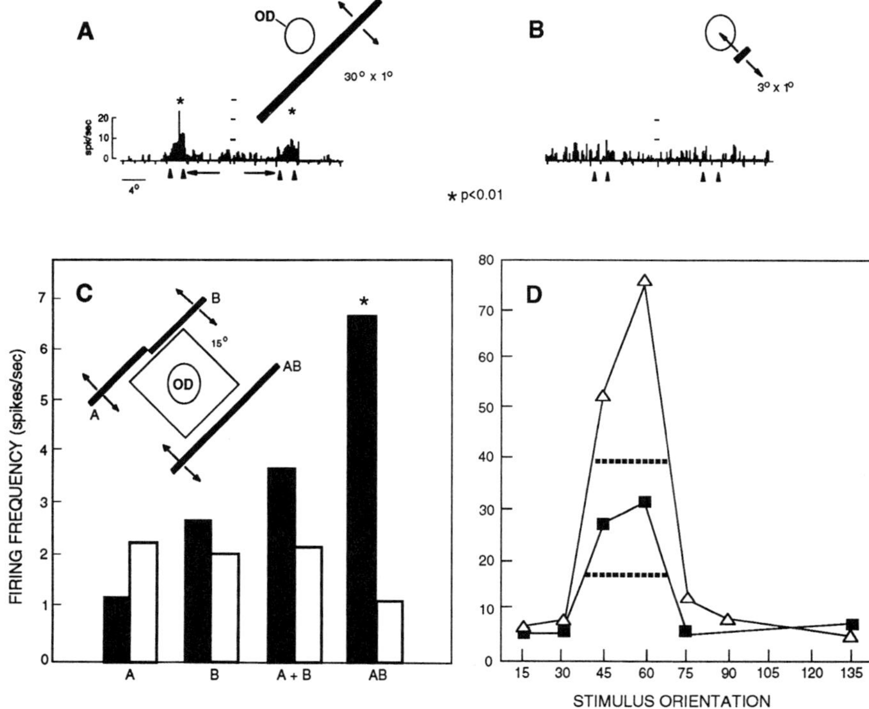

FIGURE 16.3. Characteristics of a neuron with discontinuous receptive field within the optic disk representation. Top: poststimulus histograms in response to a long (A) and a short (B) moving bar. Arrowheads under the histograms mark the borders of the interpolated field. Asterisks indicate statistically significant differences. Lower left (C): Average firing frequency in response to bilateral (AB) and unilateral (A or B) stimulation of regions surrounding the optic disk (OD). Cell response is shown in black bars and spontaneous activity in white bars. Lower right (D): Average firing frequency to different stimulus orientations (in degrees) after stimulation through the ipsilateral (open triangles) and contralateral (black squares) eyes. Dashed line: half-peak bandwidth.

masks, the contralateral responses may be stronger. The size of the surround from which effects can be produced varies depending on the orientation and direction of motion of the stimuli, but the size or location of the interpolated receptive fields do not change. Fiorani and colleagues (1992) showed poststimulus histograms under different mask sizes to different orientation and axis of movement for a neuron located in the lower field representation of the visual field, outside the representation of the optic disk. That neuron had no orientation selectivity and its response to a long (20) stimulus decreased with increasing mask size. In contrast, the location and size of the interpolated response field of the neuron remained constant. Thus, with different mask sizes the extent and the shape of the previously silent surround varied according to the direction and orientation of the stimulus, but the size and location of the interpolated receptive fields were similar to those of the receptive field mapped without mask (Fiorani et al., 1992).

The sizes of the active surrounds that we found in area V1 are larger than the aggregated receptive fields of a cortical column or of a cortical region of the size of a point image. They are much larger than the "classical" receptive field both in *Cebus* and in *Macaque*. This finding suggests that neurons in V1 can be driven from much wider regions of the visual field than expected from the size of the classical receptive field. Previous studies had already shown selective inhibitory modulations from a wider area beyond the classical receptive field (Blakemore and Tobin, 1972; Mafei and Fiorentini, 1976; Nelson and Frost, 1978).

In conclusion, by masking the classical receptive field we were able to demonstrate that other, previously unresponsive regions can be made to excite neurons in V1. Thus, the extent of the area that can drive a given cell in V1 is several times larger than the classical receptive field. If one considers the size of the active surrounds, present estimates of cortical point image size in V1 would have to be multiplied at least 5 times.

Several factors enhance the response of the active surrounds. Coherent motion and collinearity of the stimuli and the difference in color, texture, or luminance between the mask, the stimulus, and the background are important elements for the interpolated responses or the completion effect. Collinearity is important but not necessary. Two noncollinear bars with angles up to 30 elicit some response. A long bar moving behind a conspicuous mask always elicits multiunit responses. Stimulus motion and a clearly visible contrast between the artificial mask and the background are, in most cases, necessary to elicit interpolated responses. In some experiments we employed artificial masks generated on a video monitor used to present stimuli. Neutral (gray) masks made of opaque cardboard elicited better response than inconspicuous masks generated on the video monitor. The response of most cells decreased or was abolished when the mask generated on the video monitor had the same color, texture, and luminance as the background. In several penetrations, neurons showing interpolated responses were found intermixed with units showing no response outside the classical receptive field. In the experiments with the tangent screen they corresponded to 30% of the units studied and they were located from layer 3 to layer 6.

MECHANISMS OF COMPLETION

Interpolated and discontinuous receptive fields behave analogously to the perceptual completion phenomenon that "fills in" the image across naturally blind regions of the retina. The requirements for completion at the cellular level seem similar to those that elicit completion in psychophysical studies (Ramachandran and Aiken, 1991; Ramachandran and Gregory, 1991).

The dynamic changes in receptive field size can be attributed to local circuits in V1 or to a more distributed network that includes extrastriate areas located anterior to V1. The characteristics of the interpolated responses suggest that they are derived from a much broader region than the boundaries of the classical receptive field.

The large size of the active surrounds and the restricted arborization of geniculostriate fibers make it unlikely that this pathway is the only one responsible for the generation of interpolated responses. Cells in the lateral geniculate nucleus do not show the completion effect (Pettet and Gilbert, 1991). Intrinsic connections in V1 and "backward" prestriate projections may both have the necessary coverage for spanning the gaps in the image. The size of these active surrounds is comparable to the aggregated receptive field size of area V4 or MT. Thus, it would be plausible that feedback projections to V1 could play an important role in completion.

The cortical afferents of V1 in the *Cebus* were studied by Sousa and colleagues (1991) to reveal the extent of the region of prestriate cortex that project back to V1. Figure 16.4 summarizes the data of seven successful injections in V1 in a flattened reconstruction of striate and extrastriate visual areas. Injections of fluorescent tracers in V1 at different topographical locations revealed labeled cells in several areas anterior to V1. Central injections in V1 labeled neurons in V2, V3, V4, V4t, and MT, while peripheral injections also labeled areas PO and MST. Central and intermediate injections labeled an area anterior to V4 and medial to the occipito-temporal sulcus, which we call TVP and which may correspond to part of area TEO, as defined by Boussaoud and colleagues (1991). This data suggest that several areas, including V4, V4t, MT, and PO, that have large receptive fields project back to V1 and could be the likely sources of the active surrounds. In summary, the completion-like phenomenon described for a single cell in V1 is due to the existence of active surrounds that could be generated in local microcircuits in V1 *or* be the product of a more distributed network involving areas located anteriorly to V1.

Masking the visual field around a receptive field is qualitatively different from the use of a region naturally devoid of photoreceptors, but in the present data both conditions yielded comparable results. In a paralyzed preparation, like the one we used in this study, any static boundary (such as the border of a mask) gradually fades, and probably the corresponding portions of the visual field are "filled in" by the pattern present in the background as shown by Ramachandran and Gregory (1991). For the purposes of the visual system, the absence of visual input inside the statically masked region may be regarded as qualitatively similar to the lack of visual input generated by a lesion either in the retina or in the visual pathways, and therefore the filling-in process may be similar.

In several studies using stimuli of anomalous contours, Peterhans and von der Heydt have emphasized that neurons in V2 were selective for the orientation of contrast borders as well as for anomalous contours (Peterhans and von der Heydt, 1989a; von der Heydt and Peterhans, 1989) and lines defined by collinear dots (Peterhans and von der Heydt, 1991). In contrast, they found that neurons in V1 required continuous contrast borders to respond with the set of stimuli they used (Peterhans and von der Heydt, 1989b). The difference between our results and

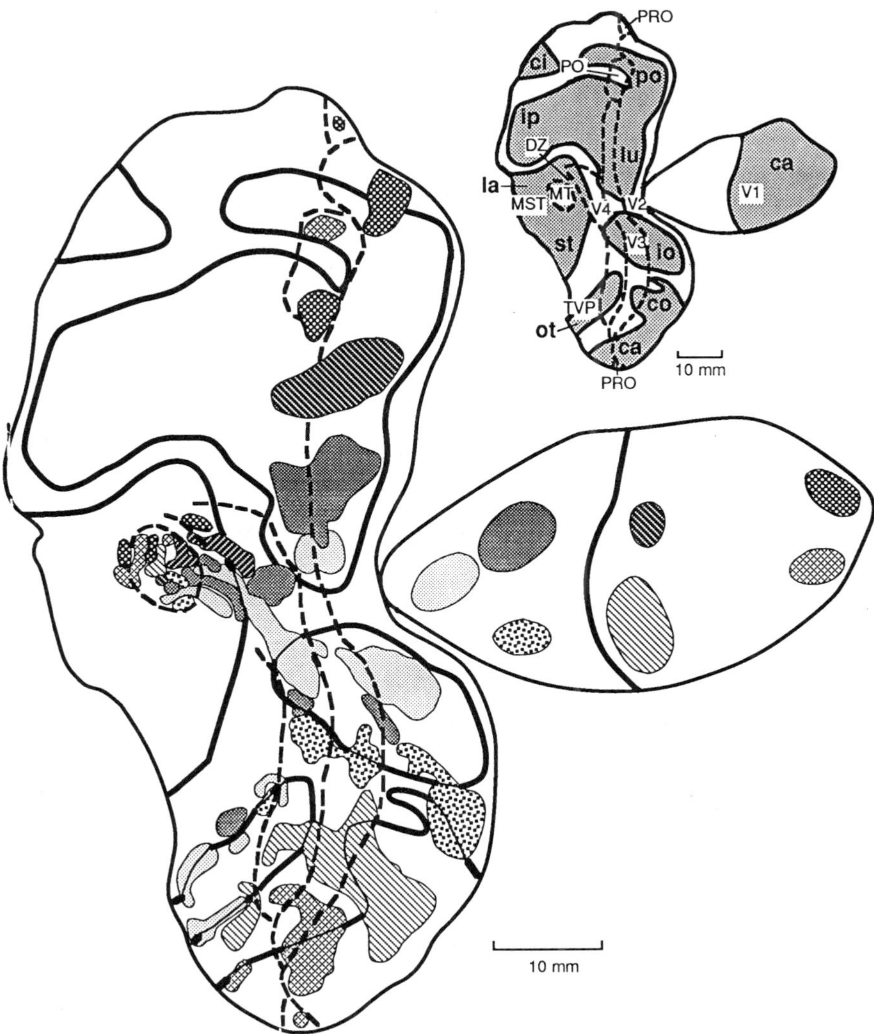

FIGURE 16.4. Topography of the afferent cortical projections to V1. Thick lines indicate the borders of the sulci. Dotted lines are myeloarchitectonic borders. The injection sites in V1 and the corresponding projecting zones are represented with the same symbols. Insert shows the location of the sulci (gray areas) and of the cortical visual areas. Abbreviations: ca—calcarine sulcus; ci—cingulate sulcus; co—collateral sulcus; DZ—area DZ; io—inferior occipital sulcus; ip—intraparietal sulcus; la—lateral sulcus; lu—lunate sulcus; MST—area MST; MT—area MT; ot—occipitotemporal sulcus; PO—area PO; po—parietooccipital sulcus; PRO—area prostriata; st—superior temporal sulcus; TVP—temporal ventral posterior region; V1—striate cortex; V2—second visual area; V3—third visual area; V3v—ventral part of 3rd visual area; V4—fourth visual area (Modified from Sousa et al., 1991).

those of Peterhans and von der Heydt could be due to the different nature of the stimulus used. In a series of experiments we used a video monitor to generate black bars moving over either a black mask or over a mask with the same color, texture, and luminance of the background. In these conditions no cell gave vigorous response to two collinear bars moving over the active surround in coherent motion. The lack of vigorous responses by the cells in V1 to this stimulus configuration was very intriguing, but it reproduces the findings of Peterhans and von der Heydt (1989b) in area V1 of the macaque.

The interpolation responses seem to be generated by a different mechanism from that proposed by Peterhans and von der Heydt for anomalous-contour responses in V2 (Peterhans and von der Heydt, 1991). Interpolation responses occur over a much wider area to moving stimuli that are bigger than those used in V2. Moreover, on psychophysical grounds, illusory contours are apparent even when subjects are free to move the eyes, while completion may require steady fixation or image stabilization. We hypothesize that the properties described here are relevant for completion, but not for the perception of anomalous contours, which may be reflected only on the activity of extrastriate neurons (Peterhans and von der Heydt, 1989b). Another alternative is that completion and perception of anomalous contours are related to different stimulus configurations.

The stimulus configuration that elicited vigorous responses in our study is different from that used by Peterhans and von der Heydt to study cells of V1 (Peterhans and von der Heydt, 1989b). When we used a stimulus configuration similar to that used by Peterhans and von der Heydt (1989b), only a very small percentage of the cells showed interpolation responses. The stimulus configuration that led to a lack of response of V1 neurons (Peterhans and von der Heydt, 1989b) conforms to the condition of *modal completion* as defined by Kanizsa (1979). In this condition the bar defined by an illusory border appears to move in the foreground, that is, one perceives a bar moving in front of the mask. In contrast, the stimulus configuration used in the study of the "artificial blind spot" conforms to the condition of *amodal completion*, as defined by Kanizsa (1979). The interpolated bar is seen in the background, moving behind a conspicuous mask. The use of masks with different texture and luminance than those of the background over a tangent screen or a video monitor enhanced the response. Thus, the difference between our results and those of Peterhans and von der Heydt in V1 could be due to the characteristics of the completion explored with the two types of stimulus condition. These differences cannot be attributed to a species difference since behavioral data obtained from cats show that these animals can see illusory contours (Bravo et al., 1988; De Weerd et al., 1990, 1991). The difference in the stimulus requirement to evoke interpolated responses in V1 and V2 suggests that these cells may underlie different perceptual functions. Interpolated responses in V1 depend on elongated moving stimuli and they may mediate perceptual completion for the blind spot, for vascular scotomas, and for hidden moving objects. In contrast, cells in V2 that signal illusory contours may contribute to the extraction of stationary or moving occluded contours.

Acknowledgments. We thank Dr. Esther Peterhans for helpful comments on the manuscript. Our thanks are also due to the Fundação Parque Zoológico de São Paulo for providing the animals used in these studies. Research supported by FINEP, CNPq, and CEPG/UFRJ.

REFERENCES

Allman J, Miezin F, McGuinness E (1985): Direction- and velocity-specific responses from beyond the classical receptive field in the middle temporal visual area (MT). *Perception* 14:105–126

Barlow HB, Blakemore C, Pettigrew JD (1967): The neural mechanisms of binocular depth discrimination. *J Physiol (Lond)* 193:327–342

Bender MB, Teuber HL (1946): Phenomena of fluctuation, extinction and completion in visual perception. *Arch Neur Psychiatr* 55:627–658

Blakemore C, Tobin EA (1972): Lateral inhibition between orientation detector in cat's visual cortex. *Exp Brain Res* 15:439–440

Boussaoud D, Desimone R, Ungerleider LG (1991): Visual topography of area TEO in the macaque. *J Comp Neurol* 306:554–575

Bravo M, Blake R, Morrisson S (1988): Cats see subjective contours. *Vision Res* 28:861–866

Desimone R, Shein SJ (1987): Visual properties of neurons in area V4 of the macaque: Sensitivity to stimulus form. *J Neurophysiol* 57:835–868

De Weerd P, Sprague JM, Vandenbussche E, Orban GA (1991): The effect of the lower and higher order visual cortical lesions on illusory contour orientation discrimination and texture segregation in the cat. *Soc Neurosci Abstr* 17:1211

De Weerd P, Vandenbussche E, De Bruyn B, Orban GA (1990): Illusory contour orientation discrimination in the cat. *Behav Brain Res* 39:1–17

Fiorani M Jr, Gattass R, Rosa MGP, Rocha-Miranda CEG (1990): Changes in receptive field (RF) size of single cells in primate V1 as a correlate of perceptual completion. *Soc Neurosci Abstr* 16:1219

Fiorani M Jr, Gattass R, Rosa MGP, Rocha-Miranda CE (1992): Dynamic surrounds of receptive fields in area V1 of primates: A physiological basis for perceptual completion? *Proc NY Acad Sci*. (in press) 1992

Gattass R, Sousa APB, Rosa MGP (1987): Visual topography of V1 in the *Cebus* monkey. *J Comp Neurol* 259:529–548

Kanizsa G (1979): *Organization in Vision. Essays on Gestalt Perception*. New York: Praeger

Mafei L, Fiorentini A (1976): The unresponsive regions of visual cortical receptive fields. *Vision Res* 16:1131–1139

Moran J, Desimone R (1985): Selective attention gates visual processing in the extrastriate cortex. *Science* 229:782–784

Nelson JI, Frost BJ (1978): Orientation-selective inhibition from beyond the classic visual receptive field. *Brain Res* 139:359–365

Peterhans E, von der Heydt R (1989a): Mechanisms of contour perception in monkey visual cortex: 2. Contours bridging gaps. *J Neurosci* 9:1749–1763

Peterhans E, von der Heydt R (1989b): The whole and the pieces—cortical neuron responses to bars and rows of moving dots. In: *Seeing Contour and Colour*, Kulikowski JJ, Dickinson CM, Murray IF, eds. Oxford: Pergamon Press

Peterhans E, von der Heydt R (1991): Elements of form perception in monkey prestriate cortex. In: *Representation of Vision: Trends and Tacit Assumptions in Vision Research*, Gorea A, Frégnac Y, Kapoula Z, Findlay J, eds. Cambridge: Cambridge University Press

Pettet MW, Gilbert CD (1991): Contextual stimuli influence receptive field size of single neurons in cat primary visual cortex. *Soc Neurosci Abstr* 17:1090

Pöppel E (1985): Bridging a neuronal gap: Perceptual completion is dependent on stimulus motion. *Naturwissenschaften* 72:599–560

Pöppel E (1986): Long-range colour-generating interactions across the retina. *Nature* 320:523–524

Ramachandran VS, Aiken WH (1991): On filling the blind spot and other medical marvels. *Soc Neurosci Abstr* 17:847

Ramachandran VS, Gregory RL (1991): Perceptual filling in of artificially induced scotomas in human vision. *Nature* 350:699–702

Roger-Ramachandran D, Ramachandran VS (1991): Phantom contours: Stimuli that selectively activate the magnocellular pathway. *Soc Neurosci Abstr* 17:848

Rosa MGP, Gattass R, Fiorani M Jr (1988): Complete pattern of ocular dominance stripes in V1 of a New World monkey, *Cebus apella*. *Exp Brain Res* 72:645–648

Rosa MGP, Gattass R, Fiorani M Jr, Soares JMP (1992): Laminar, columnar and topographic aspects of ocular dominance in primary visual cortex of *Cebus* monkey. *Exp Brain Res* 88:249–264

Rosa MGP, Gattass R, Soares JGM (1991): A quantitative analysis of cytochrome oxidase-rich patches in the primary visual cortex of *Cebus* monkey: Topographic distribution and effects of late monocular enucleation. *Exp Brain Res* 84:195–209

Sergent J (1988): An investigation into perceptual completion in blind areas of the visual field. *Brain* 111:347–373

Sousa APB, Piñon MCG, Gattass R, Rosa MGP (1991): Topographic organization of cortical input to striate cortex in the *Cebus* monkey: A fluorescent tracer study. *J Comp Neurol* 308:665–682

Tanaka K, Hikosaka K, Saito H, Yukie M, Fukada Y, Iwai E (1986): Analysis of local and wide-field movements in the superior temporal visual areas of the macaque monkey. *J Neurosci* 6:134–144

von der Heydt R, Peterhans E (1989): Mechanisms of contour perception in monkey visual cortex: 1. Lines of pattern discontinuity. *J Neurosci* 9:1731–1748

Zeki S (1983): Colour coding in the cerebral cortex: The reaction of cell in monkey visual cortex to wavelengths and colours. *Neuroscience* 4:741–745

17

Inferior Temporal Cortex: Neuronal Properties and Connections in Adult and Infant Macaques

CHARLES G. GROSS AND HILLARY R. RODMAN

In monkeys, inferior temporal cortex (cytoarchitectonic area TE) is crucial for the perception, recognition, and storage of information about visual form and color (Gross, 1973). It is the final stage in the processing hierarchy for visual pattern that begins in the retina and continues along the ventral-coursing occipitotemporal pathway (Desimone and Ungerleider, 1989). In humans, the ventral occipitotemporal cortex, particularly in the right hemisphere of right-handers, appears to play a similar role in vision (Corbetta et al., 1991; Damasio, 1989; Milner, 1968).

In the first part of this chapter we review the properties of inferior temporal (IT) neurons in mature macaques that underlie its role in pattern vision. In the second part we compare these properties in infant and adult macaques, and in the third part we describe the connections of IT cortex in infants and adults.

PROPERTIES OF INFERIOR TEMPORAL NEURONS IN ADULT MACAQUES

1. Inferior temporal (IT) cells respond only to visual stimuli. This fits with the exclusively visual effects of IT lesions (Gross et al., 1967, 1972).

2. Virtually every IT receptive field includes the fovea, and responses to stimuli at the fovea are more vigorous than elsewhere in the receptive field (Desimone and Gross, 1979). These properties mirror the crucial role of the center of gaze in form perception. Unlike striate cortex and the various prestriate visual areas, IT cortex has no visuotopic organization.

3. Receptive field size is relatively large (median about $25 \times 25°$) as compared with that in "earlier" visual areas such as striate cortex, V2, and V4 (Desimone and Gross, 1979; Gattass et al., 1981, 1985, 1988). Stimulus selectivity remains similar throughout the central portion of the visual field, thus providing a basis for perceptual equivalence across retinal translation (Gross and Mishkin, 1977).

4. Most receptive fields extend across the vertical meridian well into both visual half-fields. Thus, major portions of the two halves of visual space are represented in single cells for the first time in visual cortex (Gross et al., 1972). This unification of space is accomplished through the anterior commissure and the splenium (Gross et al., 1977).

5. Many IT cells are selective for some aspect of shape, texture, and/or color. Shape-selective cells are the most common. A small proportion of these are selective for faces and hands. Rather than acting as narrow filters for particular stimuli, IT cells tend to respond to a variety of different stimuli, often with different firing rates or firing patterns (Desimone et al., 1984; Gross et al., 1972; Perrett et al., 1982; Tanaka et al., 1991). This has led to the idea that IT cortex represents stimuli through an ensemble or population code rather than through a set of specific feature detectors (Desimone et al., 1984; Gross, 1992).

6. Selectivity of many IT neurons for shape is maintained over changes in stimulus size, contrast, and wavelength, as well as over changes in retinal location. That is, IT cells show shape constancy, both for arbitrary geometrical patterns and for natural ones such as faces (Rolls and Baylis, 1986; Schwartz et al., 1983). The response invariance of many IT neurons over size, contrast, color, and location indicates that, unlike neurons in "lower" visual areas, they are sensitive to global shape and not just local contour. Presumably these properties are built up from the convergence of the outputs of V4 and TEO cells that project to IT cortex and are themselves sensitive to local contour.

7. The six "sensory" properties listed above are found in adult macaques both when awake and behaving and when anesthetized with nitrous oxide (Gross et al., 1972, 1979). As will be discussed in the second part of this chapter, similar properties are also found in awake infant monkeys. By contrast, IT neurons show very slight or no responsiveness in animals under barbiturate anesthesia (Gross et al., 1967, 1972, 1979; Richmond et al., 1983).

8. In behaving animals, IT responses can be modulated by recent experience in delayed matching from sample and habituation paradigms (e.g., Gross et al., 1979; Mikami and Nakamura, 1988; Miller et al., 1991; Rolls et al., 1989). IT cells may also show longer term effects of experience (Miyashita, 1988).

9. The response of an IT neuron can be gated by the site of the animal's attention within the cell's receptive field (Moran and Desimone, 1985) and by the task relevance of the stimulus (e.g., Gross et al., 1979; Sato, 1988; Spitzer and Richmond, 1991).

10. IT cortex is the final purely visual stage of the ventral occipitotemporal pathway specialized for the analysis of form and color. IT cortex receives visual input from striate cortex along a route involving projections from striate to V2, from V2 to V4, and from V4 to IT both directly and via area TEO (Desimone et al., 1980). At least in anesthetized animals, this corticocortical input is necessary and sufficient for the visual properties of IT cells. However, in the awake animal, input from the tectopulvinar system may be important for modulation of attention and receptive field size (Gross et al., 1974; Moran and Desimone, 1985; Petersen et al., 1987).

As one moves along the ventral pathway from striate through V2, V4, and area TEO to IT cortex, the properties of the neurons encountered change systematically in several ways. First, receptive fields become larger (Gattass et al., 1985), affording greater opportunity for generalization across retinal translation (Gross and Mishkin, 1977). Second, the precision of the visual topography declines: in V2 it is less precise than in striate, it becomes somewhat crude in V4, slight in

TEO, and disappears entirely in IT (Boussaoud et al., 1991; Desimone and Gross, 1979; Fenstemaker et al., 1985; Gattass et al., 1981, 1985, 1988). Third, the stimulus-selective properties of the neurons become more complex (Desimone and Ungerleider, 1989). Fourth, ipsilateral influences become more pronounced starting at V4. Fifth, attentional and other extraretinal influences become clear by V4 and pronounced in IT (e.g., Moran and Desimone, 1985). Sixth, the concentration of opiate receptors and activity of F1 protein increases, perhaps reflecting the increasing cognitive function of the tissue (Lewis et al., 1981; Nelson et al., 1987). Overall, the biggest discontinuity in neuronal properties along this pathway occurs as we enter IT: now visual topography is gone, the receptive fields are usually bilateral and there is a marked increase in the complexity of stimulus selectivity (Gross et al., 1972; Tanaka et al.,1991).

What goes on beyond IT cortex is still unclear. IT projects to the hippocampus by way of synapses in perirhinal, entorhinal, and/or adjacent cortices (Squire and Zola-Morgan, 1991; Suzuki and Amaral, 1990). Whereas the storage of visual representations seems to occur only in IT cortex, the hippocampus and the adjacent cortex are crucial for that storage to occur, at least for declarative as opposed to procedural memories (Mishkin, 1982; Squire, 1987).

11. IT cortex may be partitioned into an anterior and a posterior portion on the basis of cytoarchitectonics, cortical and subcortical connections, and receptive field size. Furthermore, both the anterior and posterior portions may be further subdivided on similar criteria along the dorsoventral axis (Desimone and Gross, 1979; Felleman and Van Essen, 1991; Seltzer and Pandya, 1978; Yukie and Iwai, 1988).

12. As indicated by cross-correlation of spike trains of simultaneously recorded IT cells, nearby neurons are more likely to share common inputs and have similar stimulus selectivities than are relatively distant cells (Gochin et al., 1991). The latter finding supports earlier reports of "clustering" within IT cortex (Gross et al., 1972). In contrast, direct connections between IT cells are equally common for near and relatively distant neuron pairs.

In addition, estimates based on cross-correlation analysis indicates that the number of synaptic inputs required to activate an IT neuron is typically greater than the number required to activate a striate neuron (Gochin et al., 1991). Whereas IT requires an estimated mean of 40 inputs, the estimate for striate can range from about 2 to 10 inputs (Michalski et al., 1983; Toyama et al., 1981). The high level of convergence in IT cortex may reflect a convergence of information about stimulus parameters and thus underlie the complex stimulus-selective properties of many IT cells.

PROPERTIES OF INFERIOR TEMPORAL NEURONS IN INFANT MONKEYS

Despite extensive study of IT cortex in the past several decades, very little has been learned about its development until recently. The influential pioneering studies of Raisler and Harlow (1965) indicated that considerable sparing of visual discrimination ability follows early IT lesions. Subsequent investigations of frontal lesions in infancy (Goldman, 1971) made it clear that some regions of "association" cortex remain functionally uncommitted until well into the first year of life. However, no further investigation of the functional development of IT itself appeared until the last decade, when a series of investigations by Bachevalier and Mishkin and their colleagues provided further evidence that the contribution of IT cortex to behavior may be a delayed one. They confirmed that IT lesions in infant monkeys lead to considerable sparing of visual discrimination abilities whether tested immediately (Bachevalier et al., 1990; Hagger et al., 1985) or later in life (Bachevalier and Mishkin, 1988). In addition, they showed that the metabolic activity of IT cortex, as measured by 2-deoxyglucose uptake, is barely detectable by 3 months and does not achieve the adult pattern until about 6 months (Hagger et al., 1988). Moreover, they showed that certain types of visual learning that depend on IT in adult monkeys are not evident at all before about 3 months of life in macaques (Bachevalier et al., 1990). These lines of evidence suggested to us that the physiological properties and afferent connectivity of IT cortex might also be immature in the first half year or so of life. In the next section we describe our physiological studies of IT cortex in infant monkeys. In the following section, we report on patterns of cortical connectivity of IT in infants and compare them with IT connections in adults.

Studies in Alert Infant Monkeys

We studied the visual properties of IT neurons in six macaque monkeys ranging from 5½ weeks to 6 months of age (Rodman et al., 1991). Surgical procedures, recording methods, and behavioral training were similar to those used for prior studies of adult monkeys and are described fully in Rodman (1991). Briefly, a recording chamber, headbolt for holding the head, and scleral search coil for monitoring eye position were implanted under anesthesia and aseptic conditions. A week after the surgical procedure, the infant monkeys were trained to maintain fixation on a small spot of light while visual stimuli were presented. For each trial, 200 msec after the spot of light appeared, a stimulus appeared at the fovea or at other locations on a tangent screen and remained on for an additional 500 msec. Correct fixation within a 3° window for the duration of the trial was followed by a juice reward. For both the alert and anesthetized infant recordings, the stimuli were similar to those used previously for studies of IT cortex in adult monkeys and consisted both of three-dimensional objects and of projected images under computer control. Specifically, the stimuli consisted of 1) a standard set of complex objects including head models, monkey dolls, brushes, and plastic food objects, and 2) images projected from slides, including monkey faces, scrambled versions of faces, food items, and geometrical patterns varying in complexity. In the alert infant monkeys, we relied predominantly on projected images as stimuli.

Visual responsiveness

Overall, in the alert infant monkeys, more than three-quarters of the neurons studied responded to visual stimulation, a proportion similar to that found in

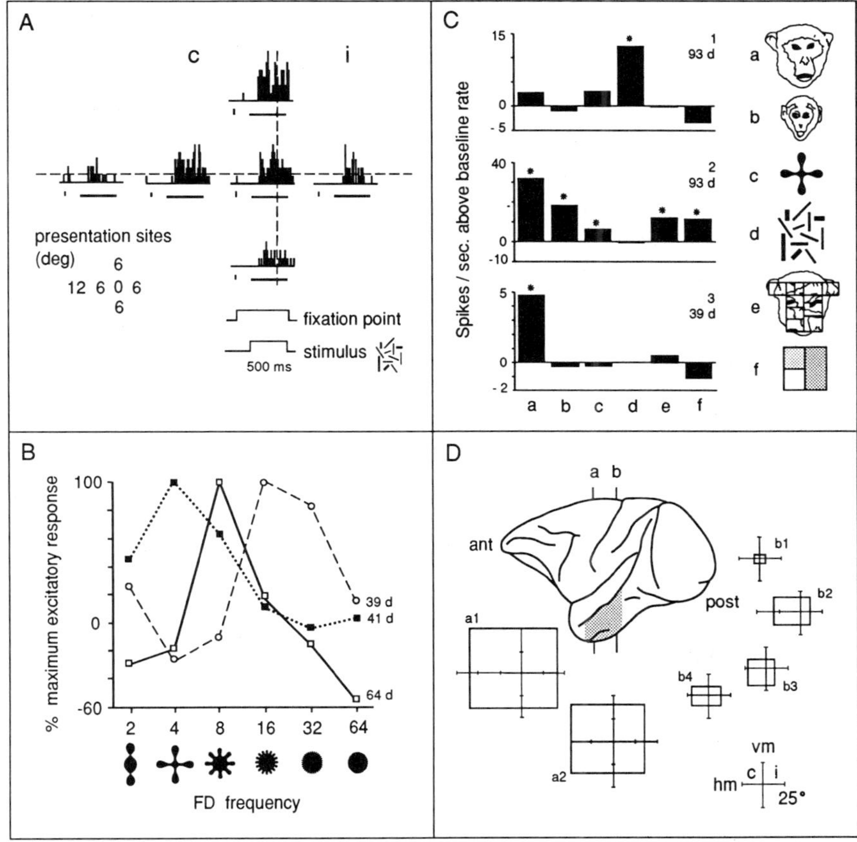

FIGURE 17.1. Receptive field properties and stimulus selectivity of inferior temporal neurons in infant monkeys (adapted from Rodman et al., 1991). Properties illustrated here are also characteristic of IT neurons in adult macaques. **A**. Poststimulus time histograms of spikes from an IT cell in a 12-week-old awake infant monkey performing a fixation task. Stimulus was a set of randomly oriented lines; c, contralateral field; i., ipsilateral field. **B**. Tuning curves for FD frequency (number of lobes around perimeter of FD) for several IT cells in awake infant monkeys; d, days of age. **C**. Bar graphs illustrating responses of several IT neurons in awake infant monkeys to the set of images shown at the right of the figure. **D**. Receptive fields of IT cells in anesthetized infant monkeys taken from levels a and b shown on the brain. Shaded area represents the zone in which recordings were made in both awake and anesthetized infant monkeys. hm, horizontal meridian; vm, vertical meridian.

previous studies of IT cortex in behaving adult animals (Gross et al., 1979; Richmond et al., 1983; H.R. Rodman, J.P. Skelly, and C.G. Gross, unpublished data). Moreover, we observed no developmental trend toward greater responsiveness with increasing age within the time window studied; robust and specific responses typical of adult IT cells were obtained in recording sessions at the youngest ages. Figure 17.1A shows poststimulus time histograms obtained from an IT neuron in a 12-week-old monkey as the stimulus was presented at various locations in the central visual field. The cell had little spontaneous activity, and a vigorous response began with a latency of approximately 134 msec. Overall, latencies for visual responses in IT cells in the alert infant monkeys ranged from 110 msec to about 320 msec, similar to the ranges reported in adult monkeys (Baylis et al., 1987). As is the case for IT neurons in adult monkeys, IT cells in the alert infants showed responses to stimulation at both contralateral and ipsilateral sites in the visual field, with the strongest responses resulting from stimulation at the fovea or nearby locations in the contralateral field. This is in contrast, for example, to cells in the overlying superior temporal polysensory area (area STP) where responses in both adults (Bruce et al., 1981) and infants were typically at least as strong in the far periphery as in the center of the field.

Stimulus selectivity

The incidence and types of stimulus selectivity observed in the alert infant monkeys were strikingly similar to those observed in both anesthetized and unanesthetized adult monkeys. As was the case for visual responsiveness, there was no indication of a trend toward more prevalent or pronounced selectivity for the stimuli we tested with increasing age. As early as the second month of life, individual IT neurons exhibited responses selective for faces, for shape, for geometrical patterns and other complex stimuli, and for color irrespective of other attributes.

Figure 17.1B illustrates patterns of responsiveness shown by several infant IT neurons tested with our standard set of shapes derived from Fourier descriptors (our so-called FD stimuli). As in adult IT (Schwartz et al., 1983), over half of the infant IT cells tested showed "tuned" responses to this stimulus set, that is, responses that were graded with the frequency in cycles per perimeter (the number of lobes of the FD stimulus).

Selectivity within a set of standard face and nonface stimuli is shown in Figure 17.1C. Some cells responded only to a single image (cells 1 and 3), whereas others, such as cell 2, showed responses to many but not all images in the set. Cell 3, which responded well to an adult monkey face, did not respond to either an infant monkey face or a scrambled version of an adult monkey face. Other "face-selective" neurons in alert infant monkeys exhibited properties described for populations of face-selective neurons in adult monkeys (Desimone et al., 1984; Perrett et al., 1982), including activation by any face stimulus in some cases and responses specific to profiles in others. As in adult monkeys, for some of these neurons it was apparent that specific features played a major role in the selectivity of the response, such as the visibility of teeth in the preferred profile views of one such cell. Although about 10% of IT neurons in the alert infant monkeys showed statistically significant responses to only one or two of a dozen or more images tested, the majority of neurons responded to a subset of the stimuli whose unifying or "critical" features were often unclear. Overall, 19% of the responsive cells responded to more than half of the stimuli tested. Similar patterns of stimulus selectivity have been found for IT cells in adult monkeys (Gross et al., 1985).

Studies in Anesthetized Infant Monkeys

In order to accurately plot receptive fields and to make comparisons with previous studies in anesthetized adult monkeys, we also recorded from IT cortex in 11 infant macaques ranging from 5 weeks to 7 months of age under conditions of nitrous oxide anesthesia and muscular paralysis with methods described in detail in Rodman (1991).

Anesthetized monkeys 4–7 months of age

Our first recordings in anesthetized infant monkeys were made in animals ranging from 4–7 months of age. In most respects, response properties of IT neurons at this age closely resembled those found in our previous studies of IT cortex in anesthetized adult animals (Desimone and Gross, 1979; Desimone et al., 1984; Gross et al., 1972). The only difference we found was a smaller proportion of responsive cells: whereas over 80% of IT cells responded to visual stimuli in anesthetized adult monkeys, only about half of the neurons we studied in the anesthetized 4–7 month olds did so.

Figure 17.1D illustrates receptive fields plotted on two penetrations through IT cortex in anesthetized infant monkeys in this age group. As in the adult monkey, about two-thirds of the receptive fields in the 4–7 month olds extended across the midline into both the contralateral and ipsilateral half-fields, and every receptive field plotted included the fovea. Median receptive field size (square root of receptive field area) was 20°, similar to our previous studies of adult monkeys. Moreover, as we have also found for adult monkeys (Desimone and Gross, 1979), the largest fields were obtained from neurons lying anteriorly within IT near its border with the superior temporal polysensory area (STP). As in both the adult animal and the alert infants, responses were strongest at the center of gaze, and we found neurons selective for shape, color, and specific complex objects, as well as a few cells selective for faces.

Anesthetized monkeys less than 4 months of age

In contrast to the adultlike properties seen in older infant monkeys under anesthesia and in alert infant monkeys at all ages, only a very few cells (10%) were responsive to visual stimulation in infant monkeys studied between 5 weeks and 4 months of age. All but one of the responsive cells encountered in this time period were found in animals over 3 months of age. Moreover, in the very young anesthetized infants, action potentials in IT tended to be small and difficult to isolate from background activity, and the spontaneous activity was low relative to that seen in adult IT and often had a bursty quality. These phenomena became less marked with increasing age in the anesthetized infants and were rare in the alert infants.

In order to ensure that the very low incidence of visual responsiveness in the youngest anesthetized infants was not due to a compromise of the animal's physiological condition by anesthesia and immobilization, we made control recordings in striate cortex and extrastriate visual area MT in several of the same sessions in which we failed to find visual responses in both IT cortex and in area STP in these animals. Cells in both these areas showed response magnitudes and selectivity typical of the adult. These results suggest that the effects of anesthesia on visual responses in IT does not reflect a general impairment due to the anesthetic regimen but instead indicates a relatively specific effect on "high-order" areas of the cortical visual pathway.

As an additional control, three of the alert recording subjects were restudied under nitrous oxide anesthesia when they were 3–3½ months old. Although the majority of IT cells in each of these animals was visually responsive when they were previously studied in the alert behaving paradigm, only 1 of a total of 30 cells studied in these animals under anesthesia gave a visual response. Moreover, responsive cells were found subsequently in these animals when they were again recorded from while awake. These results indicate that the effect of anesthesia on IT cortex was not an irreversible one produced in the initial recording sessions.

Summary of Physiological Studies

Even at the earliest ages sampled (6 weeks), the visual properties of neurons in IT cortex of alert infant macaques are strikingly similar to those found in adult monkeys. These include selectivity for shape, biologically significant images (faces), arbitrary geometrical patterns, and color. As in the adult, many IT cells in infants tend to fire to a number of different stimuli whose common features were often unclear. Similar selectivity was found in IT cortex in the anesthetized animals as early as visual responses could be elicited. In both alert and anesthetized infant monkeys, IT cells respond to stimulation in both contralateral and ipsilateral portions of the visual field. IT cells with adultlike properties thus appear to be available to the visual recognition apparatus without prolonged periods of visual experience or maturation.

Although stimulus selectivity was normal in the anesthetized infants, the proportion of visually responsive cells was reduced, and almost totally so in the youngest group, relative to unanesthetized infants and to both anesthetized and unanesthetized adults. This effect of anesthesia in infants was found both in IT and the superior temporal polysensory area but not in striate cortex or area MT, suggesting that it may be characteristic of the "higher" visual cortical areas.

CONNECTIONS OF INFERIOR TEMPORAL CORTEX IN INFANTS

In order to determine whether the organization of inputs to IT is adultlike during the age window we had examined with physiological approaches, we injected WGA-HRP into anterior IT in four infant monkeys, one at 7 weeks, one at 13 weeks, and two at 18 weeks (Rodman and Consuelos, 1992). The pattern of retrograde labeling was examined throughout the cortex in both hemispheres. The pattern of labeling in the infants was then compared to that seen after similar injections of adult monkeys, both in our own lab and in other studies.

Ipsilateral Afferents

From occipitotemporal areas

Overall, the pattern of afferents from occipitotemporal zones in the ipsilateral hemisphere, containing areas known to be mainly visual, was strikingly similar in infant monkeys of different ages and in adults. Figure 17.2 illustrates the pattern of cortical labeling found after an injection of retrograde tracer into IT cortex of an adult macaque. The pattern of labeling seen after similar injections in our youngest (7 weeks) and oldest (18 weeks) infant cases is shown in Figure 17.3. As with adult IT cortex, the strongest inputs to IT in infant monkeys arise from ipsilateral visual area TEO and from portions of IT cortex posterior to the injection site (see Figures 17.3 and 17.4). Sparse to moderately dense retrograde label was also found in visual area V4 in both infant and adult monkeys. Labeling in V4 was most pronounced in the ventral portion of the prelunate gyrus and adjacent superior temporal sulcus, which represents the central portion of the visual field (Gattass et al., 1988). In both infants and adults, inputs to IT from posterior zones originate almost exclusively in the supragranular layers in V4 and become progressively more concentrated in the infragranular layers going more anteriorly, until supragranular and infragranular label is about equally dense just posterior to the injection site in IT.

In both infant and adult monkeys, inputs to IT cortex from ipsilateral occipitotemporal regions also arise from the superior temporal polysensory area (STP), from parahippocampal areas TF and TH, the lateral bank of the rhinal sulcus, and the temporal pole (see Figures 17.2, 17.3, and 17.4A). Retrograde label was predominantly infragranular in each of these regions in both infants and adults, consistent with a feedback type of projection (Felleman and Van Essen, 1991; Rockland and Pandya, 1979). In the case of STP, pronounced labeling was seen in the infant monkeys only when the injection halo encroached on the portion of IT cortex on the lower bank of the superior temporal sulcus (compare Figures 17.3A and 17.3B); likewise, in adult monkeys, this region was heavily labeled only in cases in which the injection encroached on the lip of the superior temporal sulcus and not in cases with injections placed more ventrally within IT (Fenstemaker, 1986), supporting the suggestion of functional subdivisions within IT cortex mentioned above. Finally, a sparse projection was noted from the superficial layers of a zone on the lateral bank of the intraparietal sulcus in both adult monkeys and infants of various ages.

From anterior regions

In both infant and adult monkeys, afferents to IT cortex originate in several portions of frontal cortex. In addition, we noted projections in the infants from several additional zones not usually reported to project to IT cortex in adults,

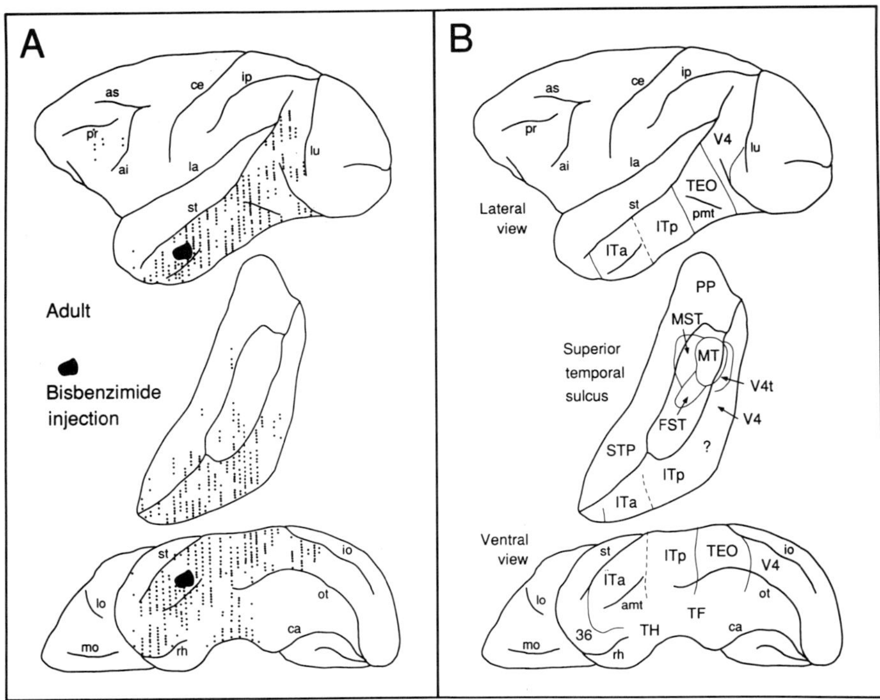

FIGURE 17.2. **A**. Reconstruction of cortical labeling shown on lateral and ventral views of the brain and on opened view of the superior temporal sulcus following injection of a retrograde tracer (bisbenzimide) into anterior IT in an adult macaque (redrawn from Fenstemaker, 1986). Each dot represents one or two retrogradely labelled neurons. **B**. Subdivisions of temporal neocortex referred to in the present study. Areal boundaries are approximate and are based on von Bonin and Bailey (1947), Gattass et al. (1985, 1988), Fenstemaker (1986), Amaral et al. (1987), Boussaoud et al. (1990, 1991), and Webster et al. (1991). ai, inferior arcuate sulcus; ce, central s.; ip, intraparietal s.; la, lateral s.; lo, lateral orbital s.; lu, lunate s.; mo, medial orbital s.; ot, occipitotemporal s.; pmt, posterior medial temporal s.; pr, principal s.; rh, rhinal s.; st, superior temporal s.

namely, the insula and anterior cingulate cortex. Moreover, labeling in these latter two regions was most pronounced in the youngest infant (7 weeks), sparse at 13 weeks, and sparse or absent at 18 weeks.

Sparse-to-moderate projections to IT in both infants and adults originate from two zones in frontal cortex (see Figures 17.2, 17.3, and 17.4). The first of these included the lateral bank of the principal sulcus, the medial bank of the inferior arcuate sulcus, and the intervening lateral surface. The second zone was located ventrally, within and around the lateral orbital sulcus. In both infants and adults, labeled cells in these regions showed a bilaminar distribution weighted toward the supragraular layers.

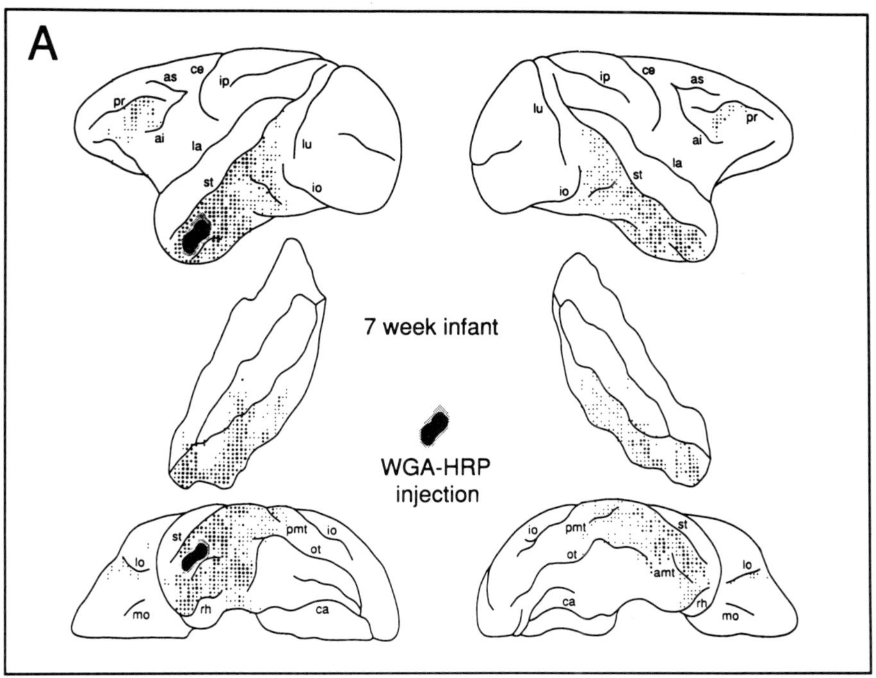

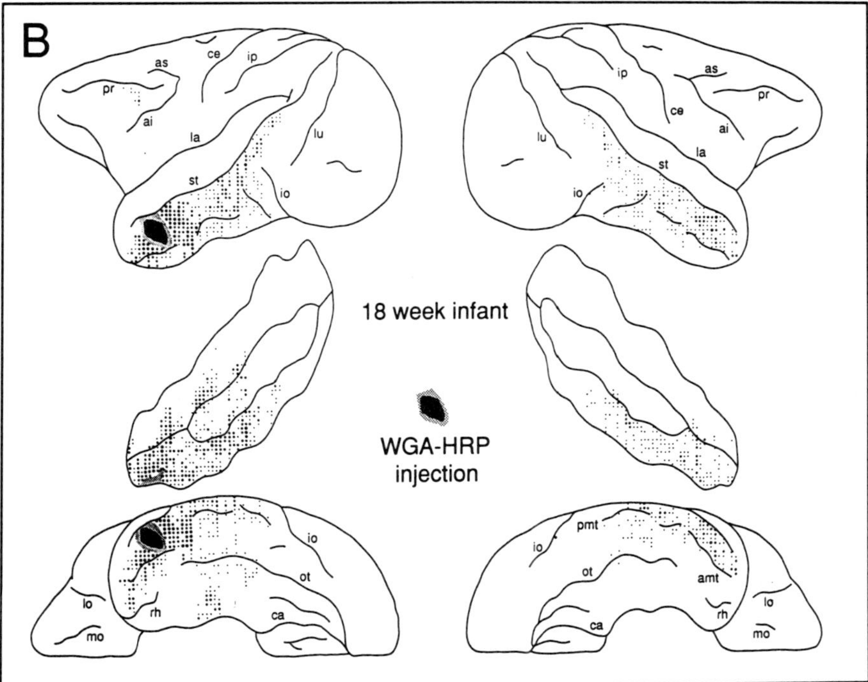

FIGURE 17.3. Reconstruction of cortical labeling following injection of retrograde tracer (WGA-HRP) into anterior IT in our youngest (A) and oldest (B) infant macaques (adapted from Rodman and Consuelos, 1992). Small, medium, and large dots indicate sparse, moderate, and dense labeling, respectively. See also legend to Figure 17.2.

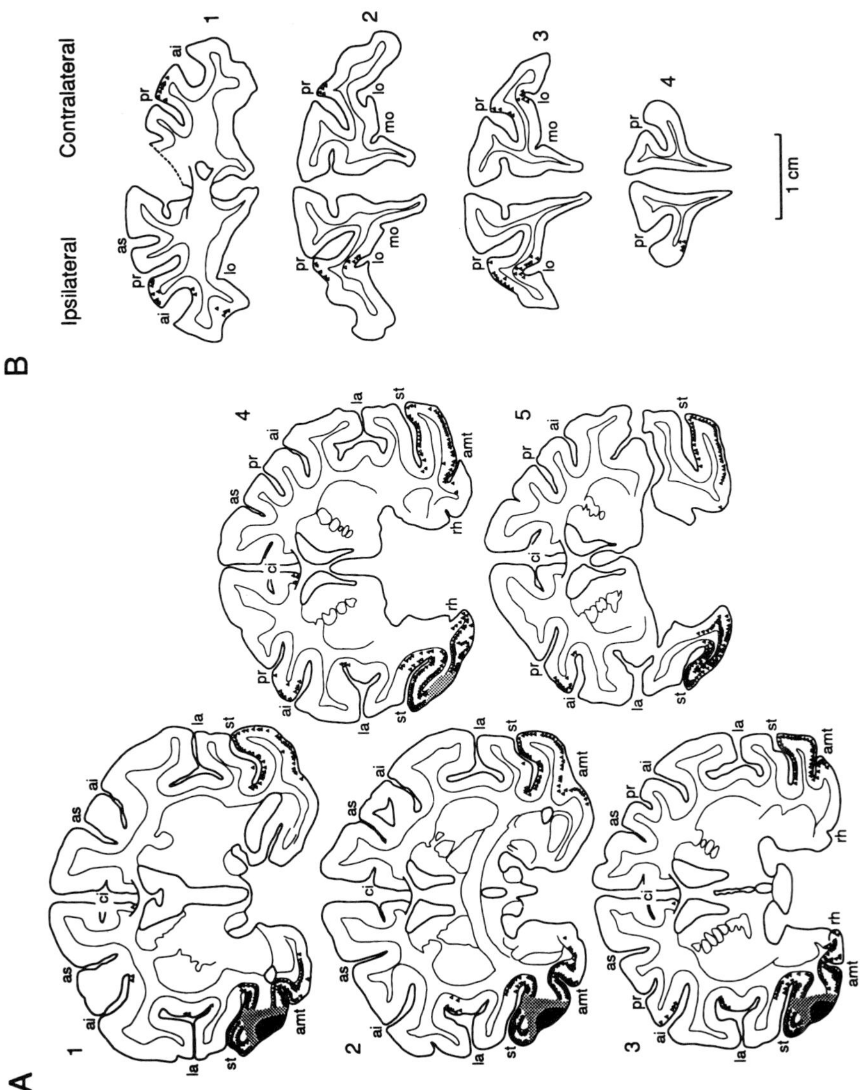

FIGURE 17.4. Cross-sections illustrating locations of retrogradely labeled cells (triangles) in anterior temporal lobe of a 13-week-old infant monkey (A) and in the frontal lobe of a 7-week-old infant monkey (B) (adapted from Rodman and Consuelos, 1992). Density of triangles reflects density of labeling. See also legend to Figure 17.2.

In all infants but one of the two oldest (18 weeks), sparse-to-moderate labeling was found within the insular cortex of the ipsilateral hemisphere (sections 1–4 of Figure 17.4A). Although the densest patch of label was located within the granular region of the insula, a few cells were also found more ventrally within the dysgranular insula (Mesulam and Mufson, 1982). Most of the labeled cells in the insula were found in the supragranular layers. Although labeled cells were not observed in the insula after IT injections in adult macaques in our own previous studies (Desimone et al., 1980; Fenstemaker et al., 1985) or indeed in most other previous studies of connections of IT cortex in adults (e.g., Shiwa, 1987), one brief report (Martin-Elkins and Horel, 1987) noted such a projection after injections of the ventral portion of anterior IT.

Moderately dense label was found in the bottom portion of ipsilateral cingulate cortex in the 7-week-old animal (see Figure 17.5B). Sparse label was found within the same region in the 13-week-old animal and in one of the 18 week olds (see Figure 17.4A). In each case, labeled cells were located predominantly in the supragranular layers. We did not see such a projection in adult monkeys. One previous study in adult monkeys (Shiwa, 1987) reported a very small number of labeled cells in the same part of cingulate cortex labeled in the infants; however, the weakness of the projection illustrated in that study contrasts with the relatively robust cingulate labeling we found at 7 weeks, lending credence to the possibility that this represents a largely transient pathway.

Contralateral Afferents

Projections from the contralateral hemisphere to IT in infant monkeys originated from homotopic portions of IT and from parahippocampal and perirhinal zones, as has also been found for adult monkeys (Desimone et al., 1980; Martin-Elkins and Horel, 1987; Webster et al., 1991a). In addition, we identified a number of contralateral pathways to IT cortex in infant monkeys which have not been investigated thus far in adult monkeys. These contralateral projections were most widespread in the youngest infant.

In all the infant monkeys, the heaviest contralateral labeling was found in a mirror-symmetric region of IT cortex. Consistent projections also arise from contralateral TEO and V4, as well as from portions of contralateral IT posterior to the level of the injection site (see Figures 17.3 and 17.4). Earlier, we had obtained similar results after large injections of IT in adult monkeys (Desimone et al., 1980). Labeled cells in contralateral V4 and TEO were found only in the supragranular layers, consistent with the exclusively supragranular origin of contralateral projections to IT in adult monkeys (Desimone et al., 1980; Webster et al., 1991a). In contralateral IT of the infants, while the majority of labeled cells were located in the supragranular layers, a few cells were also located in layers V and VI (see Figure 17.5A). Interestingly, these cells were much more numerous in the 7-week-old monkey.

Projections from contralateral zones anterior to occipitotemporal regions were identified only in the 7-week-old infant monkey. These included sparse-to-

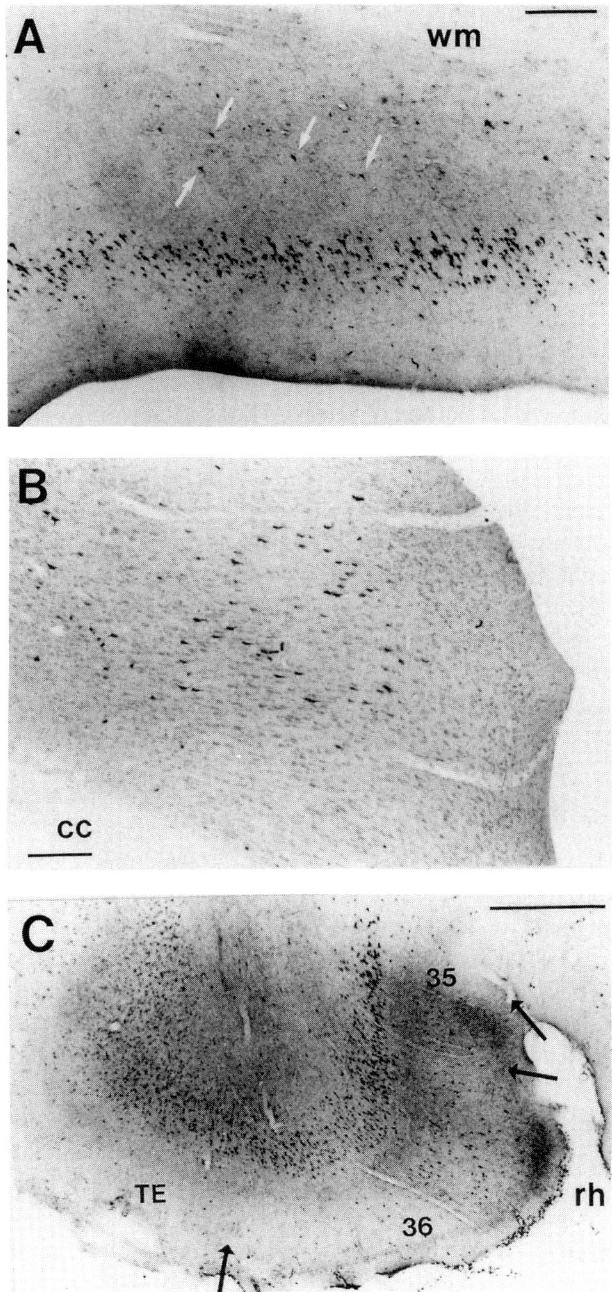

FIGURE 17.5. Brightfield photomicrographs of WGA-HRP labeling in 7-week-old infant (adapted from Rodman and Consuelos, 1992). Neutral red counterstain. **A**. Contralateral IT cortex. Although most labeled cells were located in the superficial layers, a few (white arrows) were also found in the deep layers; wm, white matter. Scale bar = 200 μm. **B**. Labeling in ipsilateral cingulate cortex. cc, corpus callosum. Scale bar = 100 μm. **C**. Labeling in ipsilateral perirhinal cortex. Note anterograde label in layers I and IV. rh, rhinal sulcus. Scale bar = 500 μm.

moderate projections from the same sectors of frontal cortex (see Figure 17.4), insular cortex, and cingulate cortex labeled in the ipsilateral hemisphere. Labeled cells in these contralateral regions were overwhelmingly supragranular, as in more posterior zones; however, a few cells were also found in deep layers in contralateral ventral frontal and cingulate cortex. To our knowledge, no such contralateral pathways to IT cortex have yet been reported for adult monkeys.

Efferent Connections

The quality of anterograde labeling observed in the infants varied across cases and cortical regions. In general, the location and laminar distribution of anterograde labeling in the infants corresponded closely to that seen in adult monkeys in previous studies. One exception of potential significance involved the pattern of labeling in the rhinal cortex (cytoarchitectonic areas 35 and 36). In a recent study, Webster and colleagues (1991) reported that projections from IT to this part of rhinal cortex are relatively restricted and concentrated in layer IV in adult monkeys, whereas in newborns they are widespread and also distribute heavily to layer I. Similarly, in our older infants anterograde label was found to predominate in layer IV of rhinal cortex, whereas in our 7-week-old animal heavy label was also found in layer I (see Figure 17.5C).

Summary of Anatomical Studies (also see Table 17.1)

The cortical areas providing afferents to anterior IT cortex from occipitotemporal regions are essentially the same in infant and adult animals. Ipsilaterally, these afferent pathways include sparse projections from area V4 and the intraparietal sulcus, moderately dense projections from area TEO, parahippocampal, and perirhinal/temporopolar areas, and dense projections from portions of IT cortex posterior to the injection site. There is also a moderately dense input to dorsal anterior IT from area STP in both infants and adults. Contralateral areas V4 and TEO provide a sparse projection to anterior IT, as do parahippocampal and perirhinal/temporopolar zones, whereas contralateral IT provides a fairly dense input. We did not see any "exuberant" projections from "lower-order" cortical visual areas in the infant monkeys. Overall, these results suggest that inputs from predominantly visual cortical regions, and zones surrounding them in the ventral temporal lobe, are adultlike at 7 weeks of age.

One exception to the adultlike pattern of occipitotemporal inputs to IT in infant monkeys is the finding of a small projection from the deep layers of contralateral IT cortex, a pathway apparently absent in adult monkeys. Since this pathway was most extensive in our youngest infant, it may be a transient projection, relatively pronounced at birth but nearly eliminated by 18 weeks. Transient corticocortical inputs deriving from layers of extrastriate cortex not providing afferents in adult monkeys have been reported by Kennedy and colleagues (1989), who described such projections from the superficial layers of MT back to striate cortex in newborn macaques.

TABLE 17.1. Cortical Afferents to Anterior IT

A. IPSILATERAL AFFERENTS

Area	Adult	13–18 Weeks	7 Weeks
V4	(+)	+	+
TEO	++	++	++
Posterior IT	+++	+++	+++
Anterior IT	+++	+++	+++
STP	++[a]	++[a]	?
Parahippocampal (TF/TH)	++	++	++
Perirhinal/Polar (35/36)	++	++	++
Lateral Intraparietal	+	+	+
Lateral Frontal	++	++	++
Ventral Frontal	+	+	++
Insula	+	+	++
Anterior Cingulate	(+)	+	++

B. CONTRALATERAL AFFERENTS

Area	Adult	13–18 Weeks	7 Weeks
V4	?	+	+
TEO	?	+	+
Posterior IT	++[b]	++	++
Anterior IT	++[b]	++	++
STP	?	+[a]	?
Parahippocampal (TF/TH)	+	(+)	–
Perirhinal/Polar (35/36)	+	+	++
Lateral Intraparietal	?	–	–
Lateral Frontal	?	–	++
Ventral Frontal	?	–	+
Insula	?	–	(+)
Anterior Cingulate	?	–	+

Summary of anatomical studies.For the adult comparisons, we relied both on our own findings (Desimone et al., 1980; Fenstemaker, 1986) and on studies performed in other laboratories (Martin-Elkins and Horel, 1987; Mesulam and Mufson, 1982; Morel and Bullier, 1990; Shiwa, 1987; Webster et al., 1991). One, two, and three pluses indicate sparse, moderate, and strong projections, respectively. Parentheses indicate projections that are very weak or uncertain. Minus signs indicate projections that have been looked for and found to be absent. Question marks denote pathways whose existence has not yet been sought.

[a]To dorsal-most portion of IT and/or IT within superior temporal sulcus.

[b]From supragranular layers only.

Inputs from anterior cortical regions (frontal, insular, and cingulate cortex) were more dense and widespread in our 7-week-old infant than in either the older infants or adults. Although frontal cortex of the ipsilateral hemisphere provides moderately dense inputs to IT in both adults and infants at various ages, projections from contralateral frontal cortex were found only at 7 weeks. Moreover, projections from insular and cingulate cortex were also bilateral at 7 weeks, and weaker, uncertain, or exclusively ipsilateral in older infants and adults. The more extensive inputs to IT in very young monkeys from contralateral cortical areas (including the deep layer projection from contralateral IT) may be related to the specification of the highly precise connectivity presumably needed to maintain stimulus equivalence across both halves of visual space within IT receptive fields.

In our infant cases, an immature pattern of anterograde labeling in the rhinal sulcus was found only at 7 weeks and not at later ages. This finding provides evidence that the cortical connectivity of IT undergoes additional maturation between 7 and 18 weeks of age in terms of its outputs as well as inputs.

RELATION TO PHYSIOLOGICAL AND ANATOMICAL DEVELOPMENT OF IT CORTEX TO BEHAVIORAL DEVELOPMENT

In awake behaving infant macaques, we found that the visual properties of IT neurons were essentially indistinguishable from those found in mature monkeys. Moreover, we found that inputs to IT from visual cortical regions are also essentially adultlike at this stage. However, visual learning and recognition, and in particular the participation of IT cortex in these functions, does not approach adult status until after at least 6 months. What might be the reason or reasons for this discrepancy?

First, although we have documented that basic adultlike response properties are present in IT cortex early in life, more work is needed to determine whether subtle quantitative differences in response magnitude or tuning (or some other unit property, such as temporal firing pattern, which we did not study) might be detectable in infants.

Another possibility is that the development of visual recognition ability may depend, at least in part, on nonvisual cortical regions that are themselves immature in infant macques. Indeed, an immature pattern of IT connectivity in infant monkeys is seen specifically with areas known to play a role in learning and memory, namely frontal, rhinal, and cingulate cortex. Furthermore, at least one of these regions, the lateral frontal cortex, is known to be functionally immature at the ages studied here (Goldman, 1971). There also appears to be a difference between infant and adult monkeys in the laminar pattern of inputs from contralateral IT. The infant pattern may be important in establishing stimulus equivalence across retinal translation, a crucial mechanism in the development of visual pattern learning.

Finally, it is possible that the connections of IT cortex with subcortical structures believed to be important in visual attention and memory, such as the pulvinar, hippocampus, and amygdala, are different in infant macaques (see also Webster et al., 1991).

Although the stimulus selectivity of IT neurons in the awake infants appeared adultlike as did the connections with visual cortical areas, the much greater sensitivity of IT neurons in infants to anesthesia demonstrates in itself that IT cortex in infant monkeys is clearly physiologically different from IT cortex in adults. How this difference is related to visual learning and recognition remains to be seen; it may involve developmental changes in cellular energy metabolism (Hagger et al., 1988) or synapse formation (Rakic et al., 1986).

SOME UNANSWERED QUESTIONS

Our findings on the physiological properties and connections of IT neurons in infant monkeys raise a number of questions for further research.

1. What are the properties of IT neurons at birth? Is visual experience necessary for the emergence, maintenance, or fine-tuning of the characteristic adultlike properties of IT cells? In particular, is early social experience essential for the development of selective responsiveness to faces, either to faces as a general class of stimuli or to particular facial views or to expressions?
2. Visual experience can alter the responses of IT neurons, both transiently and more permanently (Fuster, 1990; Miyashita, 1988; Miyashita and Chang, 1988; Miller et al., 1991). The former phenomenon may underlie short-term visual memory and the latter, long-term memory. Are either or both types of neuronal plasticity present in infant monkeys, and do they change during the course of maturation? If so, does their development parallel the development of visual recognition ability?
3. Attentional factors can also modulate the responses of IT neurons in adult monkeys (Moran and Desimone, 1985; Spitzer and Richmond, 1991). How early can these influences be detected and do they change with age?
4. What changes in connectivity between IT cortex and subcortical structures take place during development, particularly in the pattern of inputs from chemically specific systems thought to play a role in arousal and attention in adult primates? In addition, are there developmental changes in expression of postsynaptic receptors on IT neurons that alter the effectiveness of specific classes of inputs at different stages in development?
5. The development of the adult pattern of connectivity of IT cortex appears to involve the loss of "exuberant" connections with limbic and anterior cortical regions. There is some evidence that maintenance of such connections may provide a basis for functional recovery after early damage to temporal lobe structures in monkeys (Webster et al., 1992). Perhaps the transient connections reflect an immature processing system that normally disappears during development but remains functional if the mature system is damaged (e.g., Goldman, 1971). What is the relationship between the loss of these connections and the development of visual recognition abilities in normal animals?

Acknowledgments. Preparation of this chapter was supported by National Institutes of Mental Health Grant MH 19420-21, National Science Foundation Grant BNS-8919538 and James S. McDonnell Foundation Grant 90-16.

REFERENCES

Amaral DG, Insausti R, Cowan WM (1987): The entorhinal cortex of the monkey: 1. Cytoarchitectonic organization. *J Comp Neurol* 264:326–355

Bachevalier J, Brickson M, Hagger C, Mishkin M (1990): Age and sex differences in the effects of selective temporal lobe lesion on the formation of visual discrimination habits in rhesus monkeys. *Behav Neurosci* 104:885–889

Bachevalier J, Mishkin M (1988): Long-term effects of neonatal temporal cortical and limbic lesions on habit and memory formation in rhesus monkeys. *Neurosci Abstr* 14:4

Baylis GC, Rolls ET, Leonard CM (1987): Functional subdivisions of temporal lobe neocortex. *J Neurosci* 7:330–342

Boussaoud D, Desimone R, Ungerleider LG (1991): Visual topography of area TEO in the macaque. *J Comp Neurol* 306:554–575

Boussaoud D, Ungerleider LG, Desimone R (1990): Pathways for motion analysis: Cortical connections of the medial superior temporal and fundus of the superior temporal visual areas of the macaque. *J Comp Neurol* 296:462–495

Bruce CJ, Desimone R, Gross CG (1981): Visual properties of neurons in a polysensory area in the superior temporal sulcus of the macaque. *J Neurophysiol* 46:369–384

Corbetta M, Dobmeyer S, Shulman GL, Petersen SE (1991): Selective and divided attention during visual discriminations of shape, color and speed: Functional anatomy by positron emission tomography. *J Neurosci* 11:2383–2402

Damasio AR (1989): Neural mechanisms. In: *Handbook of Research on Face Processing*. Young A, Ellis HD, eds. Holland: Elsevier

Desimone R, Albright TD, Gross CG, Bruce C (1984): Stimulus-selective properties of inferior temporal neuron in the macaque. *J Neurosci* 4:2051–2062

Desimone R, Fleming J, Gross CG (1980): Prestriate afferents to inferior temporal cortex: An HRP study. *Brain Res* 184:41–55

Desimone R, Gross CG (1979): Visual areas in the temporal cortex of the macaque. *Brain Res* 178:363–380

Desimone R, Ungerleider LG (1989): Neural mechanisms of visual processing in monkeys. In: *Handbook of Neuropsychology*, vol. 2, Boller F, Grafman J, eds. England: Elsevier

Felleman DJ, Van Essen DC (1991): Distributed hierarchical processing in primate cerebral cortex. *Cerebral Cortex* 1:1–48

Fenstemaker SB (1986): *The organization and connections of visual cortical area TEO in the macaque*. Unpublished doctoral dissertation, Princeton University

Fenstemaker SG, Albright TD, Gross CG (1985): Organization and neuronal properties of visual area TEO. *Neurosci Abstr* 11:1012

Fuster JM (1990): Inferotemporal cortex units in selective visual attention and short-term memory. *Neurophysiol* 64:681–697

Gattass R, Gross CG, Sandell J (1981): Visual topography of V2 in the macaque. *J Comp Neurol* 201:519–539

Gattass R, Sousa APB, Covey E (1985): Cortical visual areas of the macaque: Possible substrates for pattern recognition mechanisms. In: *Pattern Recognition Mechanisms*, Chagas C, Gattass R, Gross CG, eds. New York: Springer-Verlag

Gattass R, Sousa APB, Gross CG (1988): Visuotopic organization and extent of V3 and V4 of the macaque. *J Neurosci* 8:1831–1845

Gochin PM, Miller EK, Gross CG, Gerstein GL (1991): Functional interactions among neurons in macaque inferior temporal cortex. *Exp Brain Res* 84:505–516

Goldman PS (1971): Developmental determinants of cortical plasticity. *Acta Neurol Exp* 32:495–511

Gross CG (1973): Visual functions of inferotemporal cortex. In: *Handbook of Sensory Physiology*, vol. 7, part 3B, Jung R, ed. Berlin: Springer-Verlag

Gross CG (1992): Representation of visual stimuli in inferior temporal cortex. *Philos Trans R Soc Lond (Biol)* 335:3–10

Gross CG, Bender DB, Gerstein GL (1979): Activity of inferior temporal neurons in behaving monkeys. *Neuropsychologia* 17:215–229

Gross CG, Bender DB, Mishkin M (1977): Contributions of the corpus callosum and the anterior commissure to the visual activation of inferior temporal neurons. *Brain Res* 131:227–239

Gross CG, Bender DB, Rocha-Miranda CE (1974): Inferotemporal cortex: A single unit analysis. In: *The Neurosciences: A Third Study Program*, Schmitt FO, Worden FG, eds. Cambridge: MIT Press

Gross CG, Desimone R, Albright TD, Schwartz EL (1985): Inferior temporal cortex and pattern recognition. In: *Pattern Recognition Mechanisms*, Chagas C, Gattass R, Gross CG, eds. Vatican City: Pontifica Academia Scientiarum

Gross CG, Mishkin M (1977): The neural basis of stimulus equivalence across retinal translation. In: *Lateralization in the Nervous System*. Harnad S, Doty R, Jaynes J, Goldstein L, Krauthamer G, eds. New York: Academic Press

Gross CG, Rocha-Miranda CE, Bender DB (1972): Visual properties of neurons in inferotemporal cortex of the macaque. *J Neurophysiol* 35:96–111

Gross CG, Schiller PH, Wells C, Gerstein GL (1967): Single-unit activity in temporal association cortex of the monkey. *J Neurophysiol* 30:833–843

Hagger C, Bachevalier J, Macko KA, Kennedy C, Sokoloff L, Mishkin M (1988): Functional maturation of inferior temporal cortex in infant rhesus monkeys. *Neurosci Abstr* 14:2

Hagger C, Brickson M, Bachevalier J (1985): Sparing of visual recognition after neonatal lesions of IT in infant rhesus monkeys. *Neurosci Abstr* 9:26

Kennedy H, Bullier J, Dehay C (1989): Transient projection from the superior temporal sulcus to area 17 in the newborn macaque monkey. *Proc Natl Acad Sci U S A* 86:8093–8097

Lewis ME, Mishkin M, Brown RM, Pert CB, Pert A (1981): Opiate receptor gradients in monkey cerebral cortex: Correspondence with sensory processing hierarchies. *Science* 211:1166–1169

Martin-Elkins CL, Horel JA (1987): Cortical areas projecting to anterior inferior temporal gyrus as demonstrated by retrogradely transported wheat germ agglutinin conjugated to horseradish peroxidase in the macaque. *Neurosci Abstr* 14:11

Mesulam MM, Mufson EJ (1982): Insula of the Old World monkey: I. Architectonics in the insulo-orbito-temporal component of the paralimbic brain. *J Comp Neurol* 212:1–22

Michalski A, Gerstein GL, Czarkowska J, Tarnecki R (1983): Interactions between cat striate cortex neurons. *Exp Brain Res* 51:97–107

Mikami A, Nakamura K (1988): Behavioral role of stimulus selective neuronal activities in the superior temporal sulcus of macaque monkey. *Neurosci Abstr* 14:10

Miller EK, Gochin PM, Gross CG (1991): A habituation-like decrease in the responses of neurons in inferior temporal cortex of the macaque. *Vis Neurosci* 7:357–362

Milner B (1968): Visual recognition and recall after right-temporal lobe excision in man. *Neuropsychologia* 6:191–209

Mishkin M (1982): A memory system in the monkey. *Philos Trans R Soc Lond (Biol)* 298:85–95

Miyashita Y (1988): Neuronal correlate of visual associative long-term memory in the primate temporal cortex. *Nature* 335:817–820

Miyashita Y, Chang HS (1988): Neuronal correlate of pictorial short-term memory in the primate temporal cortex. *Nature* 331:68–70

Moran J, Desimone R (1985): Selective attention gates visual processing in the extrastriate cortex. *Science* 229:782–784

Morel A, Bullier J (1990): Anatomical segregation of two cortical visual pathways in the macaque monkey. *Vis Neurosci* 4:555–578

Nelson RB, Friedman DP, O'Neill JB, Mishkin M (1987): Gradients of protein kinase C substrate phosphorylation in primate visual system peak in visual memory storage areas. *Brain Res* 416:387–392

Perrett DI, Rolls ET, Caan W (1982): Visual neurones responsive to faces in the monkey temporal cortex. *Exp Brain Res* 47:329–342

Petersen SE, Robinson DL, Morris JD (1987): The contribution of the pulvinar to visual spatial attention. *Neuropsychologia* 25:97–105

Raisler RL, Harlow HF (1965): Learned behavior following lesions of posterior association cortex in infant, immature and preadolescent monkeys. *J Comp Physiol Psychol* 60:167–174

Rakic P, Bourgeois JP, Eckenhoff MEF, Zecevic N, Goldman-Rakic PS (1986): Concurrent overproduction of synapses in diverse regions of the primate cerebral cortex. *Science* 232:232–235

Richmond BJ, Wurtz RH, Sato T (1983): Visual responses of inferior temporal neurons in awake rhesus monkey. *J Neurophysiol* 50:1415–1432

Rockland KS, Pandya DN (1979): Laminar origins and terminations of cortical connections in the occipital lobe in the rhesus monkey. *Brain Res* 179:3–20

Rodman HR (1991): Methods for repeated recording in visual cortex of anesthetized and awake behaving infant monkeys. *J Neurosci Methods* 38:209–222

Rodman HR, Consuelos MJ (1992): Cortical inputs to anterior inferior temporal cortex in infant monkeys. Submitted for publication

Rodman HR, Skelly JP, Gross CG (1991): Stimulus selectivity and state dependence of activity in inferior temporal cortex in infant monkeys. *Proc Natl Acad Sci USA* 88:7572–7575

Rolls ET, Baylis GC (1986): Size and contrast have only small effects on the responses to faces of neurons in the cortex of the superior temporal sulcus of the monkey. *Exp Brain Res* 65:38–48

Rolls ET, Baylis GC, Hasselmo ME, Nalwa V (1989): The effect of learning on the face selective responses of neurons in the cortex in the superior temporal sulcus of the monkey. *Exp Brain Res* 76:153–164

Sato T (1988): Effects of attention and stimulus interaction on visual responses of inferior temporal neurons in macaque. *J Neurophysiol* 60:344–364

Schwartz EL, Desimone R, Albright TD, Gross CG (1983): Shape recognition and inferior temporal neurons. *Proc Natl Acad Sci USA* 80:5776–5778

Seltzer B, Pandya DN (1978): Afferent cortical connections and architectonics of the superior temporal sulcus and surrounding cortex in the rhesus monkey. *Brain Res* 149:11–24

Shiwa T (1987): Corticocortical projections to the monkey temporal lobe with particular reference to the visual processing pathways. *Arch Ital Biol* 125:139–154

Spitzer H, Richmond BJ (1991): Task difficulty: Ignoring, attending to, and discriminating a visual stimulus yield progressively more activity in inferior temporal neurons. *Exp Brain Res* 83:340–348

Squire LR (1987): *Memory and Brain*. New York: Oxford University Press

Squire LR, Zola-Morgan S (1991): The medial temporal lobe memory system. *Science* 253:1380–1386

Suzuki WA, Amaral DG (1991): Cortical inputs to the CA1 field of the monkey hippocampus originate from the perirhinal and parahippocampal cortex but not from area TE. *Neurosci Lett* 115:43–48

Tanaka K, Saito HA, Fukada Y, Moriya M (1991): Coding visual images of objects in the inferotemporal cortex of the macaque monkey. *J Neurophysiol* 66:170–189

Toyama K, Kimura M, Tanaka K (1981): Cross-correlation analysis of interneuronal connectivity in cat visual cortex. *J Neurophysiol* 46:191–201

von Bonin G, Bailey P (1947): *The Neocortex of Macaca Mulatta*. Urbana, IL: University of Illinois Press

Webster MJ, Ungerleider LG, Bachevalier J (1991): Connections of inferior temporal areas TE and TEO with medial temporal-lobe structures in infant and adult monkeys. *J Neurosci* 11:1095–1116

Webster MF, Ungerleider LG, Bachevalier J (1992): Lesions of inferior temporal area TE in infant monkeys produce reorganization of cortico-amygdalar projections. *Neuroreport* 2:769–772

Yukie M, Iwai E (1988): Direct projection from ventral TE area of the inferotemporal cortex to hippocampal field CA1 in the monkey. *Neurosci Letters* 88:6–10

18

Reorienting Visual Spatial Attention: Is It Based on Cartesian Coordinates?

LUIZ GAWRYSZEWSKI, ROSALIA B. FARIA, TANIA G. THOMAZ, WALTER M. PINHEIRO, GIACOMO RIZZOLATTI, AND CARLO UMILTA

Selective attention is an old topic within experimental psychology (James, 1890) and most frequently refers to performance when there are conflicts between signals. Attention involves selection of higher levels of processing, while preventing access of other signals to those same high levels of processing. Selective attention plays an important role in most cognitive tasks, including pattern recognition, reading, and mental imagery (Posner, 1982).

An important aspect of selective attention is orienting to a source of visual signals. Orienting can be divided into covert and overt changes. It is clear that one frequently orients by moving head and eyes toward the stimulus event. However, it is also possible to orient without any concomitant change in eye position or postural response (Posner, 1978, 1980). The important question is how to measure such covert changes in orienting.

Cognitive studies with normal humans using visual cues to direct attention covertly to a location eccentric from the point of fixation show more efficient processing of signals at the cued location. This enhancement includes lowered manual (Gawryszewski et al., 1987; Hughes and Zimba, 1985; Posner, 1978, 1980; Rizzolatti et al., 1987; Tassinari et al., 1987) and saccadic (Posner, 1978, 1980) reaction times, reduced sensory threshold (Bashinski and Bachrach, 1980), improvement in conjoining features (Prinzmetal et al., 1986), and modulation of evoked electrical potentials recorded from the scalp (Harter and Aine, 1984; Hillyard et al., 1985).

In these studies the covert orienting of attention was induced by using either peripheral cues that are near the target, or central symbolic cues telling the subject where the target is most likely to occur (Posner and Cohen, 1984). Peripheral cues—that is, abrupt changes in the peripheral visual field—have reflexive control over attention allocation, so that, when one of these stimuli occurs, a shift of attention is automatically elicited (Jonides, 1981; Jonides and Mack, 1984). Observers also have internal control over spatial allocation of attention, so that when directed by a symbolic, centrally positioned cue, they can voluntarily shift attention.

Based on criteria such as capacity demands, resistance to suppression, and sensitivity to expectancy, peripheral cues were shown to cause automatic shifts of attention, whereas central cues were shown to cause voluntary shifts of attention (Jonides, 1981). These and other results, which have demonstrated differential time courses of orienting in response to peripheral and central cues (Müller and Findlay, 1988; Müller and Rabbitt, 1989; Yantis and Jonides, 1984) have led to a two-component model of attention shifts (Müller and Findlay, 1988; Müller and Rabbitt, 1989). Peripheral cues are thought to trigger a fast, automatic orienting mechanism, which responds to physical properties of the cue. This component is believed to be transitory, and, as it fades out, control is taken over by a second, slower, orienting mechanism which responds to the spatial information conveyed by the cue. Central cues, in contrast, are assumed to initiate this second orienting mechanism only.

Another difference between peripheral and central cues consists of the appearance in the former case of some inhibition effect. For example, a peripheral cue speeds up target detection if the cue-target interval is within 100 msec. However, this early advantage is replaced by a delay as the interval increases. The delay is explained as due to an inhibition (inhibition of return) that maximizes sampling of new locations (Posner and Cohen, 1984). This inhibition of return occurs when the target follows the cue by 200–1500 msec and appears to affect the whole hemifield where the cue was presented (Berlucchi et al., 1989; Tassinari et al., 1987, 1989).

In order to explain some features of attention reorientation, Rizzolatti and colleagues (1987) proposed a hypothesis that postulates a strict link between covert attention and the programming of ocular movements (premotor hypothesis of attention). The basic idea is that both overt and covert orienting of attention are controlled by the neural mechanisms that are also in charge of saccade programming. Covert orienting would occur when a mechanism prevents eye movements, but leaves unchanged the program for the corresponding saccade. Thus, once a directional cue is presented, a motor program for a saccade is prepared, which specifies the direction of the eye movement and the amplitude of the saccade. The motor program is prepared either when the saccade is subsequently executed (i.e., overt orienting) or not executed (i.e., covert orienting). When the imperative stimulus appears in the target position for the saccade, the manual response is emitted without further delay. When the stimulus appears in an unexpected position, a time-consuming change in the saccade program takes place before the manual response is emitted. This modification in the program produces a delay in the response, whose magnitude depends on the importance of the movement parameter that has to be changed. If amplitude must be changed, what is needed is an adjustment in activation of the already programmed set of muscles. It is reasonable to suppose that larger adjustments consume more time than smaller adjustments. If direction is the feature that must be changed, the change in the motor program is particularly time consuming, because an entirely different set of muscles must be programmed.

In this chapter, we present two experiments designed to test the premotor hypothesis of attention. In these experiments we used a central cue and examined orienting of attention that occurs after an invalid cue (Rizzolatti et al., 1987). With this procedure observers are asked to direct attention to a certain location and then, on some (invalid) trials, the target is presented at various predetermined distances from the cued location. This allowed us to estimate the time necessary for attention to be reoriented from the expected to the stimulated location.

The first experiment was designed to answer the following questions:

1. Are costs for reorienting attention across the vertical meridian equivalent to costs for reorienting it across the horizontal meridan?
2. Are costs for reorienting attention from fixation point to a peripheral position equivalent to costs for reorienting it from a peripheral position to another peripheral position requiring crossing of the vertical or horizontal meridian?
3. Are costs for reorienting attention across both the principal meridians greater than costs for reorienting it across only one principal meridian?

In terms of the premotor hypothesis, these questions can be rephrased as follows:

1. Are costs for reprogramming the horizontal vectorial component of an ocular program equivalent to costs for reprogramming the vertical vectorial component?
2. Are costs for canceling a program of active fixation and generating a program to a peripheral position equivalent to costs for canceling one vectorial component and generating the opposite one?
3. Are costs for reprogramming two vectorial components of an ocular program greater than costs for reprogramming only one vectorial component?

Having obtained affirmative answers to all the above questions we designed a second experiment in which we tested whether the greater costs observed in the first experiment when attention had to cross the two principal meridians were due to a double change in the motor program (premotor hypothesis) or to the fact that attention had to cross the fovea, a region of expanded cortical representation (see Downing and Pinker, 1985). In this experiment, we explored whether costs for reorienting attention between positions located in orthogonal meridians were equivalent to costs for reorienting attention between positions located in different hemifields. In the latter case, reorienting of attention corresponds, according to the premotor hypothesis (Rizzolatti et al., 1987), to canceling one vectorial component of the ocular program and generating one vectorial component in a direction perpendicular to the previous one.

EXPERIMENT 1

The first experiment was designed to answer questions concerning costs for reorienting attention across the vertical meridian, the horizontal meridian, across both the principal meridians, or away from a central fixation point.

Methods

SUBJECTS

Twelve male subjects were tested in this study. They were students of Universidade Federal Fluminense. All were right-handed according to the Edinburgh inventory (Oldfield, 1971), had normal or corrected vision, and were ignorant of the purpose of the study.

Apparatus and display

The experiment took place in a sound-attenuated room, under dim ambient light. The subject sat in front of a CRT screen driven by an APPLE II microcomputer, which timed the stimuli and recorded the responses. The head was positioned in a head-and-chin rest so that the distance between the eyes and the screen was approximately 67 cm. The visual display (see Figure 18.1) comprised a fixation point located inside a square box (1.1° wide) at the geometric center of the screen, and four peripheral empty boxes located one in each quadrant with an eccentricity of 4.5°. The display also contained the visual cues that indicated in which square the imperative stimulus would probably appear. The imperative stimulus was a square (0.6° wide) that was shown for 100 msec within one of the five boxes. The response was emitted by pressing a key on the computer keyboard (the character "N") with the right index finger.

Procedure

Each trial began with the presentation of the fixation point and the five boxes, which were followed, after an interval of 500 msec, by the cue. A variable interval (800, 1000, or 1200 msec) elapsed and then the imperative stimulus was presented. When the cue was a small arrowhead pointing to a peripheral box (see Figure 18.1A, 18.1B, and 18.1C), the stimulus appeared with high probability (70%) in the cued box and with low probability in the other peripheral boxes (10% in each box). When the cue was a small square (see Figure 18.1D), the stimulus appeared with high probability (60%) in the central box and with low probability in the peripheral boxes (10% in each box).

The subject was instructed to look at the fixation point while directing attention to the cued box. The importance of maintaining fixation was stressed. At the beginning of the first session eye movements were monitored, by means of a suitably oriented mirror, by an experimenter sitting behind the subject.

The subject's task was to press as fast as possible the response key to the occurrence of the imperative stimulus, regardless of its position. Response latency (simple reaction time, RT) was measured to the nearest millisecond from stimulus onset to response emission. The response ended the trial and was followed by a 1.0 sec feedback about speed and accuracy. RTs shorter than 150 or longer than 700 msec were considered errors and were discarded. The display, including the cue, stayed on until the occurrence of a response or until a 3.0 sec interval had elapsed.

Subjects attended 4 sessions each. A session consisted of a short series of training trials and 320 experimental trials, subdivided into 4 blocks with some minutes rests in between.

Following the accepted terminology, trials in which the stimulus was shown in the cued box will be referred to as "valid" and trials in which the stimulus was shown in a box different from the cued one will be referred to as "invalid." In our

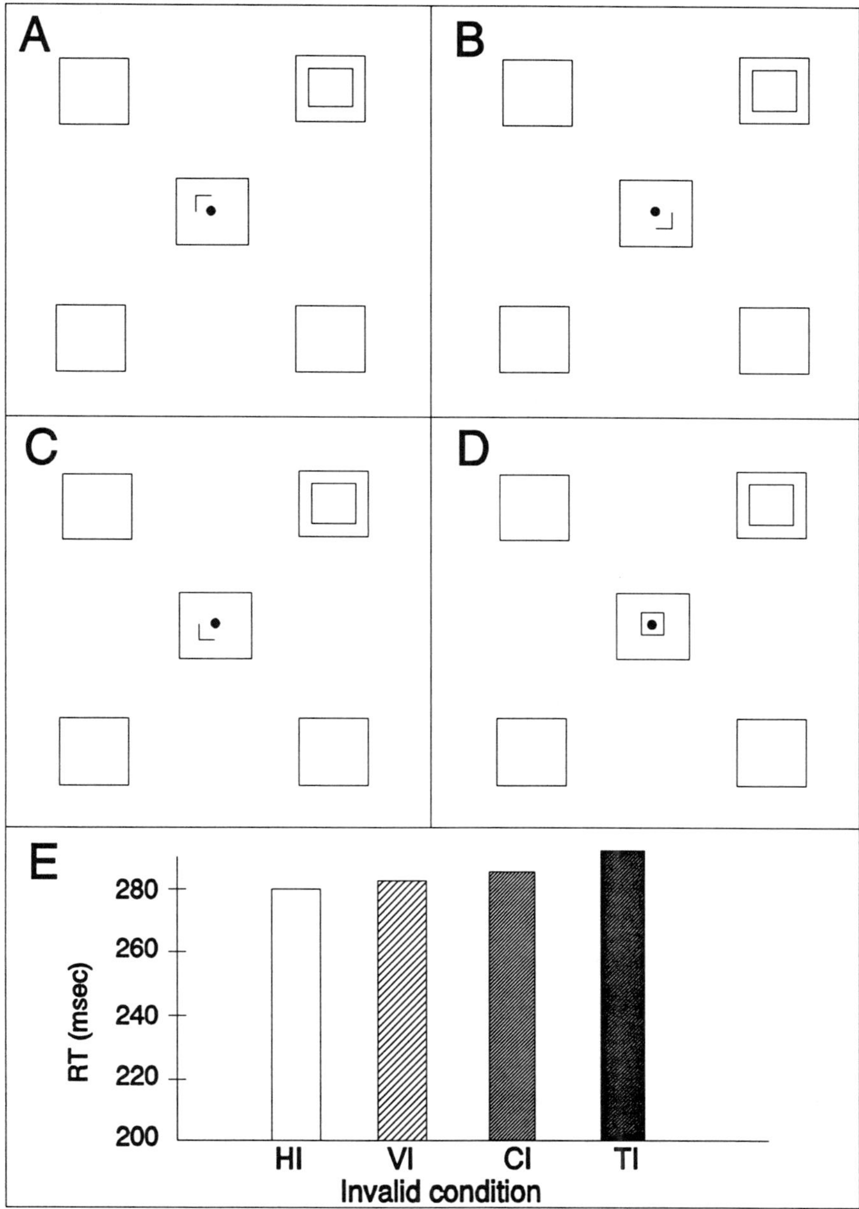

FIGURE 18.1. Experiment 1. The top panels show examples of horizontal (A), vertical (B), transfoveal (C), and central (D) invalid trials (not in scale). For details, see text. The bottom panel (E) shows reaction times for horizontal (HI), vertical (VI), central (CI), and transfoveal (TI) invalid conditions.

experimental conditions we had four types of invalid trials, namely: (1) horizontal invalid trials, in which one peripheral box was cued and the stimulus appeared in a symmetrical position on the other side of the vertical meridian; (2) vertical invalid trials, in which one peripheral box was cued and the stimulus appeared in a symmetrical position on the other side of the horizontal meridian; (3) transfoveal invalid trials, in which one peripheral box was cued and the stimulus appeared in the opposite quadrant in relation to both the vertical and horizontal meridians; and (4) central invalid trials, in which the central box was cued and the stimulus appeared in a peripheral box located in one of the four quadrants.

There was no invalid stimulus in the central box. That is, after cuing one peripheral box, the stimulus would appear in the cued box or in another peripheral box. The valid central trials were not analyzed.

At the end of the 4 sessions, for each subject, 21 medians of correct RTs were calculated for subsequent analysis: 5 valid (V), 4 horizontal invalid (HI), 4 vertical invalid (VI), 4 transfoveal invalid (TI), and 4 central invalid (CI).

Data analysis

Errors were not analyzed because they were very rare (about 3.5% overall).

Correct RTs were submitted to 2 analyses of variance (ANOVAs). In ANOVA 1, the variables that defined the treatments were stimulus *side* (left or right), stimulus *field* (upper or lower), and *condition*: valid (stimulus in the cued peripheral box) or invalid (stimulus in an uncued peripheral box). In ANOVA 2, only the invalid conditions were considered. The variables that defined the treatment were stimulus *side* (left or right), stimulus *field* (upper or lower) and *type of invalid conditions* (HI, VI, CI, or TI) (see Figure 18.1). Besides these ANOVAs, pairwise comparisons with the Newman-Keuls method were preformed whenever necessary. The level of significance was always set at .05.

Results

ANOVA 1 showed that *field* [$F(1,11) = 8.58$, $p < .02$] and *condition* [$F(1,11) = 173.37, p < .001$] were significant sources of variance. RT for lower stimuli was faster than for upper ones (248 vs. 252 msec) and RT for valid trials was faster than for invalid ones (214 vs. 286 msec).

ANOVA 2 showed that both *field* [$F(1,11) = 7.21, p < .03$] and *type of invalid condition* [$F(3,33) = 7.75, p < .001$] were significant sources of variance. RT for lower stimuli was faster than for upper ones (283 vs. 290 msec). Newman-Keuls comparisons of the means in the invalid conditions showed that RT in the transfoveal invalid condition (293 msec) was slower than RT in the other invalid conditions, which did not differ from one another (HI = 281, VI = 284, CI = 286 msec) (see Figure 18.1E).

Discussion

The purpose of the experiment was to compare the costs paid by the subjects for reorienting attention across the vertical meridian, across the horizontal meridian, across both the principal meridians, and away from a central fixation point. The main findings are as follows. First, costs for reorienting attention across one meridian or for orienting it away from fixation are equivalent. Second, costs for crossing both meridians are greater than costs for reorienting attention across one meridian or away from the fovea.

These results suggest that the orienting of attention to a visual target involves the computation of two vectorial components, one horizontal (left or right) and one vertical (up or down). According to the premotor hypothesis of attention, these components correspond to the horizontal and vertical components of the eye movement, which would direct the fovea to the target, if an overt orienting of attention were permitted. Thus, when a stimulus occurs in an unexpected position, such as on invalid trials, the reorienting of attention involves a rearrangement of that attentional vector that does not match those set up by the cue predicting the stimulus location. The present findings show that there are greater costs when attention is reoriented across two meridians than when it is reoriented across a single one, and indicate that it is more time-consuming to modify the parameters of the two vectors than those of a single vector.

There are two objections one can raise against our interpretation of the findings. The first is that the distance between the cued and stimulated box was greater in the conditions in which two meridians were crossed than in the conditions in which one meridian was crossed. Thus, the greater costs observed in the former conditions could be the effect of distance rather than of crossing an additional meridian. The second objection is that in the conditions in which both meridians were crossed, the attention moved across the fovea. It has been suggested that, because of the large foveal representation in the visual cortex, crossing the fovea would consume more time than crossing the meridians outside the fovea (Downing and Pinker, 1985). To eliminate these two possible sources of error (distance and cortical magnification), we conducted a second experiment in which we compared costs for the reorienting of attention across the fixation point with costs for reorienting attention from points located on one principal meridian to points located on the orthogonal meridian (see Figure 18.2). This experimental design, in addition to solving the above-mentioned problems, also allowed us to examine some aspects of the so-called meridian effect reported by Rizzolatti and colleagues (1987). In particular, we were interested in learning whether the extra cost observed by Rizzolatti and colleagues (1987) when the subjects had to change direction of attention (for example, from right to left), would also be observed in conditions in which the required program change is from a horizontal direction to a vertical direction (or vice versa).

EXPERIMENT 2

In this experiment we compared the costs produced by reorienting attention along the horizontal or vertical meridian with costs produced by reorienting attention from a location on one meridian (horizontal or vertical) to a location on the

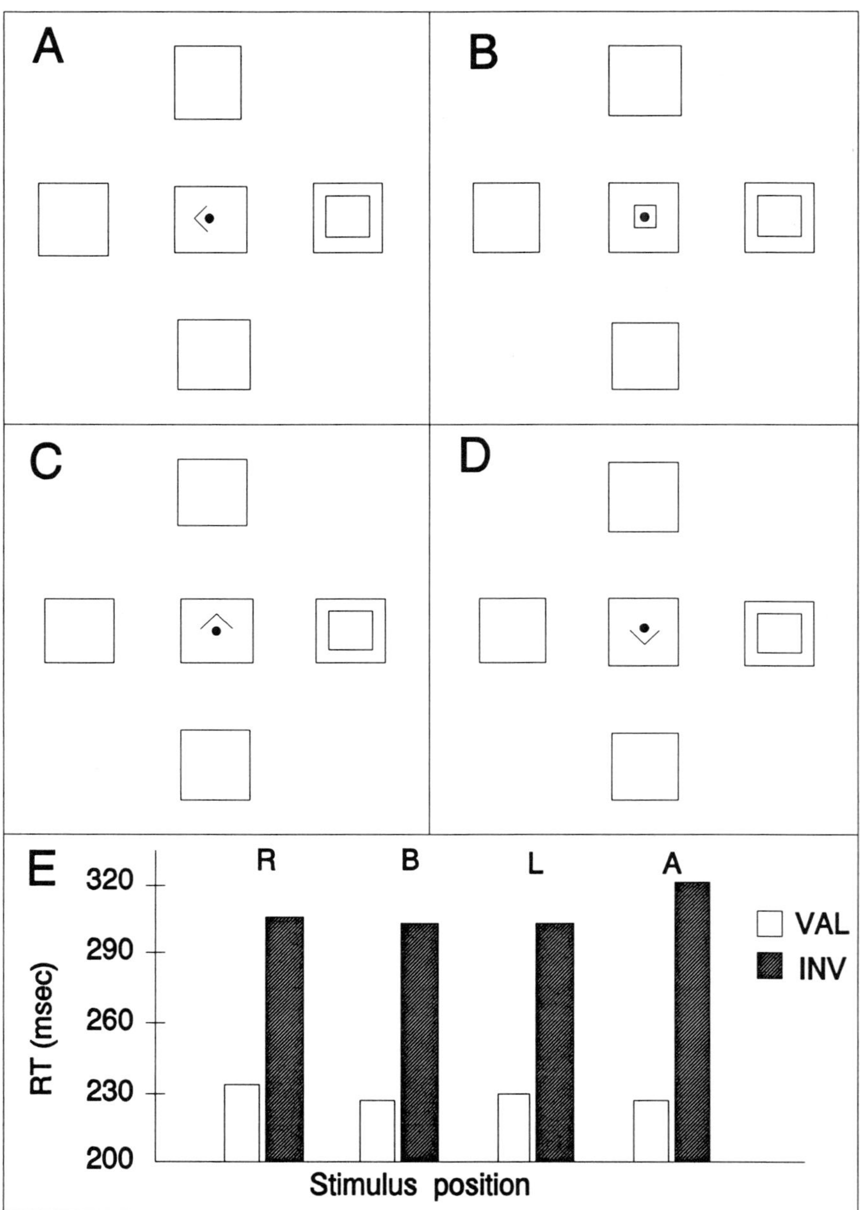

FIGURE 18.2. Experiment 2. The top panels show examples of symmetrical (A), central (B), and orthogonal (C and D) invalid trials (not in scale). For details, see text. The bottom panel (E) shows reaction times for valid (VAL) and invalid (INV) stimuli occurring to the right of (R), below (B), to the left of (L), and above (A) the fixation point.

meridian orthogonal to it (vertical or horizontal, respectively). In the former condition, attention has to cross the fovea, whereas this is not necessary in the latter.

Methods

Except for the stimulus display (see below), Experiment 2 was identical to Experiment 1. Eight male subjects of the Universidade Federal Fluminense participated in the experiment. They were right-handed (Oldfield, 1971), had normal or corrected vision, and were ignorant of the purpose of the experiment.

Apparatus and stimulus display

The stimulus display (see Figure 18.2A, 18.2B, 18.2C, and 18.2D) comprised a fixation point located inside a square box (1.1° wide) at the geometric center of the screen and four empty boxes located 4.5° above, below, to the right, and to the left of the fixation point. The display also contained the visual cues that indicated in which square the imperative stimulus would probably appear.

Procedure

In this experiment, besides valid trials, there were three types of invalid trials: (1) symmetrical invalid trials (SI), in which attention was cued to one peripheral box and the stimulus appeared in a symmetrical position on the opposite side of the same meridian; (2) orthogonal invalid trials (OI) in which one peripheral box was cued and the stimulus appeared in a peripheral box located on the meridian orthogonal to the cued one; and (3) central invalid trials (CI), in which the central box was cued and the stimulus appeared in one of the peripheral boxes.

Data analysis

Errors were not analyzed because they were very rare (2.5% overall).

As in Experiment 1, correct RTs were submitted to 2 ANOVAs. In ANOVA 1, the variables that defined the treatment were: *position* (above, below, to the right, or to the left of the fixation point), and *condition* (valid or invalid). ANOVA 2 compares the invalid conditions. The factors were *position* (above, below, to the right, or to the left of the fixation point) and *type of invalid condition* (symmetrical, orthogonal, or central).

Results

Figure 18.2E shows the results of the experiment. There was a significant effect of stimulus *position* [$F\ (3,21) = 8.25$, $p < .002$] and *condition* [$F\ (1,7) = 107.00$, $p < .001$], and a significant interaction between these factors [$F\ (3,21) = 9.69$, $p < .0001$]. RT for valid trials was shorter than for invalid trials (232 vs. 309 msec). Newman-Keuls comparisons showed that the mean RT for upper stimuli

(276 msec) was slower than RT for stimuli occurring in all other positions, which were not significantly different one from another (RTs for right, lower, and left stimuli were 270, 266, and 269 msec, respectively). The RT advantage for the upper stimulus was due to the invalid condition since post hoc comparisons showed that RT for the valid condition did not depend on position (upper = 232, right = 233, lower = 233 msec, left = 233 msec). RT for the invalid condition differed for the upper position (321 msec) versus all other positions (304, 304, 305 msec for right, lower, and left stimulus, respectively).

ANOVA 2 had two factors: *position* (above, below, to the right, to the left of fixation) and *invalid condition* (symmetrical, orthogonal, and central invalid). The main effect of *position* was the only source of variance to reach significance [$F(3,21) = 13.01, p < .001$]. Mean RT for stimuli in the upper position (321 msec) was slower than RT for stimuli in the other positions (304,304, 305 msec for right, lower, and left stimulus, respectively). The main effect of *invalid condition* was *not* significant [$F(2,14) = 1.12, p > .35$]. RT for symmetrical, orthogonal, and central invalid trials were 306, 310, and 310 msec, respectively.

Discussion

The main finding of this experiment is the absence of differences among the various invalid conditions. Cost for reorienting attention from a target located on one side (e.g., right or up) of a principal meridian to a target located on the opposite side of the same meridian (e.g., left or down) are equivalent to costs for reorienting attention from a target located on one principal meridian to a target located on the meridian orthogonal to it (e.g., right to up, or down to left). This indicates that crossing of the fovea, which is required in the former experimental conditions, but not in the latter, does not play a significant role in the cost increases when attention has to move from one visual hemifield to another. Moreover, it shows that costs for canceling one vectorial component of the ocular program and generating one vectorial component in a direction orthogonal to the previous one are equivalent to costs for canceling one vectorial component of the ocular program and generating one vectorial component in a direction opposite to the previous one.

GENERAL DISCUSSION

The premotor theory of attention proposed by Rizzolatti and colleagues (1987) is based on the notion that motor programs for eye movements control both overt and covert orienting of attention: "overt and covert orienting of attention are controlled by common mechanisms and . . . the absence of eye movements in case of covert orienting is a consequence of a peripheral inhibition, which leaves unchanged the central programming. In other words, the program for orienting attention either overtly or covertly is the same, but in the latter case the eyes are blocked at a certain peripheral stage." (p. 37)

The results of the present experiments complement the premotor hypothesis by showing that the orienting of attention to a visual target involves the computation of two vectorial components, one horizontal (left or right) and the other vertical (up or down). According to the premotor hypothesis of attention (Rizzolatti et al., 1987), these components correspond to the horizontal and vertical components of an eye movement, which would direct the fovea to the target, if an overt orienting of attention were permitted. Thus, when a stimulus occurs in an unexpected position (i.e., on invalid trials), the reorienting of attention involves a rearrangement of that attentional vector component that does not match those produced by the cue. The present findings indicate that it is more time-consuming to modify the parameters of two vectors than those of a single vector. This implies either that the two vectors are not computed in parallel or, if they are, that their simultaneous computation produces interference.

As suggested by Rizzolatti (1983; see also Rizzolatti and Berti, 1990; Rizzolatti and Camarda, 1987; Rizzolatti and Galese, 1988), it is possible to conceive of spatial attention as a property linked to the premotor activity of those various cerebral centers that program motor actions in a spatial framework. The motor plans are usually complex and involve eye movements as well as movements of the head, hand, arm, and other body parts. The attention shift toward a space sector is, then, the result of an increase of activity of neurons present in various circuits that control actions in that same space sector.

A similar hypothesis was proposed by Tassinari and colleagues (1987), according to whom the orienting of attention to a visual position corresponds to the generation of "a state of selective motor readiness which results in a general facilitation of all motor outputs potentially triggered or guided from the target area, to the disadvantage or exclusion of motor reactions to other stimulus locations." (p. 68) These authors also proposed that the segmentation of the attentional visual field by the vertical and horizontal meridians is related to the organization of motor actions in the visual space (see also Berlucchi et al., 1989; Tassinari et al., 1989).

Taken together, the spatial characteristics of various forms of visual neglect in patients and those of spatial attention in normal subjects indicate that the orienting of attention to a visual position involves the computation of the resultant of four attentional vectorial biases directed toward the right, left, upper, and lower space. In support of this contention it is worth mentioning that head movements of the barn owl are reportedly controlled by four functionally distinct neural circuits, in which upward, downward, leftward, and rightward movements are represented (Masino and Knudsen, 1990). These authors also proposed that the coding of orthogonal movement components of the head operates "as an abstract system intermediate to the sensory and motor processes that control orienting movements" and that "orthogonal coordinate systems may be a general property of intermediate stages in motor control hierarchies." (p. 3)

REFERENCES

Bashinski HS, Bachrach VR (1980): Enhancement of perceptual sensivity as the result of selectively attending to spatial locations. *Percept Psychophysiol* 28:241–248

Berlucchi G, Tassinari G, Marzi CA, Di Stefano M (1989): Spatial distribution of the inhibitory effect of peripheral non-informative cues on simple reaction-time to non-foveal visual targets. *Neuropsychologia* 27:201–221

Downing CJ, Pinker S (1985): The spatial structure of visual attention. In: *Attention and Performance*, vol. 11, Posner MI, Marin OSM, eds. Hillsdale, NJ: Erlbaum

Gawryszewski LG, Riggio L, Rizzolatti G, Umiltà CA (1987): Movements of attention in the three spatial dimensions and the meaning of "neutral" cues. *Neuropsychologia* 25:19–29

Harter MR, Aine CJ (1984): Brain mechanisms of visual selective attention. In: *Varieties of Attention*, Parasuraman R, Davies R, Beatty J, eds. New York: Academic Press

Hillyard SA, Munte TF, Neville HJ (1985): Visual-spatial attention, orienting and brain physiology. In: *Attention and Performance*, Vol. 11, Posner MI, Marin OSM, eds. Hillsdale, NJ: Erlbaum

Hughes HC, Zimba LD (1985): Spatial maps of directed attention. *J Exp Psychol (Hum Per)* 11:409–430

James W (1890): *The Principles of Psychology*. New York: Holt

Jonides JP (1981): Voluntary versus automatic control over the mind's eye. In: *Attention and Performance*, Vol. 9, Long JB, Baddeley AD, eds. Hillsdale, NJ: Erlbaum

Jonides JP, Mack R (1984): On the cost and benefit of cost and benefit. *Psychol Bull* 96:29–44

Masino T, Knudsen EI (1990): Horizontal and vertical components of head movement are controlled by distinct neural circuits in the barn owl. *Nature* 345:434–437

Müller HJ, Findlay JM (1988): The effect of visual attention on peripheral discrimination thresholds in single and multiple element displays. *Acta Psychol* 69:129–155

Müller HJ, Rabbitt PMA (1989): Reflexive and voluntary orienting of visual attention: Time course of activation and resistance to interruption. *J Exp Psychol (Hum Per)* 15:315–333

Oldfield RC (1971): The assessment and analysis of handedness: The Edinburgh Inventory. *Neuropsychologia* 9:97–113

Posner MI (1978): *Chronometric Explorations of Mind*. Hillsdale, NJ: Erlbaum

Posner MI (1980): Orienting of attention. *Q J Exp Psychol* 32:3–25

Posner MI (1982): Cumulative development of attentional theory. *Am Psychol* 37:168–179

Posner MI, Cohen Y (1984): Components of visual orienting. In: *Attention and Performance*, Vol. 10, Bouma H, Bouwhuis D, eds. Hillsdale, NJ: Erlbaum.

Prinzmetal W, Presti D, Posner MI (1986): Does attention affect feature integration? *J Exp Psychol (Hum Per)* 12:361–396

Rizzolatti G (1983): Mechanisms of selective attention in mammals. In: *Advances in Vertebrate Neuroethology*, Ewert JP, Capranica RR, Ingle DJ, eds. London: Plenum Press

Rizzolatti G, Berti A (1990): Neglect as a neural representation deficit. *Rev Neurol (Paris)* 146:626–634

Rizzolatti G, Camarda R (1987): Neural circuits for spatial attention and unilateral neglect. In: *Neurophysiological and Neuropsychological Aspects of Spatial Neglect*, Jeannerod M, ed. Amsterdam: Elsevier

Rizzolatti G, Galese V (1988): Mechanisms and theories of spatial neglect. In: *Handbook of Neuropsychology*, vol. 1, Boller F, Grafman J, eds. Amsterdam: Elsevier

Rizzolatti G, Riggio L, Dascola I, Umiltà CA (1987): Reorienting attention across the horizontal and vertical meridians: Evidence in favor of a premotor theory of attention. *Neuropsychologia* 25:31–40

Tassinari G, Aglioti S, Chelazzi L, Marzi CA, Berlucchi G (1987): Distribution in the visual field of the costs of voluntarily allocated attention and of the inhibitory after-effects of covert orienting. *Neuropsychologia* 25:55–71

Tassinari G, Biscaldi M, Marzi CA, Berlucchi G (1989): Ipsilateral inhibition and contralateral facilitation of simple reaction time to non-foveal visual targets from non-informative visual cues. *Acta Psychol* 70:267–291

Yantis S, Jonides J (1984): Abrupt visual onsets and selective attention: Evidence from visual search. *J Exp Psychol (Hum Per)* 10:601–621

Index